AF581892

10th EASN International Conference on Innovation in Aviation & Space to the Satisfaction of the European Citizens

10th EASN International Conference on Innovation in Aviation & Space to the Satisfaction of the European Citizens

Editors

Spiros Pantelakis
Andreas Strohmayer
Liberata Guadagno

MDPI • Basel • Beijing • Wuhan • Barcelona • Belgrade • Manchester • Tokyo • Cluj • Tianjin

Editors

Spiros Pantelakis
University of Patras
Greece

Andreas Strohmayer
University of Stuttgart
Germany

Liberata Guadagno
University of Salerno
Italy

Editorial Office
MDPI
St. Alban-Anlage 66
4052 Basel, Switzerland

This is a reprint of articles from the Special Issue published online in the open access journal *Aerospace* (ISSN 2226-4310) (available at: https://www.mdpi.com/journal/aerospace/special_issues/10th_EASN).

For citation purposes, cite each article independently as indicated on the article page online and as indicated below:

LastName, A.A.; LastName, B.B.; LastName, C.C. Article Title. *Journal Name* **Year**, *Volume Number*, Page Range.

ISBN 978-3-0365-4225-6 (Hbk)
ISBN 978-3-0365-4226-3 (PDF)

Contents

About the Editors

Spiros Pantelakis

Spiros Pantelakis is Prof. Emeritus, University of Patras. He has been Director of the Laboratory of Technology and Strength of Materials (2007–2020), Chairman of the Department of Mechanical Engineering and Aeronautics (2009–2013), and Vice President of the Board of Executives of the Research Committee of the University of Patras (2010–2013). He is a founding member and the Chairman of the European Aeronautics Science Network (EASN) Association since its establishment in 2008 up to 2019. In 2019, he was awarded the title of Honorary Chairman of the EASN Association. He has been Representative of the European aeronautics' academia in the Plenary of ACARE and Chairman of ACARE's Working Group on Human Resources. He was the founding member and has been a Member of the Clean Sky Academy since its establishment in 2015. He participated in the authors' team of the Clean Aviation Partnership Program and co-chaired the group, dealing with the exploration and maturation of technologies. He has been Member of the Board of Directors of Hellenic Aerospace Industry (2015–2020) and is a former Chairman of the Board of Directors of the Greek Metallurgical Society (2008–2011). He has about 45 years' experience in the field of materials and structures and has been involved in more than 100 international aeronautics related research projects. He is the author of several scientific books, chapters in international textbooks, and more than 250 scientific publications in peer-reviewed international journals and conference proceedings, and he has been the supervisor of more than 20 PhD Theses. He is chairman of about 30 International Scientific Conferences and organizer of several decades of International Workshops. Currently, he is proactively participating in the preparation of the Clean Aviation Joint Undertaking as an academia representative at the decision-making level and as a member of the core writing team of the Clean Aviation work program.

Andreas Strohmayer

Andreas Strohmayer is a professor of aircraft design at the Institute of Aircraft Design, University of Stuttgart with a research focus on manned (hybrid) electric flight ("Icaré 2" and "e-Genius") and scaled UAS flight testing. He studied Aeronautical Engineering at TU Munich and graduated in 2001 under Dr.-Ing. in the field of conceptual aircraft design. From 2002 to 2008, he was director of Grob Aerospace in Mindelheim, Germany, responsible for the design, production, and support of the Grob fleet of all-composite aircraft: the development of a four-seat aerobatic turboprop, a seven-seat turboprop, and the SPn business jet. From 2009 to 2013, he was program director for a 19-seater commuter aircraft project at Sky Aircraft in Metz, France, and then VP Programs at SST Flugtechnik in Memmingen, Germany, setting up an EASA-approved design organization, holding the TC for a six-seater all-composite turbo-prop aircraft. He then left the industry to join University of Stuttgart in 2015, teaching aircraft design and promoting electric flight. In 2016, he became a member of the Board of Directors of the European Aeronautics Science Network (EASN); since 2019, he has been an EASN Chairman. In this position, he also represents academia in the ACARE General Assembly.

Liberata Guadagno

Liberata Guadagno is Full Professor of "Foundations of Chemistry for Technologies" at the Department of Industrial Engineering of Salerno University. She held the current role in the year 2015, as an eminent scientist, according to the decision DM 276/2011. Her research activity has been mainly directed toward the design and development of smart polymeric composites (self-generating materials for autonomic damage control in thermosetting resins; multifunctional composites; self-responsive bulk and skin materials; energy-saving manufacturing processes, etc.). Her research activities are documented by 224 scientific papers in international journals, chapters of books, and 150 proceedings. She has extensive experience in public and industrial-funded projects. Recently, Prof. Guadagno has been actively involved in many industrial research and EU-funded projects, of which she has played the role of European coordinator and WPs coordinator. In regard to recent E.U. projects led by Prof. Guadagno, the European Commission's Innovation Radar highlighted excellent innovations and recognized "UNISA" team as a European 'Key Innovator' in the following topics: (1) "smart thermosetting and thermoplastic composites based on electrical conductivity properties"; (2) new mesoscopic models to simulate the electroactive and sensing behaviour of innovative materials; (3) thermoset materials with self-curing capabilities (CFRP and GFRP composites); and (4) self-deicing thermoset materials for aerospace and automotive applications. Prof. Guadagno, in collaboration with Leonardo SpA, has produced 27 international patents (with concessions in Europe and in the USA). On 15 November 2018, she was the winner of Leonardo SpA Aircraft Division 2018—"Innovation Award for the best Patent". Prof. Guadagno currently plays the role of "National Contact Point for Italy of the EASN (European Aeronautics Science Network).

MDPI

Editorial

Special Issue "10th EASN International Conference on Innovation in Aviation & Space to the Satisfaction of the European Citizens"

Liberata Guadagno [1], Spiros Pantelakis [2,*] and Andreas Strohmayer [3,*]

1 Department of Industrial Engineering, University of Salerno, Via Giovanni Paolo II, 132, 84084 Fisciano (SA), Italy; lguadagno@unisa.it
2 Department of Mechanical Engineering and Aeronautics, University of Patras, Panepistimioupolis Rion, 26500 Patras, Greece
3 Department of Aircraft Design, Institute of Aircraft Design (IFB), University of Stuttgart, Pfaffenwaldring 31, 70569 Stuttgart, Germany
* Correspondence: pantelak@mech.upatras.gr (S.P.); strohmayer@ifb.uni-stuttgart.de (A.S.)

Citation: Guadagno, L.; Pantelakis, S.; Strohmayer, A. Special Issue "10th EASN International Conference on Innovation in Aviation & Space to the Satisfaction of the European Citizens". *Aerospace* **2021**, *8*, 111. https://doi.org/10.3390/aerospace8040111

Received: 12 April 2021
Accepted: 13 April 2021
Published: 14 April 2021

Publisher's Note: MDPI stays neutral with regard to jurisdictional claims in published maps and institutional affiliations.

This Special Issue contains selected papers from works presented at the 10th EASN International Conference on Innovation in Aviation & Space to the Satisfaction of the European Citizens, which was held successfully from the 2nd until the 4th of September, 2020. Due to the COVID pandemic, it was the first time in the history of the EASN Conference series that the event took place in a virtual format. The event included 9 keynote lectures and more than 320 technical presentations distributed in close to 50 sessions. Important also to underline that 48 HORIZON2020 projects have disseminated their latest research results as well as the future trends on the respective technological field at this event. In total, the 10th EASN International Conference was attended by more than 350 participants from 37 countries worldwide.

In the present Special Issue, eleven engaging articles are contained, with more than 300 views each till now, related to aviation research. Vedernikov et al. [1] performed complex parametrical strength investigations of typical wings for regional aircraft, using an advanced four-level algorithm (FLA). The enhanced algorithm used in this work allows for an efficient strength analysis of airframes and is validated with a high-aspect-ration wing designs, also including strut-braced wings. In a systems-engineering approach, Eisenhut et al. [2] established top-level aircraft requirements (TLARs) for a 50-passenger hybrid-electric regional aircraft. Beyond performance requirements, these TLARs also include environmental factors, becoming increasingly important for sustainable aircraft designs. Furthermore, suitable reference missions are presented, as well as figures of merit that allow for an evaluation of different aircraft architectures. In the work of Bergmann et al. [3], a methodology for precise identification of the performance characteristics of a UAV test platform is described, providing an overview of the measuring system, discussing its functionality and showing flight test results. This facilitates the systematic analysis of propeller–wing interaction effects which are of interest to a synergetic aircraft configuration design with distributed propulsion. Blasi et al. [4] describe the control architecture and the control laws of an innovative Modular Iron Bird concept which aims at reproducing flight loads to test mobile aerodynamic control surface actuators for small and medium size aircraft, as well as unmanned aerial vehicles. The effectiveness of the proposed control architecture and control laws is demonstrated in numerical simulations. Pathak et al. [5] present their work on a model validation case for the distribution of the agent of an environmentally friendly fire protection system (EFFP) in the cargo hold of an aircraft. In the low-pressure vessel of the Fraunhofer Flight Test Facility (FTF), the team was able to equip an aircraft demonstrator with an EFFP and to validate refined simulation models for knockdown during different flight phases in this setup. In the work of Przysowa et al. [6], the performance and gas emissions produced from two different jet engines are compared for blends of Jet A-1 fuel with different alternative fuels in various concentrations. The acquired data serve the development of an engine emissivity model allowing for the prediction of engine emissions. The results show that under an emissions point of view blended fuels can be used to fuel gas turbines. Seitz et al. [7] present key results for the EU H2020 project CENTRELINE, undertaken to demonstrate a proof of concept for the so-called propulsive fuselage concept (PFC) with boundary layer ingestion (BLI). A performance bookkeeping scheme for the BLI propulsion is reviewed and findings from the high-fidelity aero-numerical simulation and aerodynamic validation testing in wind tunnel and BLI fan rig test campaigns are discussed. An assessment of the PFC fuel burn compared to a conventional design shows a 4.7% mission fuel benefit. Norrefeldt et al. [8] assessed the effects of an alteration of the outdoor/recirculation airflow ratio in an aircraft cabin on relative humidity, CO2 and total volatile organic compounds (TVOC) level in the cabin air. Tests were conducted in the Fraunhofer FTF facility, showing an increase of these parameters with an increased recirculation fraction, as the passenger emissions become less diluted by dry outdoor air. Kellermann et al. [9] investigated the use of ram-air based thermal management systems (TMS) for the cooling of the power train components of future hybrid electric aircraft, assessing the impact on mass, drag and fuel burn. A numerical optimization of respective TMS system was carried out for minimum fuel burn of a 180 passenger short range aircraft with a partial electric propulsion system. For a power split of 30% electric power, an additional fuel burn of 0.19% due to the TMS is reported in this work. Pavlenko et al. [10] analyzed the design requirements and operating parameters of small turbofan engines for single-use and reusable unmanned aerial vehicles (UAVs), in order to introduce alternative materials and technologies for manufacturing their compressor blades. Stress and temperature maps on compressor blades and vanes were obtained by means of thermal and structural analysis, considering the physical and mechanical properties of advanced materials and related processing technologies. It was shown that the permissible operating temperature and safety factor as well as the design requirements of the turbofan at lower manufacturing costs are met with the proposed materials. Finally, Engelmann et al. [11] analyzed the boarding procedure of a single aisle aircraft with an improved simulation methodology, increasing the level of detail in the boarding simulation. The improved model considers, for example, the passenger walking speed in dependence of surrounding objects and the location of other passengers. A validation was carried out with an Airbus A320 as a baseline, compared to an altered version with an extended aisle width and to a COVID-19 safe distance scenario. As a result, a boarding time reduction of up to 3% could be shown for the wider aisle, while the COVID-19 scenario leads to an increase of 67%.

The editors of this Special Issue would like to thank the authors for their high-quality contributions and for making this Special Issue manageable. Furthermore, the editors would like to express their gratitude to Ms. Linghua Ding and the Aerospace editorial team for their professional support.

Conflicts of Interest: The authors declare no conflict of interest.

References

1. Vedernikov, D.V.; Shanygin, A.N.; Mirgorodsky, Y.S.; Levchenkov, M.D. Strength Analysis of Alternative Airframe Layouts of Regional Aircraft on the Basis of Automated Parametrical Models. *Aerospace* **2021**, *8*, 80. [CrossRef]
2. Eisenhut, D.; Moebs, N.; Windels, E.; Bergmann, D.; Geiß, I.; Reis, R.; Strohmayer, A. Aircraft Requirements for Sustainable Regional Aviation. *Aerospace* **2021**, *8*, 61. [CrossRef]
3. Bergmann, D.P.; Denzel, J.; Pfeifle, O.; Notter, S.; Fichter, W.; Strohmayer, A. In-flight Lift and Drag Estimation of an Unmanned Propeller-Driven Aircraft. *Aerospace* **2021**, *8*, 43. [CrossRef]
4. Blasi, L.; Borrelli, M.; D'Amato, E.; di Grazia, L.E.; Mattei, M.; Notaro, I. Modeling and Control of a Modular Iron Bird. *Aerospace* **2021**, *8*, 39. [CrossRef]
5. Pathak, A.; Norrefeldt, V.; Pschirer, M. Validation of a Simulation Tool for an Environmentally Friendly Aircraft Cargo Fire Protection System. *Aerospace* **2021**, *8*, 35. [CrossRef]

6. Przysowa, R.; Gawron, B.; Białecki, T.; Łegowik, A.; Merkisz, J.; Jasinski, R. Performance and Emissions of a Microturbine and Turbofan Powered by Alternative Fuels. *Aerospace* **2021**, *8*, 25. [CrossRef]
7. Seitz, A.; Habermann, A.L.; Peter, F.; Troeltsch, F.; Castillo Pardo, A.; Della Corte, B.; van Sluis, M.; Goraj, Z.; Kowalski, M.; Zhao, X.; et al. Proof of Concept Study for Fuselage Boundary Layer Ingesting Propulsion. *Aerospace* **2021**, *8*, 16. [CrossRef]
8. Norrefeldt, V.; Mayer, F.; Herbig, B.; Ströhlein, R.; Wargocki, P.; Lei, F. Effect of Increased Cabin Recirculation Airflow Fraction on Relative Humidity, CO2 and TVOC. *Aerospace* **2021**, *8*, 15. [CrossRef]
9. Kellermann, H.; Lüdemann, M.; Pohl, M.; Hornung, M. Design and Optimization of Ram Air–Based Thermal Management Systems for Hybrid-ElectricAircraft. *Aerospace* **2020**, *8*, 3. [CrossRef]
10. Pavlenko, D.; Dvirnyk, Y.; Przysowa, R. AdvancedMaterials and Technologies for Compressor Blades of Small Turbofan Engines. *Aerospace* **2021**, *8*, 1. [CrossRef]
11. Engelmann, M.; Kleinheinz, T.; Hornung, M. Advanced Passenger Movement Model Depending on the Aircraft Cabin Geometry. *Aerospace* **2020**, *7*, 182. [CrossRef]

Article

Strength Analysis of Alternative Airframe Layouts of Regional Aircraft on the Basis of Automated Parametrical Models

Dmitry V. Vedernikov *,†, Alexander N. Shanygin †, Yury S. Mirgorodsky † and Mikhail D. Levchenkov †

Strength Department, Central Aerohydrodynamic Institute, 1 Zhukovsky Street, 140180 Zhukovsky, Russia; alexander.shanygin@tsagi.ru (A.N.S.); mirgorodskii@phystech.edu (Y.S.M.); Levchenkov.md@phystech.edu (M.D.L.)
* Correspondence: vedernikov@phystech.edu; Tel.: +7-926-255-52-13
† These authors contributed equally to this work.

Abstract: This publication presents the results of complex parametrical strength investigations of typical wings for regional aircrafts obtained by means of the new version of the four-level algorithm (FLA) with the modified module responsible for the analysis of aerodynamic loading. This version of FLA, as well as a base one, is focused on significant decreasing time and labor input of a complex strength analysis of airframes by using simultaneously different principles of decomposition. The base version includes four-level decomposition of airframe and decomposition of strength tasks. The new one realizes additional decomposition of alternative variants of load cases during the process of determination of critical load cases. Such an algorithm is very suitable for strength analysis and designing airframes of regional aircrafts having a wide range of aerodynamic concepts. Results of validation of the new version of FLA for a high-aspect-ratio wing obtained in this work confirmed high performance of the algorithm in decreasing time and labor input of strength analysis of airframes at the preliminary stages of designing. During parametrical design investigation, some interesting results for strut-braced wings having high aspect ratios were obtained.

Keywords: aircraft structure; strut-braced wing; parametric modeling; decomposition principles; strength analysis; finite element method (FEM); doublet lattice method; four-level approach

Citation: Vedernikov, D.V.; Shanygin, A.N.; Mirgorodsky, Y.S.; Levchenkov, M.D. Strength Analysis of Alternative Airframe Layouts of Regional Aircraft on the Basis of Automated Parametrical Models. *Aerospace* **2021**, *8*, 80. https://doi.org/10.3390/aerospace8030080

Academic Editor: Spiros Pantelakis

Received: 30 November 2020
Accepted: 11 March 2021
Published: 17 March 2021

Publisher's Note: MDPI stays neutral with regard to jurisdictional claims in published maps and institutional affiliations.

1. Introduction

Conditions of operations of regional aircrafts have a number of differences as compared to the middle-range and long-haul ones. The main difference is a short time of a cruise flight, meaning the absence of strong aerodynamic demands to external geometry parameters, which focused to maximize the lift-to-drag ratio. Thus, unlike the middle-range and long-haul airliners, for regional aircrafts, it is necessary to consider external geometry parameters simultaneously with the other active airframes parameters within the common design process. For this reason, designing regional airframes is quite a hard task, due to the increased number of active design parameters [1,2]. This task is similar to the initial stages of designing airframes of aircraft with non-conventional aerodynamic concept that also require simultaneous consideration of external geometry parameters together with other design parameters [3,4]. It is obvious that these tasks are impossible to solve in frame of the two-step conventional approach, when, at the first step, the values of external geometry parameters of a concept are defined, following aerodynamics demands, whereas, at the second stage, fixed values of geometry parameters are used for the definition of values of the other active airframe parameters. Thus, for successful solution of design tasks of regional aircraft airframes, a significant decrease of calculation time is needed in order to perform the preliminary design procedure. The conventional (step-by-step) design procedures within multidisciplinary design and optimization (MDO) approach are described in References [5,6].

There are several standard methods of decreasing calculation time and labor input, including automation, standardization of input and output information, high-performance calculation equipment, etc. However, one of the important matter for decreasing labor intensity is an application of decomposition of both airframe strength models and strength tasks. Development of methods and algorithms based on decomposition principles was the goal of a number of researches. In References [7,8], some examples of decomposition of a wing structure are shown. In References [9,10], decomposition of strength tasks, including decomposition of a procedure of a strength analysis, was carried out, using several different physical models, including both FEM models and simple analytical beam models.

These two abovementioned principles were realized in frame of the base version of FLA that was developed and validated in Central Aero-Hydrodynamic institute, to analyze strength of airframes and carry out preliminary design of perspective civil aircrafts with non-conventional structure concepts, including numerous different variants of layouts. The algorithm was successfully validated in frame of some European and domestic projects [11,12].

The main features of the base-version FLA are the following:

1. Global parameterization of airframe structure on the base of common special four-level database.
2. Four-level nested FE models according to detailing of the structure.
3. Full automation of initial data definition and analysis of calculations results.
4. Parallel solving strength tasks on both nested FE models and auxiliary analytical strength models.
5. Standard format of input and output data.

Full automation of the algorithm allows users to control the accuracy of a calculation, changing FE mesh size, and thus to find minimal possible nested FE model's size. Figure 1 illustrates four-level structure decomposition and relationship between FE models and analytical ones.

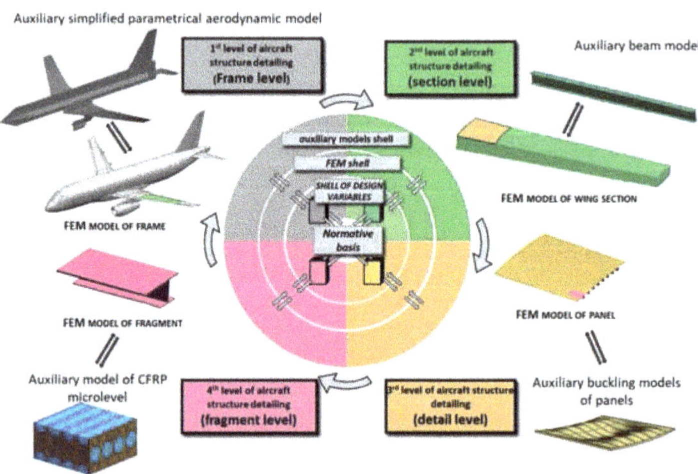

Figure 1. Scheme of calculation procedure in frame of the base four-level algorithm (FLA).

As long as, during investigation of alternative variants of non-conventional airframes, the external geometry parameters of considered aerodynamic concepts did not have serious variation, the base FLA showed high performance [13,14].

In the present work, one more decomposition principle was developed, to decrease time and labor input for procedure of analysis of aerodynamic loads. In accordance with this principle of decomposition, a new aerodynamic module for the fast choosing of critical load cases was created and validated in the frame of the new version of FLA. This module divides aerodynamic load cases into several groups in frame of the iterative process of determination of critical load cases. These groups of load cases are investigated on aerodynamic models having different mesh size. The new version of FLA keeps all advantages of the base one.

Thus, the new version of the FLA realizes the three abovementioned principles of decomposition: four-level structure decomposition, decomposition of strength tasks, and decomposition of alternative load cases during the process of determination of critical load cases. It gives the possibility to decrease no less than 20 times the labor input (as compared to the base version of FLA) of complex strength analysis, keeping required accuracy. The new version is very suitable for strength analysis and airframe designing when researchers have to consider numerous numbers of alternative aerodynamic concepts, such as a big number of alternative variants of airframes within each of an aerodynamic concept.

The structure of the new version of FLA consists of the following program modules: special four-level database modules, modules responsible for automatic building of parametrical FEM and auxiliary strength models, conventional FEM loads modules, modules for forming of parametrical aerodynamic load models, and modules responsible for relationship between four-level database and external numerical solvers. These modules were created and validated in TsAGI. In addition to the abovementioned modules, MSC Nastran program modules (SOL 101, SOL 144) were used as solvers for determination of stiffness and aerodynamic matrixes. Figure 2 shows the scheme of process of strength analysis in the FLA.

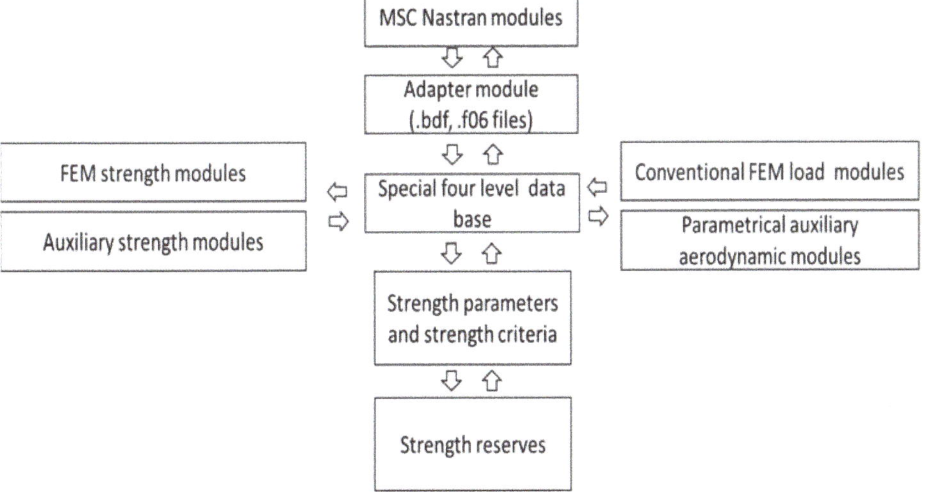

Figure 2. Scheme of process of strength analysis in the FLA.

2. Airframe Strength Analysis within FLA

The proposed FLA gives wide capabilities for reducing the time and labor input for complex strength-analysis procedures. That is very important for the preliminary design process of an aircraft airframe layout, as the check of strength constraints of numerous aircraft structure concepts is the most time-consuming part of the optimization procedure of designing.

Figure 3 shows the four-level airframe-strength models' decomposition in strength analysis, when four types of FE models are used simultaneously (FEM 1, FEM 2, FEM 3, and FEM 4) in a strength analysis. Each of these models is responsible for solution of the corresponding strength tasks. Within FEM 1, the strength tasks related to the entire aircraft structure are considered, such as estimation of global stiffness parameters, load cases definition, aeroelasticity, etc. Within FEM 2 the tasks of global buckling and post-buckling behavior of main sections are solved. Moreover, detail mass estimations of fuel, equipment, payload is performed using strength models of this level. Within FEM 3, the tasks of global stress–strain state estimation and local (panel) buckling are carried out. In addition to strength analysis, the weight of structure can be estimated on the base of the FEM 3 model. Finally, FEM 4 is used for solving local strength tasks, e.g., for local stress concentration analysis in critical zones.

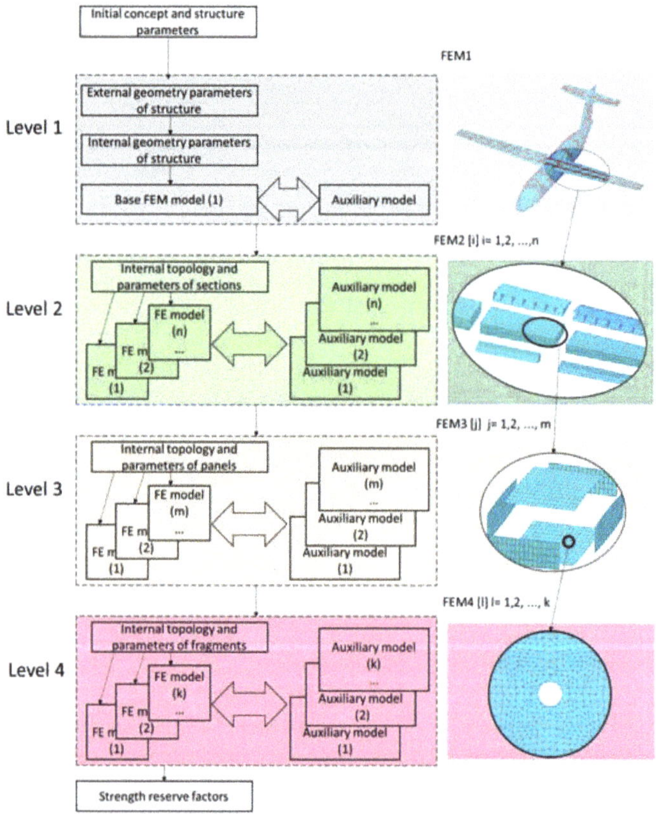

Figure 3. Block scheme of the four-level airframe strength models decomposition in strength analysis.

The FEM models of all levels have a strong connection with each other on the basis of the principle of nested models, when the models of the upper level include all nodes of the lower level models.

Thus, FEM 2 contains the main topology parameters of FEM 1 models, including the arrangement of sections into aircraft airframe. The number of FEM 2 models equals the number of sections, defined within FEM 1. Correspondingly, each of FEM 3 models include topology parameters of FEM 2 models, including arrangements of spars, ribs, panels into each section, etc.

The used principle of decomposition makes it possible to translate the correct boundary conditions from the lower level to the all the upper levels, as values of stresses and strains, defined on the base nodes (on the FEM model of the lower level), are automatically transferred to the models of upper levels. Thus, it gives possibility to solve tasks piecemeal with correct boundary conditions. It can decrease significantly time and labor input during the calculation.

In the frame of FLA, it is easy to realize decomposition-of-strength tasks because auxiliary strength models are built automatically, using the same parameters as FEMs. The auxiliary models are usually used at the first steps of strength analysis of the airframes, whereas the FEM models are used at a later stage. Auxiliary-strength models can be included into FLA, as modules working in parallel. It is very suitable for the fast solving of the tasks of aeroelasticity and determination of aerodynamic loads.

In this paper, parametrical auxiliary aerodynamic models are used, together with parametrical auxiliary beam models, in order to separate quickly non-critical load cases from the critical ones. The critical load cases are investigated by means of more detail FEM strength and aerodynamic models. In this work, only metallic airframes were considered, so no microlevel models were used in calculations. The local-level FEM models responsible for fatigue were not considered, too. Two-dimensional triangle FEs with three nodes topology and linear approximation function were used during calculations. Linear dependency between stresses and strains was assumed during the calculations.

3. Parametrical Auxiliary Aerodynamic Model for FLA

In this work, a set of modifications of the basis variant of the FLA was carried out, related to development and validation of the simplified auxiliary parametrical aerodynamic model (SAPAM).

It should be mentioned that, in the frame of the basis FLA, aerodynamic loading data were considered as an initial data that should be imported from external sources or could be calculated in the frame of a detailed aerodynamic model—that is quite time-consuming for calculations of numerous variants of a concept with various external geometry.

The modified FLA includes SAPAM based on the doublet lattice method (DLM) [15,16]. In frames of SAPAM, lifting surfaces of wing and empennage are modeled by plates, consisting of thin panels, while axisymmetric bodies (e.g., fuselage and engines) are modeled by crossing surfaces, that also consist of thin plates (Figure 4). Height and width of crossing surfaces are related explicitly with height and width of axisymmetric body by means of the parameter $k_{crossing}$. In this work, the standard value of $k_{crossing}$ parameter equal to 0.85 is taken. The typical size of panels of SAPAM is a parameter. The value of the parameter depends on the needed accuracy.

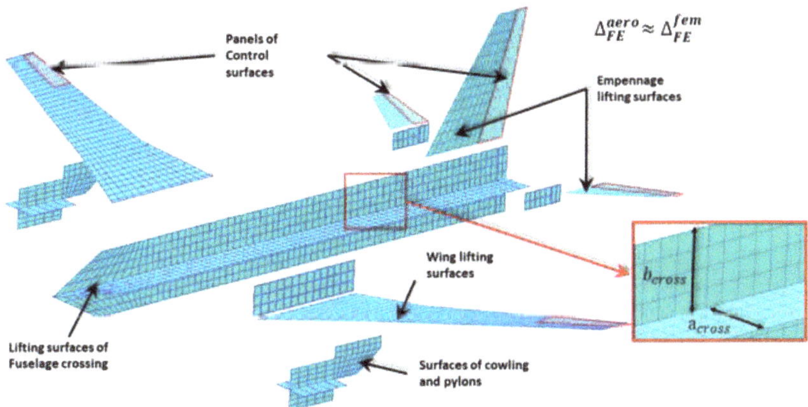

Figure 4. General view of simplified auxiliary parametrical aerodynamic model (SAPAM) of an aircraft.

The process of generation of SAPAM is as follows:

1. Building of surfaces of fuselage crossing,

2. Building of lifting surfaces of wing and empennage,
3. Building of "bridges" for correct fluid motion,
4. Building of surfaces of cowlings and pylons,
5. Setting of control surfaces from range of lifting surfaces.

SAPAM was connected with the FEM model in an automated mode. Point finite elements, which are model masses, were added into the four-level model for the balancing of aerodynamic forces and moments via inertia forces and moments. Point elements model masses were for the following:

1. Structure,
2. Fuel,
3. Payload,
4. Equipment and facility,
5. Power unit,
6. Landing gear.

In this work, SAPAM and structure-nested FEM models for regional aircraft structures have been built in frame of MSC Nastran [17,18].

In Figure 5, the modeling of inertia loads, created by fuel in a lateral wing, is shown. Forces for each point of a flight are applied to nodes of point elements, which are calculated by taking into account overloads, angle velocities, and angle accelerations. These forces are distributed to other nodes of the first level of the nested FE model via RBE3 finite elements. Using this type of finite elements make it possible to distribute loads correctly because they do not disturb the stress–strain state of the nested FE model.

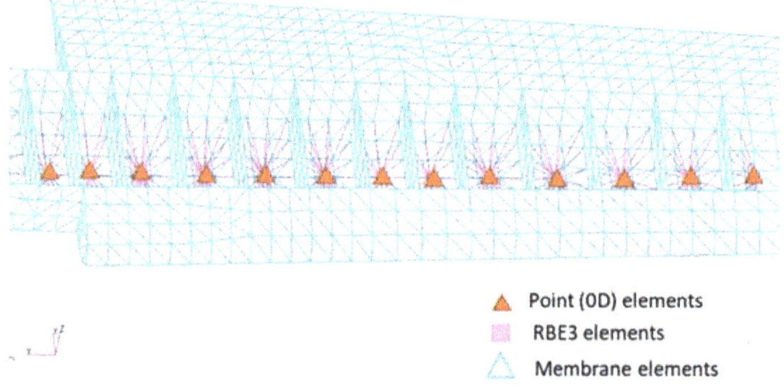

Figure 5. Modeling of inertia loads of a lateral wing.

4. Validation of SAPAM within the FLA

The main scope of the validation was to check the reliability and calculation performance of the modified FLA, namely the SAPAM module being one of the key parts of the algorithm defining its performance and the connection check between the modules of the modified FLA.

Within the validation of the modified FLA, parametrical investigations on a number of alternative variants related to the base variant of hypothetic regional aircraft were carried out. General characteristics of the base variant of the regional aircraft are listed in Table 1. The considered alternative variants differ from the base one by the following active design parameters: wing aspect ratio, wing relative thickness, wing twist, and no/yes wing strut. The wing surface area was kept constant within the investigations.

Table 1. General characteristics of base aircraft configuration.

Parameter	Value
Maximum altitude, m	10,250
Maximum speed, km/h	500
Maximum Mach number	0.445
Maximum dynamic pressure, kgf/m^2	1120
Altitude at start of cruising, m	7200
Cruise speed, km/h	400
Cruise Mach number	0.36
Maximum Take of Weight, kg	15,000
Maximum landing weight, kg	13,775
Payload, kg	2750
Fuel weight (with maximum payload), kg	2750
Minimum flight weight, kg	12,025
Maximum front gravity center, m	7.93
Maximum aft gravity center, m	8.155
Wingspan, m	26.1
Wing square, m^2	58
Fuselage length, m	19.9
Wing Aspect ratio	11.7

The structure model of the base variant of the aircraft is shown in Figure 6. It corresponds to the first level of detailing according to decomposition principles described in Section 2.

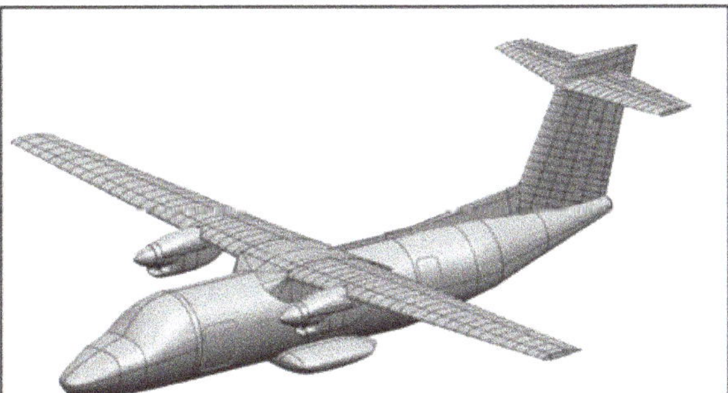

Figure 6. General view of geometry model of investigated aircraft.

The scheme of detailing of the base configuration is illustrated in Figure 7, while the scheme of detailing at the second level (sections) and the scheme of detailing at the third level (details) are shown on Figure 6a,b correspondingly. The scheme of detailing, corresponding to the fourth level (fragment), is illustrated on Figure 8.

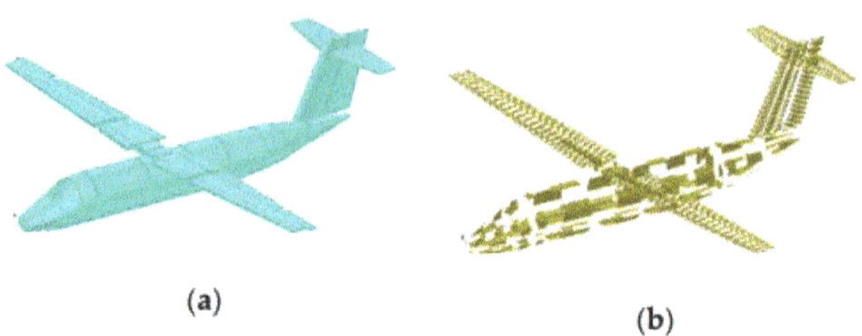

Figure 7. (**a**) Main sections of base configuration; (**b**) panels (details) of base configuration.

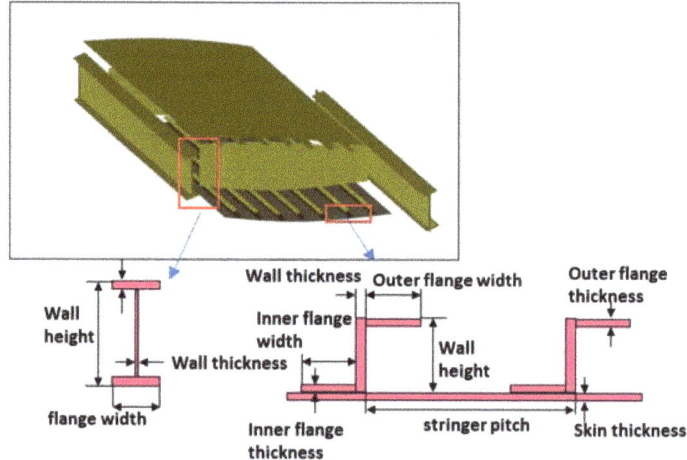

Figure 8. Parameters of the fourth level.

The first step of validation investigations was to check that the simplified model is capable of correct calculation of aerodynamic parameters for variable external geometry. In these investigations, the aspect ratio of the wing was varied at the constant wing area. Each of alternative variants was considered by means of the same automated calculation procedure. The procedure included forming the four-level FEM model, calculation of aerodynamic and inertia loads, definition of strength parameters and reserve factors, and structure optimization with preliminary weight estimation. Each variant of a structure was optimized, taking into account numerous strength constraints, including stress/strain reserves, buckling factors, and displacement constraints. Within the validation, typical aluminum alloy used in current civil aircraft structures was considered as a structure material (Table 2).

Table 2. Aluminum alloy property.

Property	Value
Young modulus, Pa	7.2×10^{10}
Poison ratio	0.3
Shear modulus, Pa	2.76×10^{10}
Allowable tensile stress, Pa	3.6×10^{8}
Allowable compress stress, Pa	4×10^{8}
Allowable shear stress, Pa	2.05×10^{8}
Density, kg/m^3	2750

Four flight loading cases (Table 3) were considered. Such load cases are the most critical for such a kind of wing structure [19]. These load cases were selected in order to compare different variants of the wing's external geometry in the same aerodynamic conditions. Structure parameters obtained during optimization for the considered load cases were used for definition of stiffness characteristics for the further strength analysis procedures.

Table 3. Flight load cases.

Case ID	Dynamic Pressure, kgf/mm^2	Ny	Mach
LC1	1100	2.5	0.45
LC2	324	2.5	0.212
LC3	1100	−1	0.45
LC4	324	−1	0.212

Variations of the wing aspect ratio were performed within the range from $\lambda_{min} = 8$ to $\lambda_{max} = 16$ (Figure 9). For each variant, wing area was a constant, as well as parameters of fuselage and empennage. Wing chords and wing thickness were decreased proportionally with the increasing of the wingspan.

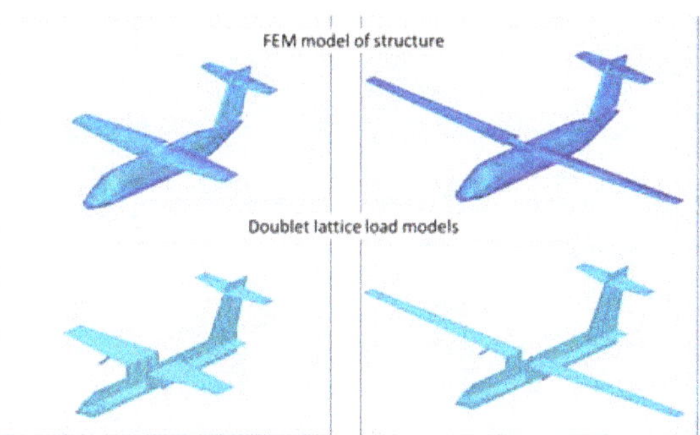

Figure 9. Numerical models of boundary configurations of the aircraft.

In addition, non-conventional structure layouts with strut were considered. The graph (Figure 10) shows dependency of mass of optimized structure of airframe vs. wing aspect ratio for two variants of structure layout with and without strut. The attachment point of wing with strut was situated at 55% of a semi wingspan.

In the graphs, weight of the wing with strut is shown relatively to the weight of the wing without strut. In the frame of validation of the modified FLA, relative wing airfoil thicknesses for the base configuration of regional aircraft were varied (Figure 11). Variations of a wing airfoil thickness were carried out within the range from 70% to 140%, as related to a wing thickness of the base configuration.

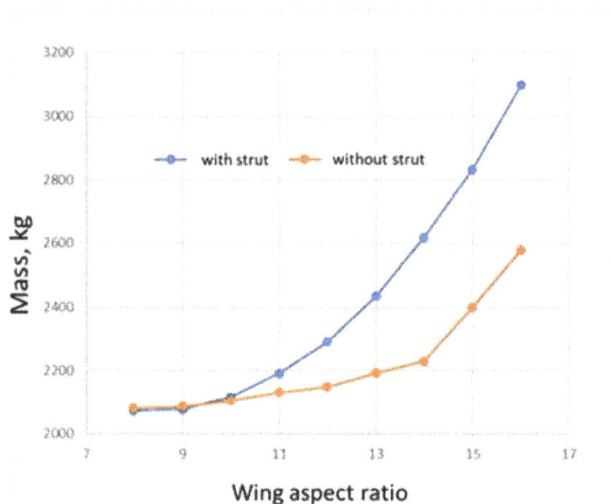

Figure 10. Dependencies of weight of frame of regional aircraft from wing aspect ratio.

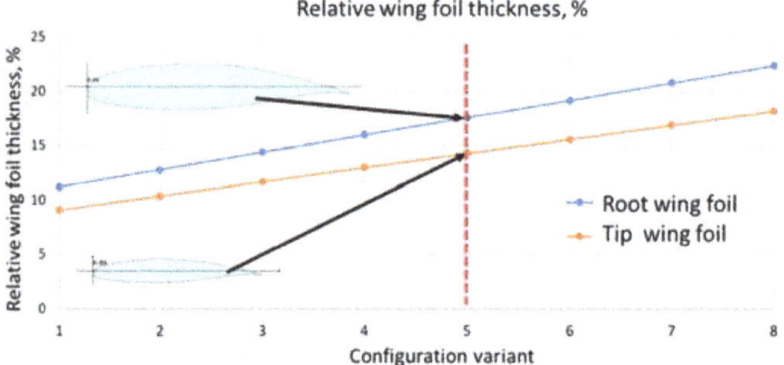

Figure 11. Variation of root and tip foils of the aircraft.

Dependencies of the bending moment applied to the wing for the most critical load case LC2 are illustrated on the graph in Figure 12a. As shown on the graph, influence of a wing stiffness on a wing loading is not essential. The weight of the airframe decreases with the increasing of wing thickness (Figure 12b).

In a similar manner, wing twist was varied (Figure 13). Angle of attack of a tip foil of the wing was varied in a range from $\alpha = -5°$ to $\alpha = 5°$. Angle of attack of a root foil of the wing was constant during this procedure, and foils between the root and the tips were formed by a linear interpolation. Dependencies of bending moment applied to the wing for the most critical load case LC2 are illustrated on the graph (Figure 13a). Increasing of the angle of attack of the tip wing foil causes increasing of bending moments applied to the wing. The mass of the airframe increases with the increasing of the angle of attack of the tip wing foil (Figure 13b).

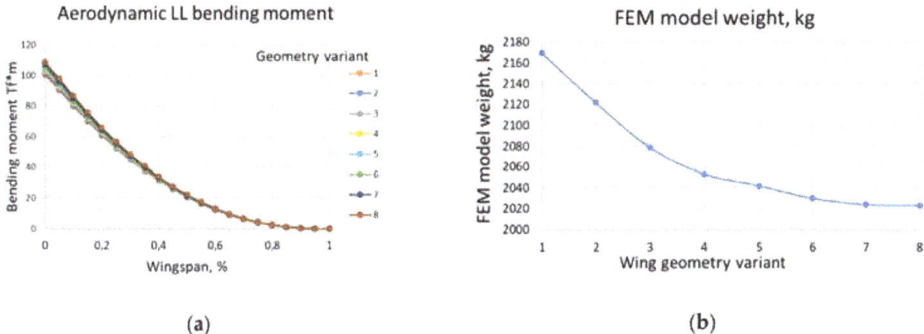

Figure 12. Bending moments (**a**) and mass of structure of frame of the regional aircraft (**b**).

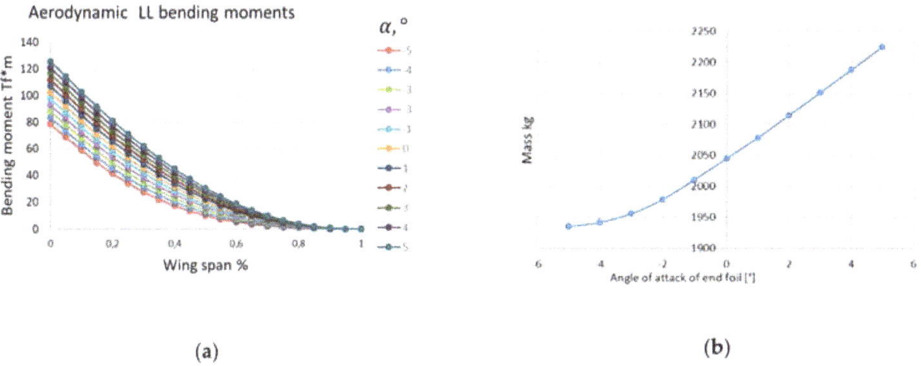

Figure 13. Bending moments (**a**) and masses of the regional aircraft (**b**).

Within the validation of the modified FLA, high performance of the SAPAM module was demonstrated.

5. Validation of the Modified FLA as a Part of Design Procedure

For validation of capabilities of the modified FLA, with a focus on the speed of calculations, numerous strength analyses of alternative variants of wing structures of the hypothetic regional aircraft were carried out.

In this investigation, three variants of wing aspect ratio were considered. In addition to the base variant with $\lambda_1 = 11.7$ (which corresponds to the base variant of the aircraft illustrated in Section 4), two more alternative variants with higher aspect ratio ($\lambda_2 = 15$, $\lambda_3 = 20$) were considered.

For these variants, parametric dependencies of weight of structure of wing from position of joint aggregate on wingspan ($\overline{Z_{st}}$) were obtained. The scheme of parametrization of the model of the regional aircraft with strut for one variant of wing aspect ratio is illustrated on Figure 14.

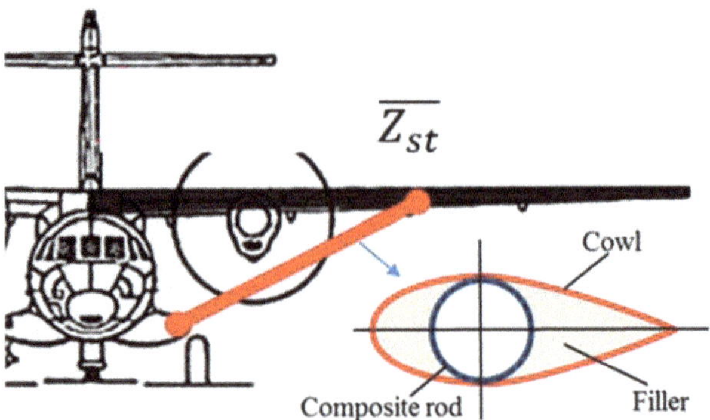

Figure 14. Scheme of parametrization of model of regional aircraft with strut.

As shown on Figure 14, wing strut is a rod with a thin wall (metal or metal-composite). In this work, local and global buckling of the strut were estimated [20,21].

Analysis of load cases, which are critical for wing structure, were carried out by using the modified methodology of load analysis. Discretization of feasible range of flight regimes was carried out in space Height–Mach–Mass (Figure 15). Load estimation was carried out in each point, for symmetric maneuvers (max and min overcharge) and gust influence. The parameters of the structure layout that were obtained were used as start parameters during the analysis of loads for each variant of wing aspect ratio. Discretization extent is also a parameter of fourth-level model (parametrized by weight, Mach, and Height step).

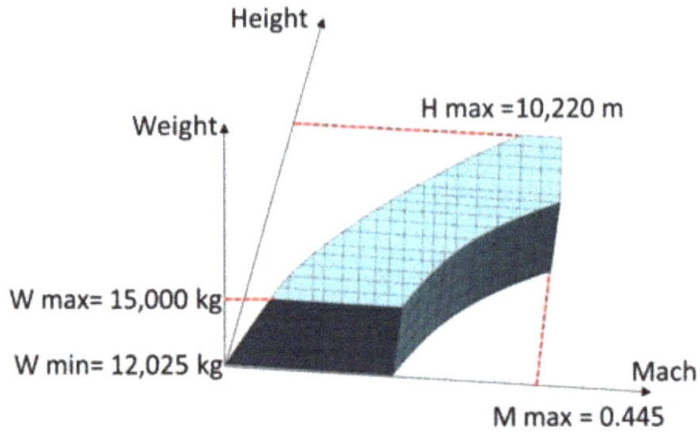

Figure 15. Discretization of feasible range of flight regimes of the regional aircraft.

Thus, about 500 points from the feasible range of flight regimes were investigated (three load cases for each point). Searching for design load cases (DLCs) from a set of load cases (LC) was carried out by the strain energy criteria. The strain energy was calculated by using auxiliary beam model of a wing. In this paper, the number of DLCs is equal to amount of control sections of the beam model (N max). LC should be considered as a DLC, when it has a maximum value of strain energy at least in one beam control section (N—index of control section) (Figure 16). Envelopes of distributions of ultimate bending and torsion moments and shearing forces for most critical load case (gust loads) are illustrated in Figure 17, for each variant of wing aspect ratio of the regional aircraft. Relative weights of the wing, as compared to the base variant of the wing without strut, for three variants of wing aspect ratio, are shown in Table 4. Dependencies of relative weight of the wing with strut on parameter ($\overline{Z_{st}}$) (for three variants of wing aspect ratio) are illustrated in Figure 18.

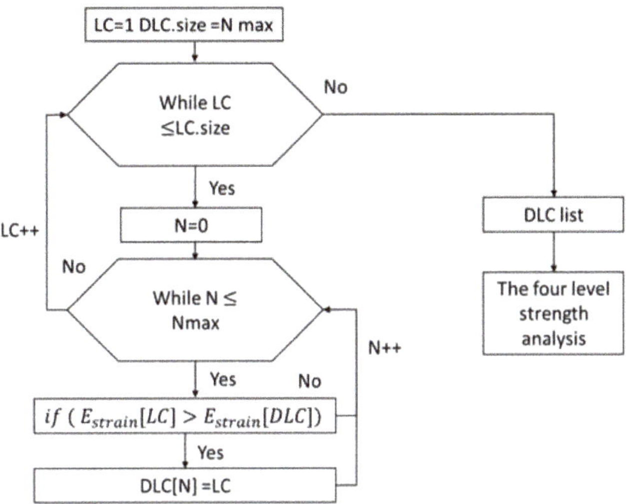

Figure 16. The scheme of searching of the wing-design load cases on the base of simple analytic beam model.

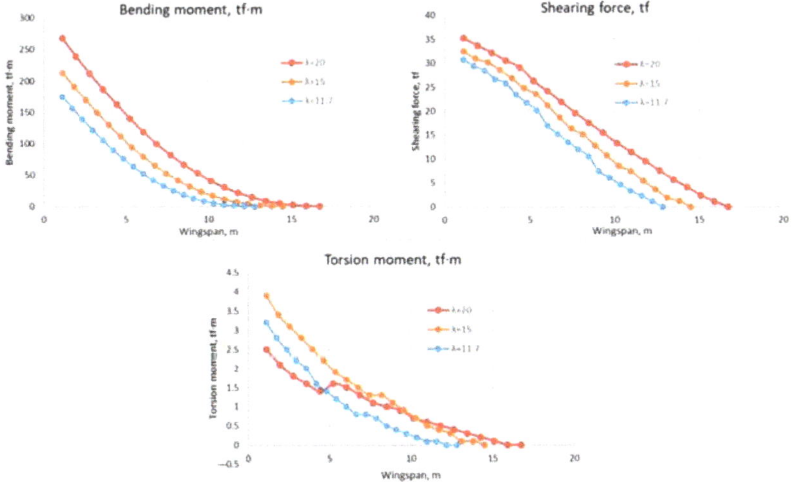

Figure 17. Ultimate bending and torsion moments and shearing forces for three variants of wing aspect ratio.

Table 4. Weight characteristics of wing, with and without strut, compared to base variant without strut.

Variant	Wing Aspect Ratio	m/m_0 without Strut, %	m/m_0 c with Strut, %
Base	11.7	100	87.7
1	15	134	93.3
2	20	183	116.5

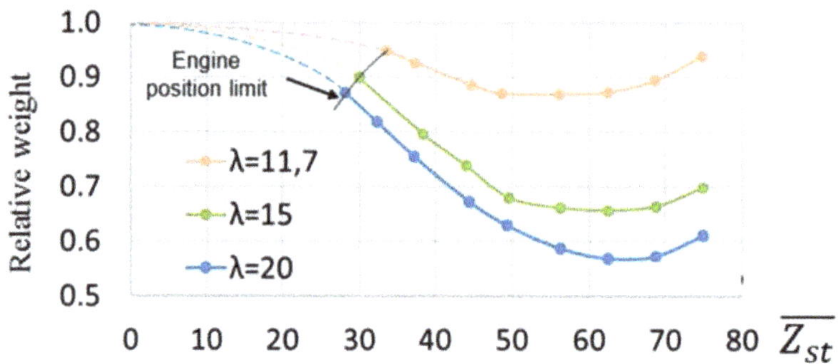

Figure 18. Dependencies of relative weight wing with strut from parameter $\overline{Z_{st}}$.

The graph (Figure 18) shows that rational value of parameter $\overline{Z_{st}}$ is equal to 50–65% of semi wingspan. Two variants of wing are rational by weight—base variant with strut (87.7% from base variant without strut) and modified variant ($\lambda_1 = 15$) with strut (93.3% from base variant without strut).

6. Time Efficiency of the FLA of Strength Analysis

The simulations were carried out on the workstation controlled by a central processor Intel I7 7700 K. An analysis of labor intensiveness of strength analysis of regional aircraft structure for the base variant configuration is shown in Table 4, for the conventional approach and the FLA.

A report for the computational capability for one iteration of strength analysis for one configuration of regional aircraft for different discretization (different size of finite element) of FEM model is illustrated in Table 5.

Table 5. Analysis of labor intensiveness of strength analysis of regional aircraft airframe structure.

Operations	The Base FLA	The Modified FLA	Labor Saving (Time)
Initial data definition, work hours	80	4	≥20
Model validation, work hours	16	2	≥8
Analysis, work hours	112	4	≥28
Post-processing, work hours	32	2	≥16
Total, work hours	240	12	≥20

In the frame of parametric investigations, the meshes of FEM models were chosen by the condition of the same allowable maximal error of calculation strain energy. Table 6 shows relations between strain energy and FE mesh size for the base variant of the regional aircraft. The error equals ~5% was chosen as the limit for all variants during parametrical investigations.

It should be noted that the parametric investigations in this work were carried out in parallel mode (each geometry configuration simulated in a separate stream).

One of the main advantages of the FLA for its implementation into the preliminary stage of aircraft design is that the labor-saving can be provided not only by means of calculation-time reduction, but also by selection of sparser mesh for initial iterations in the search for rational parameters, and more coarse mesh for getting a more accurate solution on later iterations within the chosen range of variants of aircraft concepts.

Table 6. Computational capabilities of strength analysis of a regional aircraft.

Finite Element Size, m	≤0.5	≤0.4	≤0.3	≤0.2	≤0.1
Error, strain energy of FEM model, %	11.6	7.6	5.3	3.9	2.1
Amount of nodes	9000	16,000	28,000	40,000	112,000
Memory, Gb	0.75	1.1	2.2	4	9.6
Time of FEM model generation, s	0.07	0.1	0.6	1.1	4.3
Nastran simulation time, s	16	23	48	114	480
Reading the results of simulations, s	0.3	0.61	2.2	4.3	9.63
Buckling analysis, s	5	5	5	5	5
Calculation of weight and strength reserves, s	0.13	0.5	1.2	6.3	12.4
Total time of iteration, s	21.5	29.21	57	130.7	511.33

7. Conclusions

The FLA of complex strength analysis of airframe based on the three decomposition principles, namely structural decomposition, strength task decomposition, and decomposition of alternative variant of load cases, was validated successfully on a wing structure of the hypothetic regional aircraft. The validation was focused on the developed module for fast automated calculation of aerodynamic parameters of load cases. During the calculations, high efficiency of the algorithm, both in searching rational balance between accuracy and calculation time and in choosing critical load cases for alternative concepts, was confirmed.

Parametric investigations carried out in this work have proven that the usage of the algorithm can provide no less than a 20-times decrease of calculation time, required for strength analysis of wings of regional aircrafts, as compared to conventional methods of strength analysis.

Within the validation of the modified FLA, the parametrical dependencies of weight of wing on geometrical parameters of strut-braced high-aspect-ratio wing of a hypothetic regional aircraft were obtained. The rational range of points for connection of strut and wing, corresponding to minimal weight characteristics of wing was found. The positive validation results can become a background for further development of the FLA for preliminary design procedure.

Author Contributions: D.V.V.: conceived and designed the analysis, contribute analysis tools, wrote and edit the paper, carry out numerical simulations. A.N.S.: developed the methodology and the algorithms, wrote and edit the paper. Y.S.M.: carry out numerical simulations, collect the initial data. M.D.L.: carry out numerical simulations, collect the initial data. All authors have read and agreed to the published version of the manuscript.

Funding: This research was funded within the frames of Subsidy Agreement 14.628.21.0009, Project identification number RFMEFI62818 × 0009 (Ministry of Science and High Education of the Russian Federation).

Institutional Review Board Statement: Not applicable.

Informed Consent Statement: Not applicable.

Data Availability Statement: Not applicable.

Conflicts of Interest: The authors declare no conflict of interest.

References

1. Elham, A.; La Rocca, G.; van Tooren, M.J.L. Development and implementation of an advanced, design-sensitive method for wing weight estimation. *Aerosp. Sci. Technol.* **2013**, *29*, 100–113. [CrossRef]
2. Schuhmacher, G.; Murra, I.; Wang, L.; Laxander, A.; O'Leary, O.; Herold, M. Multidisciplinary Design Optimization of A Regional Aircraft Wing Box. In Proceedings of the 9th AIAA/ISSMO Symposium on Multidisciplinary Analysis and Optimization, Atlanta, GA, USA, 6 September 2002.
3. Werner-Westphal, C.; Heinze, W.; Horst, P. Structural sizing for an unconventional, environment-friendly aircraft configuration within integrated conceptual design. *Aerosp. Sci. Technol.* **2008**, *12*, 184–194. [CrossRef]
4. Cavagna, L.; Ricci, S.; Travaglini, L. NeoCASS: An integrated tool for structural sizing, aeroelastic analysis and MDO at conceptual design level. *Prog. Aerosp. Sci.* **2011**, *47*, 621–635. [CrossRef]
5. Frediani, A.; Oliviero, F. Conceptual Design of an Innovative Large Prandtl Plane Freighter. In Proceedings of the 10th European Workshop on Aircraft Design Education, Naples, Italy, 25 May 2011.
6. Hürlimann, F. Mass Estimation of Transport Aircraft Wingbox Structures with a CAD/CAE-Based Multidisciplinary Process. Ph.D. Thesis, ETH Zürich, Zürich, Switherland, 2010.
7. Röhl, P.; Dimitri, N.; Daniel, P. A multilevel decomposition procedure for the preliminary wing design of a high-speed civil transport aircraft. In Proceedings of the 1st Industry/Academy Symposium on Research for Future Supersonic and Hypersonic Vehicles, Greensboro, NC, USA, 4–6 December 1994.
8. La Rocca, G.; van Tooren, M.J.L. Knowledge-Based Engineering Approach to Support Multidisciplinary Design and Optimization. *J. Aircr.* **2009**, *46*, 6. [CrossRef]
9. Dorbath, F. *A Flexible Wing Modeling and Physical Mass Estimation System for Early Aircraft Design Stages*; DLR: Müllheim, Germany, 2014.
10. Dorbath, F.; Gaida, U. Large Civil Jet Transport (MTOM > 40 t)—Statistical Mass Estimation. In *Luftfahrttechnisches Handbuch (LTH) Band Massenanalyse*; Luftfahrttechnisches: Handbuch, MA, USA, 2013.
11. FP7 ALaSCA Project. Available online: http://cordis.europa.eu/result/rcn/149775_en.html (accessed on 12 March 2021).
12. FP7 PoLaRBEAR Project. Available online: https://cordis.europa.eu/result/rcn/197045/en (accessed on 12 March 2021).
13. Shanygin, A.; Fomin, V.; Zamula, G. 4th-level approach for strength and weight analyses of composite airframe structures. In Proceedings of the 27th Congress of the International Council of the Aeronautical Sciences, Nice, France, 19–24 September 2010.
14. Dubovikov, E.A. Novel approach and algorithm for searching rational nonconventional airframe concepts of new generation aircrafts. In Proceedings of the 28th Congress of the International Council of the Aeronautical Sciences (ICAS 2012), Brisbane, Australia, 23–28 September 2012.
15. Albano, E.; Rodden, W.P. A Doublet Lattice Method for Calculating Lift Distributions on Oscillating Surfaces in Subsonic Flows. *AIAA J.* **1969**, *7*, 279–285. [CrossRef]
16. Rodden, W.P.; Taylor, P.F.; McIntosh, S.C. Further Refinement of the Subsonic Doublet-Lattice Method. *AIAA J.* **1998**, *35*, 720–727. [CrossRef]
17. MSC.Nastran 2017 Aeroelastic Analysis User's Guide. Available online: http//mscsoftware.com (accessed on 12 March 2021).
18. MSC.Nastran 2017 Installation and Operations Guide. Available online: http//mscsoftware.com (accessed on 12 March 2021).
19. Available online: https://www.ecfr.gov/cgi-bin/text-idx?node=14:1.0.1.3.11#sp14.1.25.c (accessed on 12 March 2021).
20. Fomin, V.; Mareskin, I.; Sidorova, I. Method of buckling analysis of hybrid rods at preliminary designing of wing caisson with truss structural layout. In Proceedings of the 7th EASN International Conference on Innovation in European Aeronautics Research, Warszawa, Poland, 26–28 September 2017.
21. Mareskin, I. Multidisciplinary optimization of pro-composite structural layouts of high loaded aircraft constructions. In Proceedings of the 30th Congress of the International Council of the Aeronautical Sciences, Daejeon, Korea, 25–30 September 2016.

Article

Aircraft Requirements for Sustainable Regional Aviation

Dominik Eisenhut [1,*,†], Nicolas Moebs [1,†], Evert Windels [2], Dominique Bergmann [1], Ingmar Geiß [1], Ricardo Reis [3] and Andreas Strohmayer [1]

1 Institute of Aircraft Design, University of Stuttgart, 70569 Stuttgart, Germany; moebs@ifb.uni-stuttgart.de (N.M.); bergmann@ifb.uni-stuttgart.de (D.B.); geiss@ifb.uni-stuttgart.de (I.G.); strohmayer@ifb.uni-stuttgart.de (A.S.)
2 Aircraft Development and Systems Engineering (ADSE) BV, 2132 LR Hoofddorp, The Netherlands; evert.windels@adse.eu
3 Embraer Research and Technology Europe—Airholding S.A., 2615–315 Alverca do Ribatejo, Portugal; rjreis@embraer.fr
* Correspondence: eisenhut@ifb.uni-stuttgart.de
† These authors contributed equally to this work.

Abstract: Recently, the new Green Deal policy initiative was presented by the European Union. The EU aims to achieve a sustainable future and be the first climate-neutral continent by 2050. It targets all of the continent's industries, meaning aviation must contribute to these changes as well. By employing a systems engineering approach, this high-level task can be split into different levels to get from the vision to the relevant system or product itself. Part of this iterative process involves the aircraft requirements, which make the goals more achievable on the system level and allow validation of whether the designed systems fulfill these requirements. Within this work, the top-level aircraft requirements (TLARs) for a hybrid-electric regional aircraft for up to 50 passengers are presented. Apart from performance requirements, other requirements, like environmental ones, are also included. To check whether these requirements are fulfilled, different reference missions were defined which challenge various extremes within the requirements. Furthermore, figures of merit are established, providing a way of validating and comparing different aircraft designs. The modular structure of these aircraft designs ensures the possibility of evaluating different architectures and adapting these figures if necessary. Moreover, different criteria can be accounted for, or their calculation methods or weighting can be changed.

Keywords: hybrid-electric propulsion; regional air travel; alternate airports; top-level aircraft requirements; figures of merit; aircraft design

Citation: Eisenhut, D.; Moebs, N.; Windels, E.; Bergmann, D.; Geiß, I.; Reis, R.; Strohmayer, A. Aircraft Requirements for Sustainable Regional Aviation. *Aerospace* **2021**, *8*, 61. https://doi.org/10.3390/aerospace8030061

Academic Editor: Ola Isaksson

Received: 11 December 2020
Accepted: 18 February 2021
Published: 26 February 2021

1. Introduction

In the past, aviation has been a main driver for economic wealth by not only connecting millions of people, but also by providing a fast option for trade between different continents. Furthermore, it also enables tourism, which allows people to experience different countries and cultures to broaden their minds. This generates economic wealth for the destination region, which is especially valuable for developing countries with a remote tourism market [1]. Apart from the many benefits created by aviation, there are also downsides, primarily concerning environmental aspects. According to different studies, aviation is currently responsible for 1–2% of human-made CO_2 emissions [1,2]. Aviation's impact on the environment is not only limited to CO_2, but also includes other forms of emissions like NO_X or noise. While emitted greenhouse gases impact the climate and contribute to climate change [3], noise is expected to influence the health and general well-being of residents in the vicinity of airports [4]. Many efforts are being pursued by governments and the aviation sector itself to reduce negative impacts, with an ultimate goal of achieving sustainable aviation.

In 2011, the European Commission presented the EU Aviation Sector's vision for the future of aviation, called Flightpath 2050 [5]. This report not only set goals for emission reductions, but also built a framework for the sector's market direction. Therefore, the vision pursued the aims of streamlining procedures and improving overall passenger experience. This included facilitating fast boarding, arriving at destinations on time regardless of weather conditions, and enabling 90% of European travelers to reach their destination from door to door within 4 h. Apart from increasing the convenience of air travel for passengers, arriving on time and improving air traffic management also contribute to reducing emissions by causing fewer deviations and less otherwise "wasted" holding time and fuel.

Roughly eight years later, at the end of 2019, the European Union (EU) introduced the next step towards a sustainable future—the European Green Deal [6], which aimed to ensure that Europe would become the first "climate-neutral" continent on Earth. Similar to Flightpath 2050, the aims of this Green Deal included setting emission targets and enforcing a circular economy. These actions were not specifically aimed at aviation, but instead aimed to cover all of the sectors impacting the climate.

In addition to governmental policies, the aviation industry is also striving to become more sustainable in its own right. With the Carbon Offsetting and Reduction Scheme for International Aviation (CORSIA) initiative, the International Civil Aviation Organization (ICAO), as a subsidiary organization of the United Nations (UN), is also setting emission targets for the aviation industry. From 2021 onwards, carbon-neutral growth should be achieved, as well as an increase in energy efficiency by 2% per year until 2050 [2].

In addition, the UN in 2015 defined an agenda including 17 sustainable development goals and 169 targets to be achieved by 2030: "They are integrated and indivisible and balance the three dimensions of sustainable development: the economic, social and environmental" [7]. Moreover, the UN stated that a global effort is required to end poverty, respect human rights, create equality, and other essential goals, all while protecting the environment where "humanity lives in harmony with nature and [where] wildlife and other living species are protected" [7].

According to the Air Transport Action Group (ATAG) and their 2020 Aviation Benefits report [1], aviation plays a major role in supporting that agenda. To name just a few ways in which the aviation plays a main role: Before the COVID-19 pandemic, the aviation sector supported nearly 88 million jobs across the world, including direct, indirect, induced, and tourism related jobs. In addition, aviation supported 4.1% of the global gross domestic product [1].

Aviation has many benefits compared to other transport modes. The most important involves travel times, which are usually the lowest of all available modes of travel due to the straight flight paths that are possible without deviations being needed. Even in regional markets, this often holds true, despite the short distances that need to be covered to travel across these areas. This benefit is further increased if air travel is used to get across natural barriers like mountains, lakes, or the seas between islands. Air travel allows for connecting smaller cities to the already existing transport networks with lower infrastructure costs, thereby reducing development and maintenance spending. The only requirements to cover an airline route are airports at the departure and destination locations, each with sufficient infrastructure. No roads or rails are necessary in between these airports to facilitate air travel. Furthermore, more remote regions can be reached by air travel where a fast connection by other means of transport is not economically feasible or even where the route itself is not economically feasible at all.

According to an ATR market forecast, 58% of the worldwide regional air routes network was created between 2003 and 2018 [8]. To support this growth and add more remote regions to the global transport system in a sustainable way, the vision of a regional 50-passenger aircraft will allow for connecting smaller routes or simply destinations with reduced passenger and/or cargo volumes. To fulfill this purpose, it is expected that the size of the aircraft is reasonably defined. In addition, a smaller plane allows for more connections every day, if the demand is high enough, and reduces the risk of unprofitable flights due to low passenger volume at the same time. This will lead to more travel options for European citizens and a denser transport network interconnecting Europe. However, the higher number of connections must be cost efficient, and the aircraft must have competitive direct operating costs (DOCs) in comparison to other transport modes. In regional aviation, the impact of fuel consumption is amplified. As a result of the remoteness of most regional airports, the fuel transportation costs are increased. Compared to hub airports, the average fuel price at regional airports is higher by one third worldwide, and in some parts of Europe, even by a factor of two or more [8]. In addition, the transport further increases overall emissions.

Following the vision of the European Commission within Horizon 2020, an entry into service (EIS) for 2035/2040 for a hybrid-electric 50-seat regional aircraft seems challenging but feasible.

Electrifying the powertrain offers many benefits over conventional architectures. Electric motors provide power without any local emissions. The only emissions might result from producing the electric energy required to power the motors. Furthermore, electric motors scale almost linearly with regard to power and mass. This might enable the use of novel propulsion concepts like, for example, distributed (electric) propulsion, where efficiency can be increased, synergistically exploiting the interaction between the wing and propulsors.

Assuming green electric power, a fully electric aircraft would offer the lowest emissions during operations. For now, the range of such an aircraft would be limited due to the low specific energy of current battery technology. To tackle this flaw, a hybrid-electric concept combines the advantages of both worlds, i.e., the range provided by conventional fuel, and the emission reduction achieved by an electrified powertrain with higher efficiency. In addition, this concept enables new operation strategies where emissions can be managed based on different factors. In flight phases where the most harmful pollutions occur, using only the battery will allow for lower emissions of greenhouse gases and noise. Different hybrid-electric architectures offer further options for the design and operation of the aircraft. A parallel architecture allows for a power boost in high load scenarios like take-off or go-around. A serial architecture can increase redundancy or offers to replace a gas turbine with a battery pack as long as certification and technology requirements are met.

Apart from architectures and operation strategies related to local emissions, an additional degree of freedom is created by different energy management strategies. Having multiple energy/power sources allows for controlling the ageing of different powertrain components. With operating hours of a gas turbine building up, the battery could provide power originally produced by the turbine and vice versa. This will increase wear on the boosting component, but might help the operator to optimize the maintenance schedule.

In contrast to these benefits, the hybrid-electric system will add additional components to the aircraft which increase mass and complexity. The mass penalty has to be outweighed by a higher overall efficiency of the aircraft and other benefits. One is the required emission-free taxiing within Flightpath 2050, which can be fulfilled easily by the hybrid-electric aircraft (HEA). Furthermore, the added complexity will require a higher effort for certification. The HEA must achieve similar safety levels to a conventional aircraft. The different operating and energy management strategies might lead to additional redundancy. However, analyzing the overall architecture and different failure scenarios is challenging.

In this context, the project FUTPRINT50, which consists of 14 international partners, was formed. To address all these different challenges and to grasp the bigger picture, the project partners adapted a systems engineering approach [9] with a dedicated mission statement, which will be explained in the following paragraph.

In Figure 1, the V-model of systems engineering is shown. It advises to break the overall design process into different levels and, by the iterative process, each level is revised whether expectations and requirements are fulfilled or not.

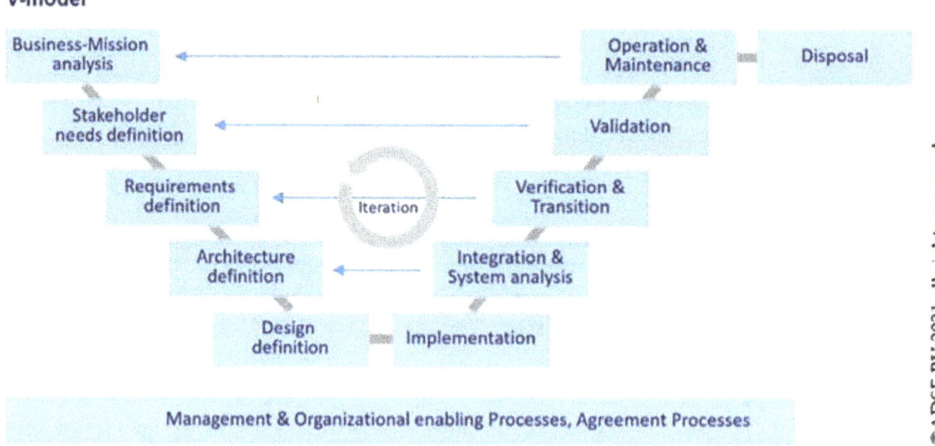

Figure 1. V-model of systems engineering.

The left part of the V starts with the need definition and works through to the definition of the design. The bottom side of the V-model contains the actual design, including systems and sub-systems design. The right side starts with the implementation and integration. After this, the design will be tested and validated if all goals are achieved. Finally, actual operation will commence.

The first conceptual design is created by defining and implementing the systems at a higher level, i.e., general layout, propulsion type, disruptive technologies, etc. The starting point is defined by using existing data and a comparative aircraft analysis. The results create the basis for the top-level aircraft requirements (TLARs) and initial architecture and design definition. Consequently, the first concept is verified and validated, and a first concept of operations can be determined.

During the conceptual design phase, iterations are made to identify if the initial requirements are met, as is shown in the figure. At the end of the design, the entire conceptual product is evaluated to identify if it still satisfies the initial expectations. Requirements can be re-evaluated, and stakeholders can refine their needs, as one can learn from the first concept. This iteration is important when developing new disruptive designs and technologies, as one can never create the perfect starting point during the business and stakeholder analysis. Each overall iteration brings the design to a more detailed level, eventually converging to one final design, aligned with the stakeholders.

For the FutPrInt50 project, the following stakeholders have been identified: EU citizen, authorities, operator, airport, air traffic management, supplier of energy, and the passenger. The stakeholders and their needs define the goals and requirements to make the project a success. Stakeholder interaction and analysis is required to define if the product or system to be designed fits in the overall picture. To better capture this overall picture, the product can be seen as a system within a system which is influenced by its surroundings (Figure 2a).

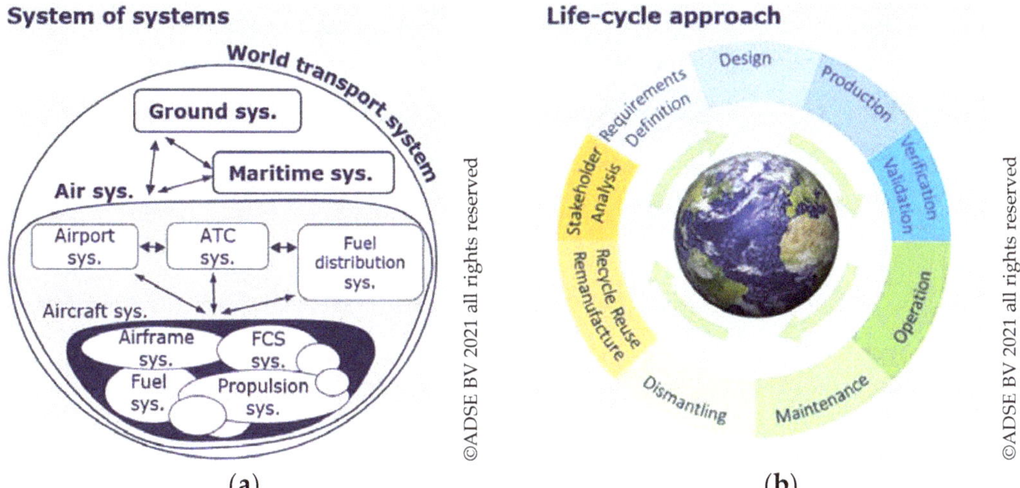

Figure 2. (**a**) System of systems and (**b**) life-cycle approach.

Taking this holistic view is important to grasp all aspects of the "larger" system and the stakeholders involved. The system of systems is one part of identifying the "larger" system, while another is the life-cycle approach shown in Figure 2b. Especially for sustainability, this shows that one should not only look at the operational period, but also all other sections of the life cycle to build a "green" solution. The needs and the requirements definition shall drive the design process to achieve a sustainable product in its life cycle, whilst still satisfying the stakeholders' needs within the systems of systems. The design decisions will drive emissions during all parts within the life cycle, which includes production, the operational phase, as well as end of life and recycling.

After formulating the needs and goals for the system, these could be summarized into a mission statement, used to define the system that is to be designed. The following is the current mission statement for a regional HEA:

"To develop a synergetic aircraft design for a commercial regional hybrid-electric aircraft up to 50 seats for entry into service by 2035/2040, to identify key enabling technologies and a roadmap for regulatory aspects. The clean sheet aircraft design shall help accelerate and integrate hybrid electric aircraft and technologies to achieve a sustainable competitive aviation growth, as well as acting as a disruptor to regulators, air traffic management and energy suppliers."

The clean sheet aircraft design shall:

- have class-leading emissions and noise,
- include technologies that ensure (operational) safety,
- offer a competitive operational cost,
- offer operational improvements during exploitation compared to current regional aircraft,
- not enforce expensive changes to the current infrastructure.

This statement not only describes the system, but also highlights the benefits for the stakeholders. Within the V-model, it represents the first two levels, the business mission analysis and the stakeholder needs definition. These two provide the framework for the requirements definition represented by the TLARs.

The TLARs help to translate the abstract needs of the stakeholders to more manageable requirements on a system level. This includes specifications that should be precisely met, like the range, for example, or ones like emissions, where overachieving the goal is desirable.

The V-model requires us to validate if the requirements are fulfilled by the design. Therefore, figures of merit have been defined in the process and are shown in Section 4. They not only check if the requirements are fulfilled, but also add the possibility to compare different design approaches and to identify superior concepts. This is especially necessary for complex interwoven requirements like emissions and DOCs. On the one hand, capital cost as part of the DOCs might increase due to the higher development and manufacturing cost of a more environmentally friendly airplane. On the other hand, when emissions are taxed, this might increase the DOCs for aircraft with higher emissions.

2. Top-Level Aircraft Requirements

All defined TLARs are grouped in different categories, relating to environment, market, operations, performance, and regulations [9]. This helps to evaluate the design with respect to different aspects. All requirements within these categories and their reasoning are explained in the following sections. It is important to note that these requirements are the result of a comparative aircraft analysis, evaluated for our hybrid-electric configuration. The purpose is to capture the requirements for an aircraft that has comparable or better characteristics than the current state-of-the-art, considering potential limitations regarding a realistic timeline for technology readiness.

2.1. Environment

The environmental requirements are mainly derived from Flightpath 2050, and all adopted requirements are shown in Table 1. Defined emission reductions for CO_2, NO_X, and noise are directly implemented into our TLARs.

Table 1. Top-level aircraft requirements (TLARs) with regard to environmental aspects.

TLAR (Environment)	Value
Reduction of CO_2 emissions	$\geq$75% vs. ATR-42
Reduction of NO_X emissions	$\geq$90% vs. ATR-42
Reduction of Noise emissions	$\geq$65% vs. ATR-42
Emissions during ground operations	No emissions of CO_2 and NO_X
Materials used in the design	Sustainable end of life solution
Use of alternative propellants	Yes

In addition to the emissions reduction during flight, all ground operations should be performed emission-free, including taxiing. In general, there are two options available: Either an emission-free system is integrated into the aircraft, or the aircraft is towed by an electric truck on the ground. The first option is easily fulfilled by the HEA. In comparison, a conventional aircraft would require a system solely for the purpose of emission-free taxiing. During all other mission sections, this system is dead mass that penalizes the overall aircraft performance. The second option, an electric tow truck, requires infrastructure on the ground at every serviced airport. Especially for smaller regional airports, this might limit the profitability of either the aerodrome itself or the desired connection.

Furthermore, the goal is not only to reduce overall emissions during the operational phase, but also during the whole aircraft life cycle. Therefore, the materials used should allow a sustainable solution like recycling or reusing at the end of life. This connects to the life-cycle approach described earlier.

The final criterion is the usage of alternative propellants. This is not only limited to drop-in sustainable aviation fuels (SAF) that replaces conventional jet fuel, but also specifically allows for hydrogen.

These defined environmental requirements would not only incorporate the Flightpath 2050 goals, it would also account for further political initiatives, like the European Green Deal [6] and its aim of being climate neutral in 2050. This can be achieved by SAF or green hydrogen in combination with a life-cycle approach that also considers recyclability in a circular economy. In this case, there is only a minor difference for aviation between the Flightpath 2050 goals and the Green Deal.

2.2. Market

Table 2 shows all aspects related to market requirements, including operator demand and passenger comfort. Aiming at the regional air traffic market, the number of passengers is set to 50.

Table 2. Top-level aircraft requirements with regard to market.

TLAR (Market)	Value
Number of passengers	≤50
Cargo capacity	≥500 kg
Luggage bins	≥0.06 m^3/passenger
In-flight entertainment	Seamless connectivity
Cabin altitude	≤2000 m (6560 ft)
Cabin ventilation	≥0.25 kg/min fresh air per passenger
Cabin temperature	23 °C
Cabin humidity	10%
Lavatory	≥1
Galley	≥1
Direct operating costs	Competitive with ground transport
Dispatch reliability	≥98%

As the world is coming closer together, not only fast connections for passengers are required, but also for cargo. Therefore, the cargo capacity is set to at least 500 kg if maximum passengers are on board. In addition, it should be possible to quickly convert the plane from passenger to cargo and vice versa. Furthermore, a combi-version would allow for trading some passengers for additional cargo. This allows flexibility for the operators and gives them the ability to react to different needs in different markets/regions.

In regional aviation, mostly carry-on luggage is used. This also helps with the turn-around time defined in the previous section. To allow for enough space in the cabin, the luggage bins are sized that each passenger has 60 L available. This is comparable to the current Embraer E-Jet E1 family, which is larger than the bin size of current turboprop aircraft, for improved traveler experience. Furthermore, the volume of regular hand luggage (55 cm × 40 cm × 20 cm) is only 44 L.

No conventional in-flight entertainment will be integrated, but seamless connectivity like Wi-Fi should be provided. The flight duration is rather short, and connection to the internet is expected to be more essential, especially for business trips where the passenger is able to work during the flight. To share flight information, an app could be provided for example.

Cabin pressure is set to 2000 m (6560 ft), which complies with certification specifications (CS) that require a cabin pressure altitude of less than 2438 m (8000 ft) (CS 25.841 (a)) [10]. For the short flight time of the regional aircraft, a lower altitude is not needed. New airplanes for long routes like the Airbus A350 and the Boeing 787 are already using a cabin altitude of 6000 ft, so the trend and market requirements for regional aviation have to be closely monitored.

To comply with CS-25, cabin ventilation must be designed to provide every passenger with at least 0.25 kg of fresh air per minute. The exact required ventilation capacity is closely connected to the temperature and humidity requirements. Every passenger on board generates heat and moisture. To keep the temperature at 23 °C and humidity at 10%, the air inside the cabin must be reconditioned, which in the end defines the exchange rate for cabin ventilation.

For passenger comfort, at least one lavatory has to be installed. Despite the trend going to serve less warm food on-board, at least one galley is foreseen. This ensures the possibility to offer hot beverages like coffee or tea during flight.

As already described, the goal of the aircraft is not to replace other transport modes. Therefore, the DOCs should be competitive with public ground transport. A metric for comparison might be the expected ticket price for each transport. Furthermore, a comparison to car travel might be of interest, even more so as the majority of cars is expected to be electric by 2040. Another important aspect which drives the DOCs of the airplane is the dispatch reliability, which should exceed 98%. This will increase utilization of the airplane for the operator and thus the potential profit.

2.3. Operations

To ensure operational flexibility for an aircraft capable of servicing many airports and achieving high utilization, operational aspects have to be considered as top-level requirements (see Table 3).

Table 3. Top-level aircraft requirements with regard to operational aspects.

TLAR (Operations)	Value
Wingspan	<36 m
Weather operations	All weather
Turn-around time	$\leq$25 min

ICAO Annex 14 [11] defines categories by which an aerodrome is rated. The number is related to the reference field length and the letter is based on the wingspan of the airplane. For the HEA, a category 2C aerodrome is required by the TLAR definitions. The "lower" the category, the more airports are available to be serviced by the aircraft. The letter C refers to the maximum wingspan of less than 36 m. This is similar to the ATR 42, which has a wingspan of just more than 24 m. A lower airport category seems to be unfeasible, as the limit for category B is 24 m. Future aircraft will likely have higher aspect ratios than the ATR 42, with an expected higher maximum take-off mass (MTOM) for an HEA, and also the wing area will be increased to end up with a similar wing loading. In combination overall, an increase in wing span is expected.

Another aspect of particular interest with respect to utilization is the weather conditions the aircraft is capable of operating in. As the goal is to design an aircraft for commercial operations, the general dispatchability should be high. Therefore, obviously the aircraft must be capable of operating in all weather conditions including icing, for example.

To further increase productivity, a turn-around time of no more than 25 min should be obtained. This is comparable to current regional aircraft like the EMB-145 or the De Havilland DHC-8-300 [12,13].

2.4. Performance

In Table 4, the TLARs with regard to performance are given. Because range and reserve policy are crucial aspects, they are analyzed separately in more detail.

The design cruise speed is set to 450–550 km/h, which is comparable to current turboprops. Reducing this speed would harm productivity of the aircraft by affecting the number of dispatches per day. Increasing it will raise the number of dispatches, but more excess thrust or power is required to accelerate the aircraft. Therefore, the aircraft mass might increase which limits the benefits gained. The design payload of 5300 kg is calculated by the number of passengers multiplied with their mass including luggage, which is estimated to be 106 kg. The maximum payload of 5800 kg is defined by the design payload plus 500 kg of cargo mass.

Table 4. Top-level aircraft requirements with regard to performance.

TLAR (Performance)	Value
Design cruise speed	450–550 km/h (Mach 0.40–0.48)
Design payload	5300 kg
Maximum payload	5800 kg
Design range (design cruise speed, design payload)	400 km + reserve
Maximum range (design cruise speed, design payload)	800 km + reserve
Reserve fuel policy	185 km + 30 min holding
Range from hot and high airports (design payload, international standard atmosphere (ISA) +28, 5400 ft)	450 km + reserve
Range from cold airports (design payload, ISA −30)	450 km + reserve
Take-off field length (maximum take-off mass (MTOM), sea level (SL), ISA, paved)	≤1000 m
Take-off field length STOL (SL, ISA, paved, > 80% pax)	≤800 m
Landing field length (SL, ISA, paved)	≤1000 m
Rate of climb (MTOM, SL, ISA)	≥1850 ft/min
Rate of Climb at top of climb	≥1.5 m/s (300 ft/min)
Time to climb to FL 170	≤12.7 min
Maximum operating altitude	7620 m (25,000 ft)
Service ceiling for one engine inoperative (OEI) (or equivalent) (95% MTOM, ISA +10)	4000 m (13,125 ft)

The take-off field length (TOFL) is characterized by two values. The first one is for standard operation with full capacity, and here the TOFL is defined as 1000 m. The second one is defined for short take-off and landing (STOL) to allow access to smaller airports—here a TOFL of 800 m is required for a passenger load factor of more than 80% compared to full capacity. The exact achievable load factor for this requirement will be a result of the final design and the grade of optimization for standard operations. Additionally, both field lengths are the criterion for landing as well. This means that for the two different kinds of operations, the landing field lengths must be no greater than 1000 m and 800 m, respectively.

All these field lengths depend on the maximum lift coefficient, where a "certifiable" coefficient should be reached. "Certifiable" describes that, for example, in case of high-lift distributed propulsion, the aircraft must still reach an acceptable sink rate in descent. If the thrust generated for the required lift would exceed the drag of the configuration, the aircraft would climb instead of losing altitude. In this case, energy harvesting could be beneficial for the overall aircraft. In general, if a novel high-lift system is used, it has to have equivalent safety compared to current conventional technology in all possible conditions, e.g., crosswinds. To address this, the National Aeronautics and Space Administration carried out initial analyses for the Maxwell X-57 and made suggestions for the design of high-lift distributed propellers [14].

The minimum rate of climb at MTOM in international standard atmosphere (ISA) conditions at sea level (SL) is set to 1850 ft/min, which is comparable to the ATR 42 today [15]. This ensures operational capabilities and allows the operators to service equivalent routes than current turboprop aircraft.

For the rate of climb at the top of climb, 1.5 m/s or 300 ft/min is defined. This ensures the aircraft's ability to perform maneuvers to avoid collisions when cruising at the top of climb. This value might change in the future if airspace rule dynamics change.

The overall climb performance is defined by the time to climb to FL 170. This criterion makes the climb comparable to other aircraft (ATR 42-600: 12.7 min [15]) because power, and therefore the rate of climb, depends on altitude for air-breathing propulsion. In hybrid-electric aircraft, these losses could be countered by the battery.

A maximum operating altitude FL 250 or 7620 m is defined for a few reasons. Firstly, above that altitude, a redundant air management system is required (CS 25.841) [10]. This would result in a higher system complexity and an increase in total aircraft mass. Secondly, due to the typical turboprop mission duration of one hour, a higher cruise altitude is often not required, as it will result in a longer climb segment. This will not only require more power, but also more fuel, which might not be offset by a more efficient cruise in higher altitude. Furthermore, the passenger comfort is increased when the cruise segment takes up a significant portion of the overall flight. Lastly, it is expected that the formation of aviation induced clouds and contrails at that defined altitude, and thus the climate impact of the aircraft is significantly lower [16].

The service ceiling with one engine inoperative (OEI) is set to 4000 m, or 13,125 ft, to overfly mountain ranges in most regions of interest to air traffic and is comparable to current turboprops [15]. This criterion is set for 95% of MTOM and ISA +10. For the HEA, the term OEI must be adapted in relation to the selected power train architecture and propulsor configuration.

2.4.1. Range

The design range of the regional aircraft is a major point concerning its future mission and purpose. The aim of regional air travel is a closer connection of rural areas and to realize a fast connection between cities. This is connected to the Flightpath 2050 goal that "90% of travellers within Europe are able to complete their journey, door-to-door within 4 hours" [5].

The regional flights should enable an easy and fast way to also travel on routes with lower passenger volume, similar to remote bus and train connections. Thereby traveling by air should work in synergy with other modes of transportation, enabling routes which are difficult or not cost-effective to realize with ground-based transport systems or add a significant and useful time differential.

At first, the coverage of airports in the EU plus Great Britain, Norway, and Switzerland is analyzed (see Figure 3). This leads to a number of 414 airports with a field length of more than 1000 m (430 airports for 800 m, respectively) that have a volume of more than 15,000 passengers per year according to the data of the reported airports by Eurostat in 2019 [17], excluding overseas departments. It is expected that these airports are commercially in use. Under the assumption of a 100 km catchment area around the airports, major parts of the EU are already covered. We expect that a larger catchment area does not result in any advantages for the traveler, since no time advantage is expected. Furthermore, we assumed that no emission savings can be achieved for a catchment area of more than 100 km.

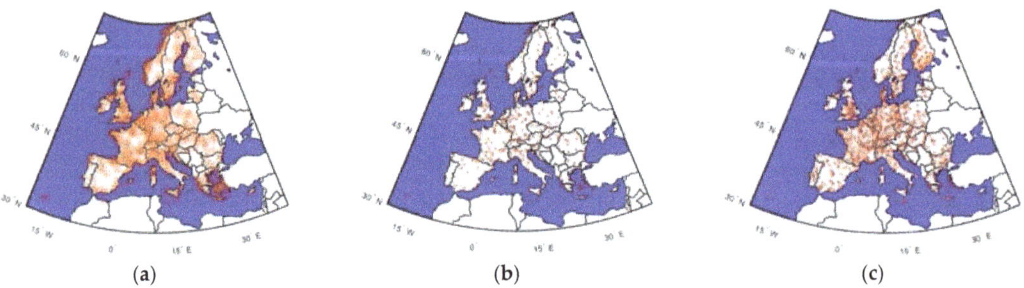

Figure 3. Coverage area of European commercial airports with an assumed catchment radius of (**a**) 100 km, (**b**) 50 km, and (**c**) 50 km considering all airfields of at least 1000 m runway.

For regions with low transport infrastructure (no highway, no railway, mountainous regions) the catchment area has to be adjusted because of a slower travel speed. Here, a more suitable radius of the catchment area for regional air traffic would be 50 km or even less from an airport. Considering this smaller catchment area, there is only sufficient coverage for metropolitan areas. These urban areas are often already connected to the regional traffic network. However, air travel can realize new and faster connections between these areas as an alternative means of transport.

However, if we consider airports or airfields in general (commercial or not) with a runway (concrete and asphalt) of at least 1000 m, a higher degree of coverage can be achieved. A study based on EU airports provided in the database of OpenAIP [18] has shown that 1112 airports and airfields fulfill this requirement (the study also includes military airfields). Taking these airfields into account, nearly all regions of the EU can be connected to a regional air traffic network. Assuming that these regions are often barely connected to a fast transport infrastructure network, this will greatly contribute to the European vision presented in Flightpath 2050.

To allow regional travel in the EU with an aircraft in respect to the defined goals of Flightpath 2050, as well as worldwide defined climate objectives, the design range is set to 400 km and the maximum range to 800 km, respectively. It is assumed that for this range, it is possible to develop an aircraft with EIS 2035/2040 reaching the global climate goals and to enable a regional air transportation system as an extension of the current traffic infrastructure. To connect different regions and cities in the EU, the existing commercial airports with a field length of 1000 m or more in operation already enable 12,467 routes (distance between 350–800 km). For a field length of 800 m or more, this would result in 13,214 routes. In both cases, these connections of interest to regional flight cover approximately 15% of the flight connections in the EU with a distance of more than 250 km (Figure 4).

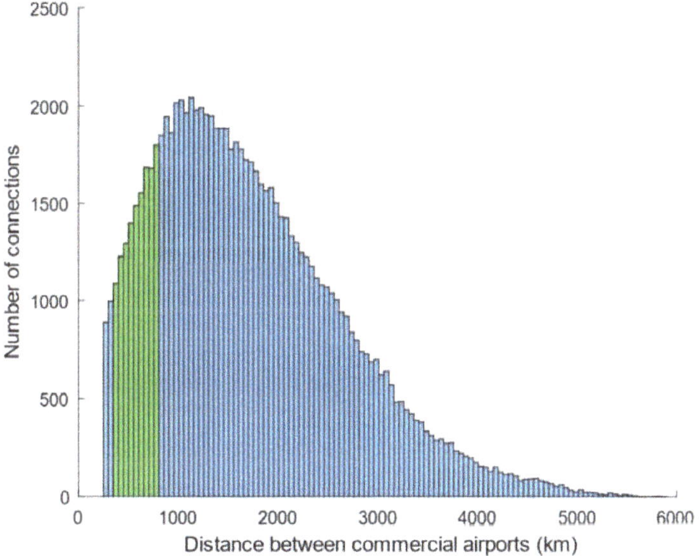

Figure 4. Number and distance of flight connections in the EU (plus Great Britain, Norway, and Switzerland) concerning commercial airports in operation (blue + green)—feasible connections for regional air traffic (green). All connections of more than 6000 km are not displayed.

Due to the lower passenger capacity of the planned aircraft, less populated regions could also be connected to the EU traffic network even if there is a lower passenger and cargo volume demand. The availability of these air connections has been shown to play a role regarding regional development [19]. Figure 5 shows the coverage of different ranges for existing commercial airports with a field length of at least 1000 m. It can be seen that nearly all states of the EU can be covered, excluding some islands and overseas departments.

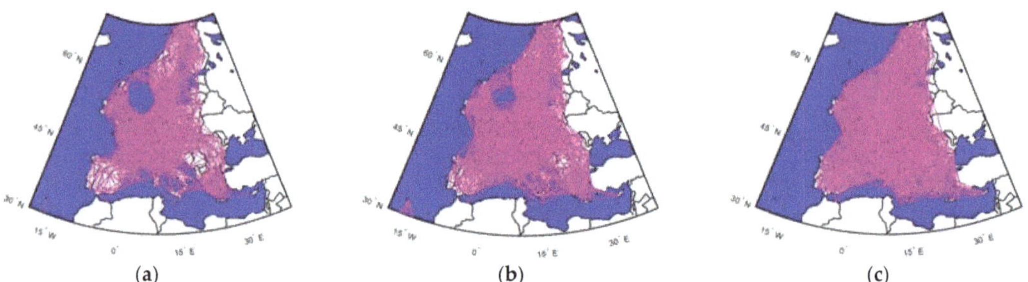

Figure 5. Connections between commercial airports in operation in different distances (**a**) 350–500 km, (**b**) 500–650 km, and (**c**) 650–800 km.

The two examples in Figure 6 showcase possibilities within this regional air traffic system. The first is Strasbourg, France, in the heart of the EU and the seat of the European Parliament. Only considering a distance between 350–800 km, 113 connections can be realized. The figure on the right presents domestic air traffic between the Greek mainland and islands. This could be an addition to current ferry connections for faster travel between the islands with a bigger distance. Therefore, the minimum flight distance was reduced to 150 km.

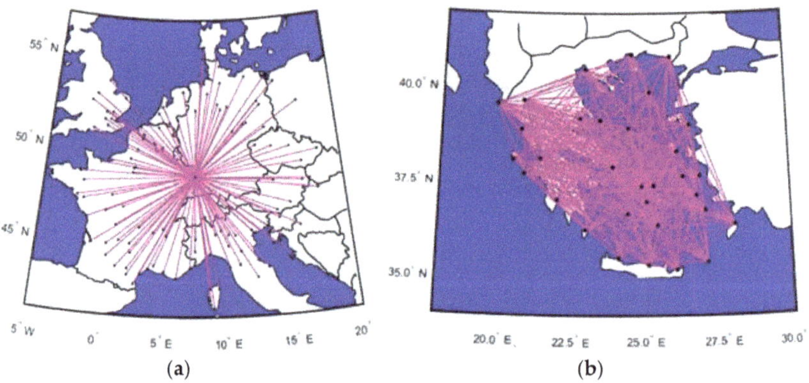

Figure 6. Example (**a**) Connections from Strasbourg (SXB) between 350–800 km, Example (**b**) Feasible domestic flight routes of Greece between 150–800 km.

2.4.2. Reserve Policy

Within ICAO Annex 6 [20], topics concerning the operation of aircraft are covered. This includes fuel requirements for a safe execution of a complete flight. It defines that the minimum amount of usable fuel should account for the following parts: Fuel required for taxiing and the overall trip, including a contingency fuel for unforeseen factors. The reserves are split into fuel required to reach the alternate airport and a final fuel reserve, which for a turbine driven aircraft is set to 30 min of holding at holding speed, 1500 ft above ground level.

The distance to the alternate, and therefore required, fuel is specific to the planned route. For this project, it is defined as 185 km in general, which is equal to about 23% of the maximum range. In many regions, this is expected to be enough for most regional flights. Two main events can happen that require diverting to an alternate. First, having already reached the planned destination and being required to divert to another airport, for example, because the airport was closed due to an accident. Second, being en route and the weather at the planned destination worsens, requiring the flight to divert to an en route alternate.

For the first case, having already reached the destination and being required to divert, Figure 7 shows the number of potential alternates for airports in the EU (including Great Britain, Norway, and Switzerland). Out of the 414 commercially operated airports with a field length above 1000 m, 3 have no alternate within the defined reserve distance. This yields more than 99% of airports adequately covered. Over 90% have more than one alternate within reach. To correctly define airports with missing alternates, airports in Serbia, Bosnia, and Herzegovina, Montenegro, Kosovo, Albania, North Macedonia, Belarus, and Ukraine were added in all reserve analysis. Here, the same criteria for the airport selection have been applied (field length of at least 1000 m and yearly passenger volume of at least 15,000).

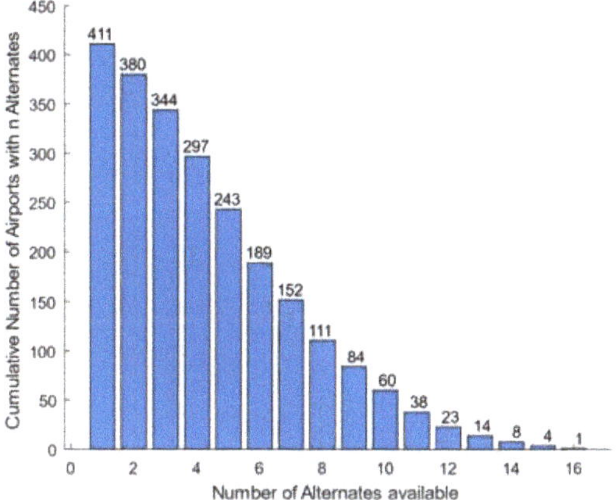

Figure 7. Cumulative number of airports that have n available alternates within reach by the defined reserve distance.

The three airports not included are highlighted in Figure 8, with a circle representing the defined reserve distance. We can see that the airports of Lisbon (LIS) in Portugal, Flores (FLW) on the Azores, and Longyearbyen (LYR) on Svalbard lack an adequate alternate airport.

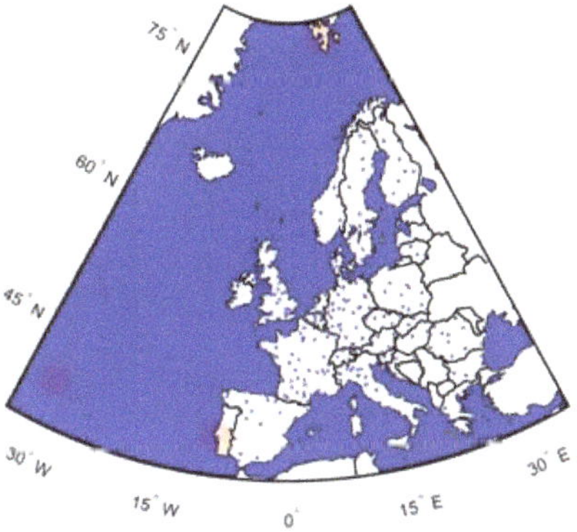

Figure 8. Airports that have no alternate in reach with the specification given in the TLARs.

The farthest distance to an available alternate is 945 km for Longyearbyen, which is longer than the maximum range of the aircraft; therefore, the airport is not relevant for the analysis. This leaves a number of 413 airports investigated in total. For the other two airports, the distance to an alternate equals 234 km for Flores and 201 km for Lisbon.

To better account for operational aspects, two safety factors are introduced for a second study. First, a safety margin should be applied in order to derive from the theoretical range of the great circle to a practical range. For this, a distance increase by 5% was assumed. Second, a safety factor should consider adverse headwind conditions, which increase the energy required to reach the diversion airport. If a "fresh breeze" on the Beaufort Scale is assumed as the reference value with wind speeds up to 38 km/h, the flight time of an aircraft with a cruise speed of 550 km/h would be increased by around 7%, and consequently the required energy for the flight is increased accordingly.

Figure 9 shows that by including safety factors, the number of airports without any alternate within reach is increased by ten. With 3%, the fraction of airports with no alternate is still small. For these twelve airports, the distances required vary between slightly over 185 km up to 263 km, again with Flores on top. This shows that even with unfavorable conditions, the selected reserve policy is suitable for the vast majority of airports in Europe.

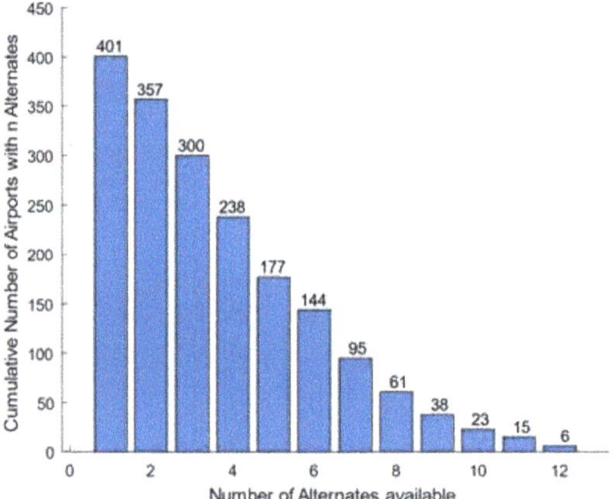

Figure 9. Cumulative number of airports that have n available alternates within reach by the defined reserve distance including safety factors for practical range and adverse headwind.

It must be mentioned that some airports do have potential alternates within the defined range without being recognized in this calculation. That is because they do not meet the requirements previously set, e.g., within Lisbon's range, Beja airport (BYJ) is not serviced by any regular scheduled flights at the moment and has a yearly passenger volume of less than 15,000 passengers.

For the second case, the distance to an en route alternate is expected to be lower than it is after having already reached the planned destination. Even when the distance to an alternate en route is higher than the defined reserve distance, the aircraft still has the energy for the remaining trip, which then could be used to reach an alternate. In case of an HEA with different energy sources like fuel and batteries, this might require that there is no single point of failure within one system part. Otherwise, if only one combustion engine and generator are installed and one component fails, the battery might not be able to provide the required energy to reach any airport.

In order to determine the distance to the nearest suitable en route alternate airport, a mesh of possible aircraft positions over land was created, expressed in latitude ϕ and longitude λ. Aircraft positions over sea and islands, except for Great Britain and Ireland, were excluded at this stage. The mesh applied in the calculation consists of positions with a spacing of $1/60°$ in latitude $\Delta]\phi$ and longitude $\Delta\lambda$, which corresponds to a maximum spacing of approximately 1 nautical mile. In Figure 10, an exemplary mesh is shown with a spacing of $1°$ in latitude and longitude. Consequently, the shortest distance of each aircraft position to the suitable en route alternate airports is determined by calculating the length of the great circle to each individual airport and determining the minimum value. With the data obtained, a cumulative frequency plot was created.

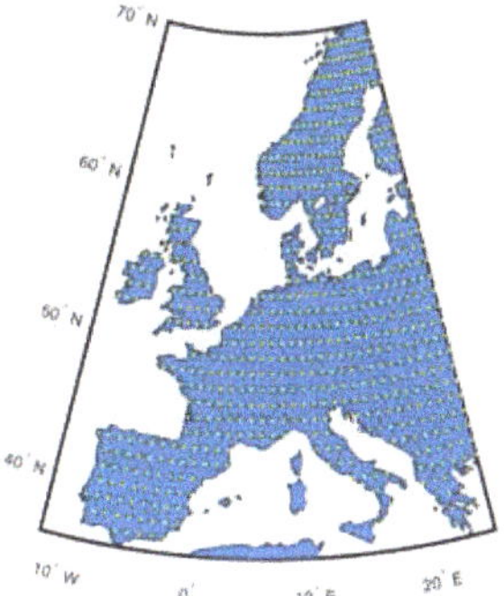

Figure 10. Exemplary visualization of investigated aircraft positions in Europe with a spacing of 1° in latitude and longitude—the calculation was carried out with a spacing of 1/60°, which results in a maximum spacing of approximately 1 NM.

Figure 11 shows the database of alternate airports on the left and the resulting curve for airports within the investigated parts of Europe featuring a concrete or asphalt runway with a field length of more than 1000 m on the right. Suitable diversion airports from 95% of all investigated aircraft positions can be reached within 120 km. A percentage of 99% can be reached within 148 km, and the farthest distance is 198 km southeast of Madrid (38.8° N, 2.75° W).

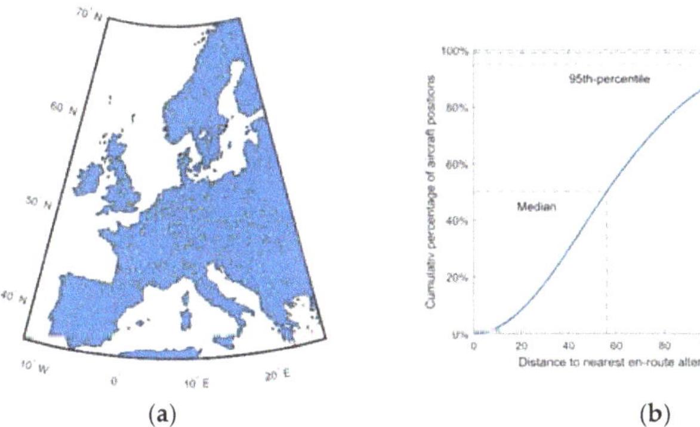

Figure 11. Possible alternate airports (**a**) and plot of cumulative frequency of the distance to a suitable en route alternate with a concrete or asphalt runway (**b**).

The maximum error of the calculation is evaluated by considering the distance of a possible aircraft position, which is located exactly in between the mesh of evaluated aircraft positions and can be determined to be 1.26 km [21]. For this evaluation, similarly to the first analysis, safety factors for deviations from the great circle and headwind conditions could be applied.

2.5. Regulatory Aspects

In general, this type of aircraft has to comply with CS-25. So, all requirements from this CS have to be fulfilled. An HEA is not fully certifiable under the current regulations in place. Some gaps have already been identified. For example, a battery could be used as redundancy within a serial architecture, and therefore one gas turbine might be enough for a safe design.

Looking at novel propulsion concepts like distributed propulsion, a new definition of the OEI condition is needed. Not limited to distributed propulsion, the overall propulsion architecture will need analysis. Depending on the interdependencies of different systems and their redundancies, a similar definition might emerge. Due to the complexity of the HEA architecture, this could be individual for every aircraft or architecture.

Other aspects might be electromagnetic interference or electromagnetic compatibility due to the required power of the overall aircraft. Voltage levels, and in connection the breakdown voltage of the insulator, are also relevant. On the other hand, the required power and the voltage level define the current, which in the end drives the cross section of the conductor and therefore mass.

3. Reference Flight Missions

The requirements and evaluation criteria serve as a framework for the output data that will be generated in the aircraft design. Those parameters depend, e.g., on weather (ambient temperature in particular), on energy management strategies, and also on the flight route and profile. Therefore, various reference flight missions were developed in the frame of FUTPRINT50, so potential markets for the regional aircraft can be analyzed:

- Design Range
- Maximum Range
- Cold Operations
- Extreme Cold Operations
- Hot and High Operations
- Mountainous Terrain
- Island Operations
- STOL Operations

For these missions, an advantageous position on the market may be anticipated for the regional HEA developed within FUTPRINT50. In regions with excellent railway systems, the regional HEA will not be able to prevail due to its limited seat capacity. However, specific markets come into play where no other means of transport can compete. This concerns routes where rough terrain or water lies in between, or infrastructure like railway or road networks are less developed.

The design range criterion can be represented by the route from Edinburgh, United Kingdom to Dublin, Ireland (EDI–DUB). The great circle distance for this route is 338 km. However, all distances of the presented routes have been increased by 7% to account for possible headwind and 5% for airways and navigation. This leads to a design range of 380 km, which represents the definition by the TLARs very well. The alternate airport for Dublin is Belfast (BFS), which is located ca. 140 km north. After adding all contingency reserves, this route represents the defined distance to an alternate airport (185 km) very well. Figure 12 shows the flight route and its altitude profile. This also includes the flight to the alternate airport and thirty minutes of loiter at 175 knots. It becomes apparent that the reserve is a significant part of the whole flight mission and must not be neglected.

For the investigation of low temperature impacts, a flight route in Iceland is selected. The flight from Reykjavík to Egilsstaðir (RKV–EGS) is a common route serviced by a De Havilland Canada DHC-8. The average temperature at both airports is about ISA −10.

In order to account for the effects to the powertrain in even colder temperatures, the flight route from Yakutsk to Vilyuisk (YKS–VYI) in Russia is examined. The average winter temperature can reach up to ISA −55 (−40 °C).

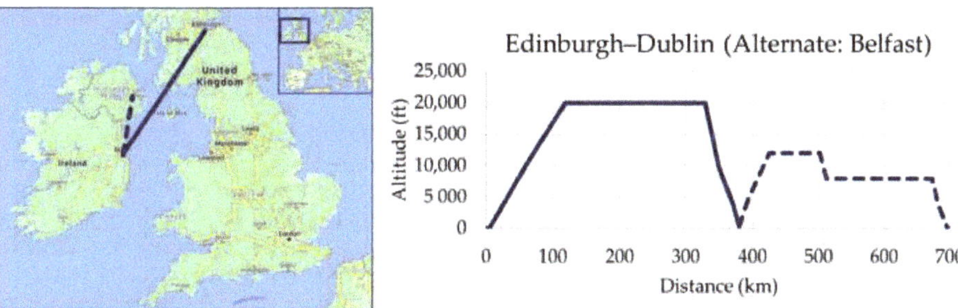

Figure 12. Exemplary flight route and profile of the design range mission EDI–DUB (Edinburgh, United Kingdom to Dublin, Ireland) (including the flight to the alternate BFS (Belfast) and a holding of 30 min).

As mentioned earlier, the regional HEA will be able to compete on routes where other means of transport lack required infrastructure. This is also the case in the mountainous terrain from Lima to Ayacucho (LIM–AYP) in Peru. This flight route requires a steep climb right after take-off to reduce detours; this will challenge the climb performance of the aircraft design. Furthermore, a Missed Approach Procedure at an airport elevation of 2720 m is included on this route.

In order to test the TLAR for hot and high missions, the flight route from Mexico City to Acapulco (MEX–ACA) in Mexico was selected. The elevation of the airport of Mexico City is 2230 m, which translates into a 20% reduced air density compared to SL. Furthermore, Mexico City shows an average daytime temperature that lies 10° above ISA conditions.

Island operations describe a mission which restricts refueling to only take place at one airport. This can be simulated by flight operations from São Tomé to Príncipe (TMS–PCP) and back. It must be noted that the distances of each leg may not be simply added up to one round trip. The need for two take-offs and climbs is expected to play a major role on the total energy required.

To account for the STOL capabilities of the aircraft, the flight route from Corvo to Horta (CVU–HOR) in Portugal was included in the list of reference missions. With a flight distance of about 270 km, both airports are located on separate islands in the Azores. The runway in Corvo offers an available take-off run distance of only 800 m. Table 5 shows a summary of the flight mission profiles selected along with each of their specific reasoning.

Table 5. Selection criteria for chosen flight missions.

Flight Mission	Route	Special Characteristic
Design Range	EDI–DUB	ca. 400 km, competing transport via 2 h-ferry
Maximum Range	TJM–NJC	ca. 800 km, 13 h by car, 17 h by train
Cold Ops	RKV–EGS	Temperature ISA −10
Hot and High	MEX–ACA	MEX: Elevation 2230 m, Temperature ISA +10
Extreme Cold Ops	YKS–VYI	Winter temperatures around ISA −55 (−40 °C)
Mountain Ops	LIM–AYP	After take-off: 3000 m at 80 km, 4500 m at 150 km
Island Ops	TMS–PCP	Two 200 km-legs without refueling
STOL Ops	CVU–HOR	CVU: Runway length 800 m

4. Figures of Merit

Within the aircraft design process, numerous unique aircraft configurations are generated, especially in the conceptual design phase. In order to assess those models as objectively as possible, various evaluation criteria, the so-called figures of merit, must be defined. While the TLARs described in Section 2 establish a framework of the aircraft design, the figures of merit quantify how well the design performs within the given evaluation criteria. In general, all quantifiable output parameters of the aircraft design process can serve as figures of merit. They are then utilized to rate and compare the configurations on the basis of those parameters. The more design parameters are considered, the better the assessment of the aircraft design might be. However, using a lot of parameters can also make an evaluation less transparent and more complex. Hence, a selection process for meaningful, quickly graspable figures of merit is important. This is described in Section 4.1.

While the separate parameters can compare isolated disciplines of the aircraft design, a combination of these parameters can form one aircraft-level figure of merit which rates the entire aircraft design. By means of a calculation method described in Section 4.2, the aircraft-level figure of merit is constructed as a number between 0 and 1. On that scale, a better design—based on the evaluation criteria set—translates into a higher score.

4.1. Identification of Figures of Merit

To find the most suitable figures of merit, all parameters generated in the aircraft design are processed through different criteria [22]. These criteria form a filter that excludes figures of merit that do not affect the overall evaluation of the aircraft in a significant way and are, therefore, discarded. The filter comprises four criteria:

1. Limitations that are defined by TLARs
2. Design parameters that are fixed
3. Parameters that do not change significantly between different designs
4. Parameters that are expressed through higher-level ones.

The first criterion discards figures of merit that do not influence the assessment because of limitations by the TLARs. An example for this criterion is the parameter of aviation induced cloudiness. As shown in Section 2, a maximum operating cruise altitude of 25,000 ft was chosen for the regional aircraft developed within the FutPrInt50 project. Since aviation induced cloudiness is mainly expected to form in altitudes higher than that [16], this figure of merit can be discarded.

Fixed design parameters act as second criterion. Some parameters must remain unchanged during a design process. For example, the maximum range is a hard requirement stated by the airlines. Falling short of it is not an option. However, exceeding it will make the design not necessarily better either; the design may become less efficient on shorter ranges because of an increased operating mass empty.

The third criterion filters out parameters that do not change significantly. This may be valid for parameters that are estimated by engineers or parameters that only vary in case of major technology changes. Switching from conventional jet fuel to hydrogen will impact the ground handling and servicing for every aircraft design similarly. This is why it can be neglected.

The last criterion discards any figure of merit that can be expressed by a higher-level one. As an example, cruise speed is often defined as a figure of merit because it can have an influence on the number of possible dispatches. However, it is also interconnected with fuel consumption and many other characteristics. All of those aspects can be combined within the DOCs.

Additional parameters in regard to the entire life cycle, like CO_2 emissions of the production or end-of-life phase, are not considered. However, they can be easily included afterwards due to the flexibility of the calculation method of the figure of merit.

This selection process leads to seven figures of merit listed in Table 6, where they are divided into three categories. These seven meaningful parameters evaluate the quality of the regional HEA developed within FutPrInt50 and can be used as a basis for the assessment of future concepts in this aircraft category. The separate figures of merit will be explained in more detail in the following.

Table 6. Summary of figures of merit selected in the frame of FUTPRINT50.

Environmental Aspects	Airline Desirability	Introduction of Hybrid-Electric Aircraft
CO_2 emissions	Direct operating costs	Development risks
NO_X emissions		Certification challenges
Noise emissions		Production aspects

The Environmental Aspects that are chosen for the figures of merit compare to the goals set in Flightpath 2050. As the CO_2 emissions of an aircraft strongly depend on fuel flow, energy management strategies have a major influence on this part of the figures of merit. Another environmental figure of merit is the emission of NO_X. It can mainly be lowered by improving the combustion characteristics of the engine. While its atmospheric effect is important to consider because of ozone forming, it is also a relevant pollutant in the vicinity of airports [23].

The same issue corresponds as well to the noise emissions. Noise is a very important aspect in order to ensure the health and well-being of citizens living close to airports. Although noise is difficult to analyze in the early design process, engineers can roughly estimate it by focusing on components protruding from the aircraft profile like landing gear or flaps. Another key element is the positioning of propulsion components and potential shielding like a ducted fan would offer.

Major contributions to these aspects can be provided by an optimum energy management strategy which is adaptive to the current state of operation. This offers the potential to reduce emissions overall or just in particular flight segments close to the ground.

Operators are the main drivers for aircraft specifications, and appealing to them is of major importance for every design. Since cost is an essential parameter in the airline industry, the DOCs are the main driver of the Airline Desirability.

The DOCs are divided into five segments which consist of capital costs, maintenance costs, energy costs, crew costs, and (navigation) fees. Capital costs are expected to rise for a new HEA because of higher development efforts. On the other hand, energy and maintenance costs are expected to decrease because of lower energy costs for HEA and less complex maintenance in the long run [24]. The energy costs are highly dependent on many parameters like cruise speed or energy management strategies, all of which are defined by the reference flight missions. While crew costs are expected to be comparable to the ones on conventional aircraft, the navigation and landing fees might decrease because of environmental rewards.

Beside operating costs, the potential of generating income is an important aspect for airlines as well. Here however, it is assumed to stay the same as seat capacity and passenger comfort remain unchanged. Though, it may be included in further enhanced models.

The figures of merit in the category Introduction of Hybrid-Electric Aircraft cover the risks in development as well as the expected challenging certification process and overall production aspects. The first item represents the risks for the manufacturer developing the HEA. This includes challenges that might occur in the design process of not only the overall aircraft, but also of all required components and subsystems. To address these topics, an estimation is made in relation to the technology readiness level of identified key technologies. Furthermore, if required, infrastructure needs are also considered. This could be the case for hydrogen fuel as well as required infrastructure to either swap batteries and recharge them at the airport or charge the batteries within the airplane.

Certification is the second part within this category. Currently, the CS-25 from the European Union Aviation Safety Agency (EASA) offers no guidelines on how to certify hybrid-electric aircraft. By assessing the complexity of the system, an estimation can be made about the amount of testing required for certification. Furthermore, adding novel propulsion concepts like distributed electric propulsion will increase complexity even more and therefore require additional verification.

Lastly, the selection of material and the application of processes also impose risks on different designs. Again, the newer and more complex the technology, the higher the overall risk for unforeseen difficulties in production. All these criteria are difficult to quantify and must oftentimes be objectively estimated by the engineer.

4.2. Calculation of the Aircraft-Level Figure of Merit

With the selected figures of merit, a new process becomes necessary: The separate figures of merit presented before must be merged into one single aircraft-level figure of merit which evaluates the entire aircraft configuration [22]. This helps by comparing different concepts and identifying which of the designs is the most promising one for further development. The scheme of this method is shown in Figure 13.

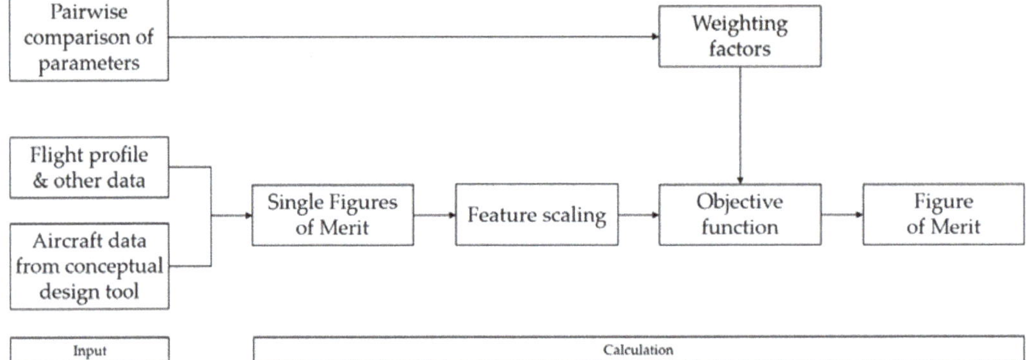

Figure 13. Calculation method for the figure of merit.

The first step of the calculation method consists of combining two major input data. The design tool provides all required aircraft data like dimensions, masses, and aerodynamics. Further input contains data about the flight profile, energy management strategies, ambient conditions, etc.; this depends on the chosen reference mission. The following module, Single Figures of Merit, calculates the seven single figures of merit that were previously selected. As those separate figures of merit are all given out in their individual unit, they need to be feature scaled so they can be combined into one. Feature scaling is a method which removes all units and resizes all the different values of the figures of merit onto a rating scale. This is done by analyzing all values of each parameter and assigning the scores 0 and 1 to the lowest or highest value. The remaining values are scaled correspondingly. However, it is important to note that the lowest score does not necessarily coincide with the lowest value. This is because a lower value in emissions, for example, will lead to a higher score on the assessment scale.

Furthermore, the resulting equation also consists of weighting factors. They are mostly generated by a pairwise comparison in a table. In the case of the different emissions of greenhouse gases, however, the parameters are weighted by their different impacts on the environment. This constructed equation forms the so-called objective function. It consists of weighted and normalized parameters which are then combined into the aircraft-level figure of merit. The flexibility of this weighting allows us to identify designs with different focusses, for example, prioritizing environmental aspects over costs or vice versa. In the end, various sets of weighting factors can be applied, where each set represents a specific focus that shall be analyzed. This process leads to a single number, which represents the figure of merit on the aircraft level. This number will then be used to assess all the aircraft designs that need to be compared.

Dividing the parameters into different categories and sticking to that strict workflow has the advantage that the structure of the figures of merit is established in a modular way.

For example, a function of power, specific fuel consumption, and time quantifies the CO_2 emissions, which merge into the environmental aspects. So, the modular approach allows for building a figure of merit that consists of others, merging into the single aircraft-level figure of merit. This means that if any outcome of a figure of merit is unclear, it is always possible to go one level deeper to understand the characteristics of that specific figure of merit. This provides transparency, traceability, and plausibility. Another advantage of the modular approach is the simple implementation of modifications. When an additional figure of merit is identified, it can be added by changing only one module of the entire workflow. This allows an easy integration of figures that were previously discarded—it is also thinkable to select a modular approach within the code. For example, the maximum altitude or cruise altitude could be analyzed and based on this value, and contrails are included or discarded in the environmental aspects.

5. Conclusions

Within this work, requirements for sustainable regional aviation and their reasoning within the FUTPRINT50 framework have been presented. Moreover, the figures of merit were displayed which validate to what extent the defined requirements are fulfilled. The V-model of systems engineering builds the framework for the overall project and ensures a successful "end product". A mission statement defines the value that is added for the stakeholders and outlines the overall system to be designed. Here, the focus was set on the European citizen and improvements to their life quality by reducing emissions of pollutants and noise.

With the help of the mission statement, the top-level aircraft requirements were defined. These requirements set the criteria that must be fulfilled by the system to be designed. Considering not only environmental aspects, but also the performance of the system should satisfy the operators and stakeholders. These specifications ensure that the aircraft to be developed can occupy its planned niche within the system of systems. In addition, we took a closer look on the defined reserve distance and range requirements. The study for the reserve policy shows that—including a safety factor—more than 95% of the investigated commercial airports have at least one alternate within 185 km. En route alternates are expected to be less problematic if the hybrid-electric aircraft is sized properly with enough redundancy for the different energy sources. The study has shown that most parts of the EU can be covered, and more than 10,000 connections are possible; this represents around 15% of possible connections within Europe above 250 km. In the future, further analyses could be performed for other regions of the world: Mountainous regions or island operations might be of particular interest. Examples could be the Andes or the Himalaya for the former, and the Caribbean or Southeast Asia for the latter.

After stating the requirements—following the iterative process of the V-model—the figures of merit were defined. Besides verifying whether stakeholder needs and requirements are met, they aim to identify the best design from many. This is done in three key categories that are most influential to the goals of the project. Seven parameters divided into environmental, economic, and technical aspects are rated for all investigated aircraft designs. This analysis reveals the quality of the aircraft in comparison to other designs. Furthermore, a method was created that provides a modular structure of all separate figures of merit. This ensures flexibility and allows easy changes to the composition of the aircraft-level figure of merit.

The TLARs and the figures of merit must be tested in consistent scenarios. To achieve this, different reference missions were defined, each one specific to test a key aspect of the requirements. With this, a focus on different scenarios can be set in the future depending on the expected size or relevance of potential and existing markets.

Author Contributions: Conceptualization, D.E., N.M., E.W., and R.R.; software, D.E., D.B., and I.G.; investigation, D.E., N.M., E.W., D.B., I.G., and R.R.; writing—original draft preparation, D.E., N.M., and D.B.; writing—review and editing, D.E., N.M., E.W., D.B., I.G., R.R., and A.S.; supervision, A.S.;

project administration, D.B.; funding acquisition, D.B., R.R., A.S. All authors have read and agreed to the published version of the manuscript.

Funding: The research leading to these results has received funding from the European Union's Horizon 2020 Research and Innovation program under Grant Agreement No 875551.

Institutional Review Board Statement: Not applicable.

Informed Consent Statement: Not applicable.

Data Availability Statement: Results not available, raw data is cited in text.

Acknowledgments: We thank all researchers and collaborators of the entire FUTPRINT50 Consortium, namely University of Stuttgart, Cranfield University, Airholding S.A., TU Delft, ADSE, CEA, EASN, Unicusano, Embraer, TsAGI, GosNIIAS, CIAM, NRC, and MAI, for their valuable input.

Conflicts of Interest: The funders had no role in the design of the study; in the collection, analyses, or interpretation of data; in the writing of the manuscript, or in the decision to publish the results.

Abbreviations

ATAG	Air Transport Action Group
CORSIA	Carbon Offsetting and Reduction Scheme for International Aviation
CS	Certification Specification
DOC	Direct Operating Costs
EASA	European Union Aviation Safety Agency
EIS	Entry Into Service
EU	European Union
FL	Flight Level
HEA	Hybrid-Electric Aircraft
ICAO	International Civil Aviation Organization
ISA	International Standard Atmosphere
MTOM	Maximum Take-Off Mass
OEI	One Engine Inoperative
SAF	Sustainable Aviation Fuel
SL	Sea Level
STOL	Short Take-Off and Landing
TLAR	Top-Level Aircraft Requirements
TOFL	Take-Off Field Length
UN	United Nations

References

1. Air Transport Action Group. Aviation: Benefits Beyond Borders. 2020 Report. 2020. Available online: https://aviationbenefits.org/media/167143/abbb20_full.pdf (accessed on 10 November 2020).
2. International Civil Aviation Organization. 2019 Environmental Report. Aviation and Environment. 2019. Available online: https://www.icao.int/environmental-protection/Documents/ICAO-ENV-Report2019-F1-WEB%20(1).pdf (accessed on 9 March 2020).
3. Lee, D.; Fahey, D.; Skowron, A.; Allen, M.; Burkhardt, U.; Chen, Q.; Doherty, S.; Freeman, S.; Forster, P.; Fuglestvedt, J.; et al. The contribution of global aviation to anthropogenic climate forcing for 2000 to 2018. *Atmos. Environ.* **2020**, *244*, 117834. [CrossRef] [PubMed]
4. Sparrow, V.; Gjestland, T.; Guski, R.; Richard, I.; Basner, M. Aviation Noise Impacts White Paper: State of the Science 2019: Aviation Noise Impacts. In *2019 Environmental Report: Aviation and Environment*; ICAO: Montreal, QC, Canada, 2019; pp. 44–61.
5. European Commission. *Flightpath 2050. Europe's Vision for Aviation; Maintaining Global Leadership and Serving Society's Needs; Report of the High-Level Group on Aviation Research*; Publications Office of the European Union: Luxembourg, 2011; ISBN 978-92-79-19724-6.
6. European Commission. The European Green Deal COM(2019) 640. 2019. Available online: https://eur-lex.europa.eu/resource.html?uri=cellar:b828d165-1c22-11ea-8c1f-01aa75ed71a1.0002.02/DOC_1&format=PDF (accessed on 20 November 2020).
7. United Nations. Transforming Our World: The 2030 Agenda for Sustainable Development. A/RES/70/1. Available online: https://sustainabledevelopment.un.org/content/documents/21252030%20Agenda%20for%20Sustainable%20Development%20web.pdf (accessed on 10 November 2020).
8. Avions de Transport Régional. ATR Turboprop Market Forecast 2018–2037. Available online: https://1tr779ud5r1jjgc938wedppw-wpengine.netdna-ssl.com/wp-content/uploads/2020/07/1011_makretforecast_digital_151.pdf (accessed on 1 February 2021).

9. Eisenhut, D.; Windels, E.; Reis, R.; Bergmann, D.; Ilário, C.; Palazzo, F.; Strohmayer, A. Foundations towards the future: FutPrInt50 TLARs an open approach. *IOP Conf. Ser. Mater. Sci. Eng.* **2021**, *1024*, 12069. [CrossRef]
10. EASA. Certification Specifications and Acceptable Means of Compliance for Large Aeroplanes CS-25. Amendment 25. 2020. Available online: https://www.easa.europa.eu/sites/default/files/dfu/cs-25_amendment_25.pdf (accessed on 11 November 2020).
11. International Civil Aviation Organization. *Annex 14 to the Convention on International Civil Aviation. Aerodromes*, 8th ed.; Aerodrome Design and Operations: Montreal, QC, Canada, 2018; Volume 1, ISBN 978-92-9258-483-2.
12. Embraer S.A. EMB-145. Airport Planning Manual. 2019. Available online: https://www.flyembraer.com/irj/go/km/docs/download_center/Anonymous/Ergonomia/Home%20Page/Documents/APM_145.pdf (accessed on 10 December 2020).
13. Bombardier Inc. DHC-8-300. Airport Planning Manual. 2001. Available online: https://eservices.aero.bombardier.com/wps/portal/eServices/Public/AirportEmergencyPublication (accessed on 10 December 2020).
14. Patterson, M.D.; Borer, N.K. Approach Considerations in Aircraft with High-Lift Propeller Systems. 2017. Available online: https://ntrs.nasa.gov/api/citations/20170005869/downloads/20170005869.pdf (accessed on 30 November 2020).
15. Avions de Transport Régional. ATR 42-600 Fact Sheet. Available online: http://1tr779ud5r1jjgc938wedppw-wpengine.netdna-ssl.com/wp-content/uploads/2020/07/Factsheets_-_ATR_42-600.pdf (accessed on 11 November 2020).
16. Kärcher, B. Formation and radiative forcing of contrail cirrus. *Nat. Commun.* **2018**, *9*. [CrossRef] [PubMed]
17. Eurostat. Air Transport Measurement: 2019 List of Reporting Airports Covered by Commission Regulation 1358/2003. Available online: https://ec.europa.eu/eurostat/cache/metadata/Annexes/avia_pa_esms_an5.docx (accessed on 18 November 2020).
18. Garrecht Avionik GmbH. Airport List | openAIP. Available online: https://www.openaip.net/airports (accessed on 1 December 2020).
19. Halpern, N.; Bråthen, S. Impact of airports on regional accessibility and social development. *J. Transp. Geogr.* **2011**, *19*, 1145–1154. [CrossRef]
20. International Civil Aviation Organization. *Annex 6 to the Convention on International Civil Aviation—Operation of Aircraft—Part I—International Commercial Air Transport—Aeroplanes*, 11th ed.; ICAO: Montreal, QC, Canada, 2018; ISBN 978-92-9258-473-3.
21. Geiß, I.; Strohmayer, A. Operational Energy and Power Reserves for Hybrid-Electric and Electric Aircraft. *Deutscher Luft- und Raumfahrtkongress 2020* **2021**. [CrossRef]
22. Moebs, N.; Eisenhut, D.; Bergmann, D.; Strohmayer, A. Selecting figures of merit for a hybrid-electric 50-seat regional aircraft. *IOP Conf. Ser. Mater. Sci. Eng.* **2021**, *1024*, 12071. [CrossRef]
23. Derwent, R.; Collins, W.; Johnson, C.; Stevenson, D. Global ozone concentrations and regional air quality. *Environ. Sci. Technol.* **2002**, *36*, 379A–382A. [CrossRef] [PubMed]
24. Hoelzen, J.; Liu, Y.; Bensmann, B.; Winnefeld, C.; Elham, A.; Friedrichs, J.; Hanke-Rauschenbach, R. Conceptual Design of Operation Strategies for Hybrid Electric Aircraft. *Energies* **2018**, *11*, 217. [CrossRef]

Article

In-flight Lift and Drag Estimation of an Unmanned Propeller-Driven Aircraft

Dominique Paul Bergmann [1,*], Jan Denzel [1], Ole Pfeifle [2], Stefan Notter [2], Walter Fichter [2] and Andreas Strohmayer [1]

1 Institute of Aircraft Design (IFB), University of Stuttgart, Pfaffenwaldring 31, 70569 Stuttgart, Germany; jan.denzel@ifb.uni-stuttgart.de (J.D.); andreas.strohmayer@ifb.uni-stuttgart.de (A.S.)
2 Institute of Flight Mechanics and Controls (iFR), University of Stuttgart, Pfaffenwaldring 27, 70569 Stuttgart, Germany; ole.pfeifle@ifr.uni-stuttgart.de (O.P.); stefan.notter@ifr.uni-stuttgart.de (S.N.); walter.fichter@ifr.uni-stuttgart.de (W.F.)
* Correspondence: dominique.bergmann@ifb.uni-stuttgart.de; Tel.: +49-711-685-60342

Abstract: The high-power density and good scaling properties of electric motors enable new propulsion arrangements and aircraft configurations. This results in distributed propulsion systems allowing to make use of aerodynamic interaction effects between individual propellers and the wing of the aircraft, improving flight performance and thus reducing in-flight emissions. In order to systematically analyze these effects, an unmanned research platform was designed and built at the University of Stuttgart. As the aircraft is being used as a testbed for various flight performance studies in the field of distributed electric propulsion, a methodology for precise identification of its performance characteristics is required. One of the main challenges is the determination of the total drag of the aircraft to be able to identify an exact drag and lift polar in flight. For this purpose, an on-board measurement system was developed which allows for precise determination of the thrust of the aircraft which equals the total aerodynamic drag in steady, horizontal flight. The system has been tested and validated in flight using the unmanned free-flight test platform. The article provides an overview of the measuring system installed, discusses its functionality and shows results of the flight tests carried out.

Keywords: unmanned aircraft; thrust determination; flight testing; e-Genius-Mod; free-flight wind tunnel

Citation: Bergmann, D.P.; Denzel, J.; Pfeifle, O.; Notter, S.; Fichter, W.; Strohmayer, A. In-flight Lift and Drag Estimation of an Unmanned Propeller-Driven Aircraft. *Aerospace* **2021**, *8*, 43. https://doi.org/10.3390/aerospace8020043

Academic Editor: Roberto Sabatini
Received: 14 December 2020
Accepted: 1 February 2021
Published: 6 February 2021

Publisher's Note: MDPI stays neutral with regard to jurisdictional claims in published maps and institutional affiliations.

1. Introduction

While investigating the flight performance of new aircraft configurations or technologies, knowledge about aerodynamic and flight mechanical parameters is required to understand their impact on the aircraft. From the perspective of aircraft design, one main focus is the impact on flight performance when investigating for example the effects of new propulsion technologies or aircraft configurations. To investigate and assess the flight performance, the determination of lift and drag is of major importance. Especially the determination of drag from flight tests with an unmanned propeller aircraft is a challenging task.

Flight tests with unmanned aircraft systems (UAS) have become increasingly important in recent years. Scaled plattforms are not only used as payload carriers, but also for the analysis of novel aircraft configurations or for the assessment of unconventional propulsion systems.

The lower costs as well as the reduction of risks especially in the area of unconventional configurations militate in favor of using unmanned systems. Typical examples for demonstrators used to study flight dynamic effects are the AlbatrossONE [1] of Airbus and the platform of the project FLEXOP (flutter free flight envelope expansion for economical performance improvement) [2]. In the case of the AlbatrossONE, gust loads are minimized by structural interventions in the area of the wing tip, while aeroelastic effects on the wing are observed with the help of the carrier platform FLEXOP. To investigate the impact on flight performance of new technology on an aircraft configuration, a direct comparison between base line configuration and modified aircraft is of interest, as demonstrated by NASA with its Area-I Prototype Technology Evaluation and Research Aircraft (PTERA) [3]. In the research, the PTERA platform was modified to investigate for example combined circulation control [4] and a spanwise adaptive wing [5].

Unmanned aircraft are an important tool to investigate new aircraft configurations and novel aviation technologies at University of Stuttgart, [6]. New technologies or concepts can be flight-tested with little effort, low costs and manageable project risks after initial theoretical investigation. This way, research concepts can reach a much higher feasibility, of particular interest to industry, to close the gap between upstream research and industrial exploitation. This can build a bridge, demonstrating technologies at a higher technology readiness level (TRL). Innovative ideas can be demonstrated and validated in a relevant environment which corresponds to a TRL 5/6. However, there are limitations due to the degree of scaling of the technologies under consideration.

A disadvantage for the investigation of flight performance with scaled unmanned aircraft is the limited data available for the different aircraft system components. When investigating the flight performance, propeller characteristics and efficiency of the individual components of the propulsion system are of specific interest.

The Institute of Aircraft Design (IFB) at the University of Stuttgart has developed the "e-Genius-Mod" test platform based on this background. The platform is modelled on a scale of 33.3% of the electrically powered "e-Genius" aircraft [7], which is also designed, manufactured and operated by the institute. In this particular case, the UAS was realized as Froude-scaled version [6].

Due to the modular design of the testbed, different configurations can be easily realized, which allows an adaptation to different applications. Furthermore, the geometry of the fuselage offers ideal conditions for accommodating various payloads.

For the purpose of identifying the aircraft and measuring flight performance, it is essential to map the aerodynamic parameters of the system. The most important parameters are the aircraft drag and the lift polar. From the correlation of the coefficients for lift and drag of the aircraft, a direct statement about the performance and energy consumption can be derived. The gliding characteristics, the required power installed and the potential range are important parameters that are required to compare different propulsion concepts and identify their impact for a prospective aircraft design.

In order to obtain reliable drag and lift polar, manned test aircraft often would carry out comparative flights with calibrated systems. Especially in the field of gliding, where a precisely identified system is very important, such survey flights are still state of the art today.

In the unmanned area, such survey flights are difficult to carry out. For UAS, a method for a glider model is described by Edwards [8]. Due to the induced drag of the propeller system, this method is not useful for the proposed identification of propeller-driven UAS. A method to identify drag of a propeller aircraft was developed by Norris and Bauer [9]. Alternatively, if available, a model for the propeller wind-milling drag or thrust could be used to apply this method. For the e-Genius-Mod testbed, a different approach was taken. In order to record the performance data, a measurement system was developed that is capable of recording the thrust values required during the flight. Such systems have not yet been installed in large UAS, but have only been used sporadically on manned aircraft. For the aerodynamic characterization of smaller UAS, on-board thrust measurements have been performed in the past, such as presented by M. Bronz and G. Hattenberger [10]. However, for the identification of the flight performance and to show the feasibility of novel technologies in aviation the measurement of drag is essential.

Although the thrust values for the operation of a propeller-driven system can be measured without any problems when stationary, the thrust values during flight are not comparable due to the induced flow to the propeller without prior identification of the propeller via wind tunnel experiments. The efficiency of the drive also changes significantly under different flow conditions. Therefore, a direct, mechanical thrust measurement in flight is of significant advantage.

1.1. Challenge of Current Research Using Unmanned Aircraft as Free-Flight Test Platform

A current research topic at the University of Stuttgart is how to use unmanned propeller-driven aircraft as a test platform to determine their flight performance without knowing specific data of propeller and powertrain. This ability is for example required to compare different propulsion configurations like Wing Tip Propellers (WTP) to a basic configuration in flight. Based on the collected flight data, the estimated benefits of this technology in flight should be validated and demonstrated. For this purpose, the basic in-flight measurement system of the test platform is extended with a system to measure the thrust of the propeller in-flight.

For this, the assumption was made that in steady horizontal flight the thrust performed by the propeller corresponds to the aerodynamic drag acting on the aircraft.

1.2. The Modular Test Platform e-Genius-Mod

The free-flight platform e-Genius-Mod (Figure 1) is established as a technology test bed to demonstrate new technologies for future aircraft design in a relevant environment. The UAS test platform [6] is used for academic and innovative research projects to investigate scaling similarities of free flight models and to demonstrate new technologies up to TRL 6. The modular design of the test bed is ideally suited for the investigation of new aircraft configuration solutions for distributed electric propulsion systems. The size of the aircraft with a wingspan of 5.62 m, maximum take-off weight of 40 kg and a payload capacity of more than 10 kg [6] is suited to perform prospective investigations. The maximum possible flight time is up to 100 min, depending on the particular battery capacity and the payload weight. An overview of the technical data of the e-Genius-Mod is summarized in Table 1.

Figure 1. Free-flight test platform e-Genius-Mod.

Table 1. Technical data of the e-Genius-Mod [6].

Aircraft Parameter	Value
Wing span	5.62 m
Wing area	1.56 m^2
Aspect ratio	20.2
Length	2.95 m
Take of mass *	34.9 kg
Electric drive power	5 kW (Plettenberg Terminator 30/8 Evo)
Max. thrust	156 N
Propeller	RASA 24 × 12
Design speed	24.8 m/s
Battery capacity *	42 Ah (44.4 V)

* flight test characteristic.

The test bed e-Genius-Mod is an extension of the full-scale aircraft for further investigation of electric flight and new aviation technologies. For this reason, the test platform is equipped as a free flight wind tunnel. In order to perform measurements in steady, horizontal flight, an autopilot is used to steer the aircraft along a predefined path of constant altitude and velocity, as described in [11]. The measurement system ensures the synchronous logging of all relevant variables (Table 2), as further detailed in Section 3.

Table 2. Basic measurement equipment of the free-flight wind-tunnel.

Variable	Sensor Type	Unit
linear accelerations rotation rates	IMU IMU	Board computer (Pixhawk4)
position attitude velocity	(IMU, GPS, magnetometer, barometer) estimation via Kalman filter	
true air speed angle of attack angle of sideslip air density	5 hole-probe temperature sensor	AirDataBoom (Vectoflow/VectoDAQ Air Data Computer)
deflection angle of the control surfaces	magnet sensor rotor sensor	Actuator (Volz DA15N)

The basic free flight measurement equipment can be easily expanded by connecting additional sensors to the dedicated bus system used solely for measurement data.

Electric propulsion systems allow for new design alternatives in terms of propulsion integration and aircraft design. To investigate and demonstrate new concepts, a modular testbed like the unmanned scale model is a very useful and flexible tool. The demonstration of innovative concepts like distributed propulsion on a manned aircraft would be expensive and time intensive and not useful for research with open-ended findings at this early stage. With its modular airframe design, the e-Genius-Mod is the basis for an efficient and systematic research of the various effects of distributed propulsion.

Onboard thrust measurements on a UAS pose a particular challenge. On the one hand, it must be possible to carry out reliable, calibrated measurements, and on the other hand, the sensor systems are subject to narrow limits in terms of dimensions and weight. For this purpose, sensor systems specifically designed for the e-Genius Mod were developed, calibrated and tested in wind tunnel experiments before installation.

The thrust values are in many respects informative for the evaluation of the flight controller itself, as well as for the assessment of the efficiency of an engine. In general, they establish a direct relationship to aerodynamic quality and, in case the electrical power consumption is known, allow a direct statement about the overall efficiency of the corresponding powertrain. Especially when considering several distributed engines, it is possible to make a reliable statement about the overall energy balance onboard the platform.

The aim of the thrust measurements is to prove the expected positive effects of the distributed engines in terms of quality, and, in further steps, to make statements about where on the aircraft they can be used most efficiently. Furthermore, previously performed simulations are to be validated with these measured values.

2. Research Objective

To assess and validate the impact of a novel configuration or technology in flight, knowledge of the flight performance is of vital importance. Required basic information in aircraft design are the aerodynamic coefficients for lift (C_L) and drag (C_D) of an aircraft with respect to the angle of attack (AoA). Therefore, the objective of this research is the identification of the $C_L - C_D$, $C_L - AoA$ and $C_D - AoA$ polars in-flight. For the investigation of these coefficients in-flight the following well-known approach is proposed as a starting point.

While the investigations are carried out under cruise conditions, it is assumed as a basis for the investigations that the measurements are carried out in a non-accelerated (steady) horizontal flight. This allows to establish the balance of forces in flight direction and perpendicular to it respectively:

$$\text{DRAG (D)} = \text{THRUST (T)} \tag{1}$$

$$\text{LIFT (L)} = \text{WEIGHT (W)} \tag{2}$$

Thrust will be directly measured between engine and engine mount. The measured force corresponds to the force acting on the aircraft caused by the propeller thrust. With

$$T \approx D = \frac{\rho}{2} * v^2 * C_D * S \tag{3}$$

and the knowledge about the true air speed (v) and air density (ρ), we can directly determine the drag coefficient (C_D) of the aircraft by assuming drag (D) and thrust (T) as balanced. S describes the reference wing area of the test platform. True air speed and air density will be measured with the air data boom installed in the nose of the aircraft (Table 2). A small deviation has to be considered by the frictional forces of linear bearings in the thrust measurement unit which can be eliminated by a calibration.

The lift coefficient (C_L) can be determined by the following equation:

$$W = L = \frac{\rho}{2} * v^2 * C_L * S \tag{4}$$

Since the aircraft is electrically powered by a battery system, the weight is constant throughout the flight. To identify the corresponding angle of attack, the air data boom is used. Angle of attack and angle of sideslip are measured with a five-hole probe. As the measured values of the different sensors are synchronised, the coefficients of lift and drag can be described directly in relation to the AoA. The measured AoA has only to be corrected by its installation position in relation to the wing. In this way, the direct connection between AoA and the corresponding lift and drag values should be representative in flight. Larger values of lift and drag are to be expected with an increasing of the AoA.

3. Approach for the Flight Test Scenario

The approach for the flight test scenario is based on the assumptions made in Section 2 to investigate the flight performance under the condition of a steady horizontal flight and with a zero-wind condition for the atmosphere.

However, considering the reality of free-flight tests, perfect atmospheric conditions will never be achieved. To meet these requirements to some degree, the flight tests are carried out in the early morning on days with calm atmosphere. The impact of atmospheric disturbances was estimated in [12]. To realize a statistical accuracy of the data measured in flight, the data for a single measurement point is collected over a minimum of four legs with two different flight directions.

To determine the thrust during flight, a measuring system is installed between the electric engine and the engine mount to measure the tensile forces. The occurring tensile forces correspond to the thrust generated by the propulsion system.

The thrust measurement is particularly relevant due to its direct correlation with the aerodynamic drag. The investigations are carried out under cruise conditions, i.e., steady, horizontal flight. For the investigation, a flight track is chosen which allows the longest possible horizontal legs. The limiting factor for the tests is the current regulatory framework, which specifies a maximum flight envelope for the tests in the permitted airspace.

The measurement flights are performed with an autopilot in control to achieve steady, horizontal flight and high repeatability. For the test flight a circuit with maximized straight segments (legs) in-between is chosen. Given the airspace restrictions, the leg on which the measurements can be made is nearly 1000 m long. Figure 2 represents the flight path of a measurement flight at 300 m altitude.

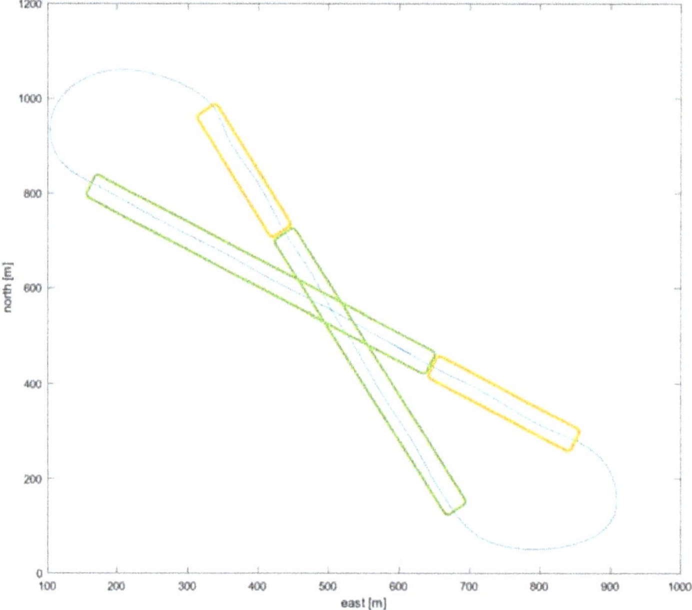

Figure 2. Flight track—sections of measurement (**green**); level off sections (**orange**).

The legs are divided into two segments. The first segment is the "level off section". After the turn, airspeed, altitude and attitude are stabilized in this section. Data in this section will not be considered. In the section of measurement, a stabilized flight attitude is expected and the data will be used for the analysis of lift and drag. As there will be no perfect steady horizontal flight with constant altitude and airspeed, limits for the deviation in roll angle, airspeed and altitude are set, to assign a confidence rating to the single leg that is considered in the evaluation. To optimize the steady horizontal flight, the autopilot controls a constant velocity and altitude.

4. Measurement Unit

The thrust measurement unit consists of two parts. The sensor unit which contains the sensor itself for direct measurement of the thrust and a self-developed data acquisition unit that contains the amplifier module and the interfaces to connect further sensors. The data acquisition unit is also the interface to the data storage.

Recorded values at the propulsion system during flight:

- Thrust
- RPM of the propeller
- Engine current
- Engine voltage

Besides collecting the data with high precision, the complete measurement system is required to feature a lightweight design for use on the UAS. This ensures that the use of the testbed for further studies is not limited by the weight of the measurement unit.

4.1. Sensor Unit

The sensor unit is based on a standard tension/pressure sensor (HBM U9C 100N). For the sensor unit to be used on the unmanned test platform, the main focus is on a robust and compact lightweight design. The unit is designed to allow for the sensor to be installed without torque and bending moments introduced by the engine.

An engine connection unit is being developed for this purpose. This consists of an engine connection plate, a force introduction bolt and linear guide pins to absorb torsion and bending loads. These loads are directed from the linear guiding pins into the linear bearings installed in the load transmission frame. For the inflight thrust measurement, the sensor unit is installed directly between the engine mount and the engine (Figure 3).

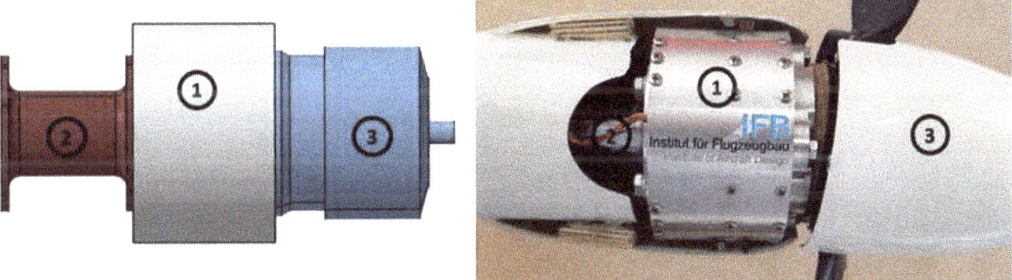

Figure 3. Integrated thrust measurement sensor unit; 1—Sensor Unit; 2—Engine Mount; 3—Electric Engine.

Due to its design, the sensor unit allows two separate load paths (Figure 4). Loads will be introduced in the sensor unit via the engine connection plate. The plate is linearly mounted and transmits the force induced by the propeller thrust as tension force directly to the sensor. The torque and the bending loads resulting from the angular and linear acceleration and vibrations of the engine are derived via the linear bearings into the structure of the sensor unit (housing and frames). This way the thrust can directly be derived from the measured tensile force.

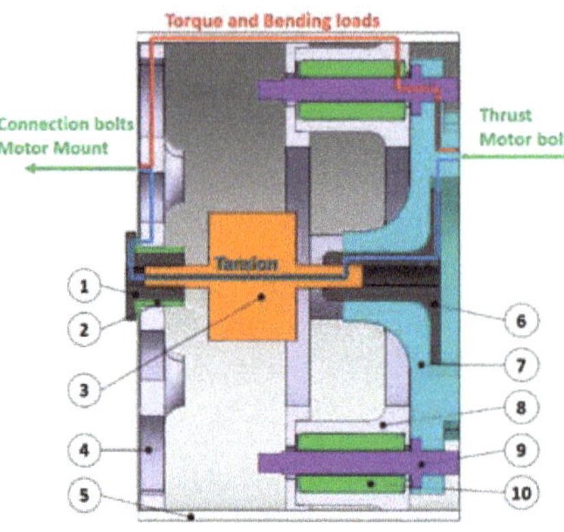

tension rod (traction stop)
friction bearing
tension sensor
connecting frame
housing
force introduction bolt
engine connection plate
load transmission frame
linear guide pins
linear bearing

Figure 4. Sensor Unit (sectional view).

4.2. Data Acquisition Unit

The task of the data acquisition unit (Figure 5) is the acquisition and processing of data from the propulsion system and the transmission of the data to the board computer for synchronous logging. An STM32 microprocessor is used to convert analog signals, to encode and time stamp the data and to handle transmission to the central board computer via an RS485 bus (Figure 6). While this is sufficient to handle the sensor input from magnetic rpm measurements and current and voltage sensors, the force sensor outputs need to be amplified first. This is done using a miniaturized amplifier module (GSV-6CPU) installed on the Data Acquisition Unit. Additional interfaces on the data acquisition board allow to connect further sensors that might be required by future studies using the e-Genius-Mod test platform. With a weight of only 28 g, the data acquisition unit (board without cable) is perfectly suited for use in UAS without restricting the payload capacity.

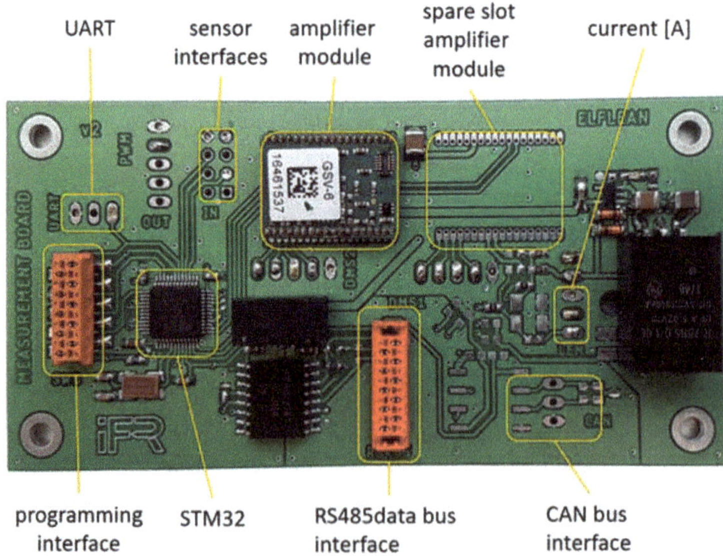

Figure 5. Data Acquisition Unit.

Figure 6. Information flow of the Data Acquisition Unit.

5. Test Flight Analysis

In order to obtain repeatable measurement flights, the flights are automatically guided along a previously planned flight path. The autopilot system [11] adjusts the flight condition to different, constant flight speeds at an altitude above mean sea level of 650 m. This results in stable conditions in stationary horizontal flight, during which the corresponding input variables of the drive system are recorded. The available endurance of 85 min for one test flight allows for more than 68 measuring sections, which are segmented into phases with different flight velocities. With this, all measurements required for a characterization of the aerodynamic parameters of a configuration can be carried out in one single flight Figure 7 provides a synopsis of the legs and Table 3 presents the commanded velocities to identify the coefficients for lift and drag in flight.

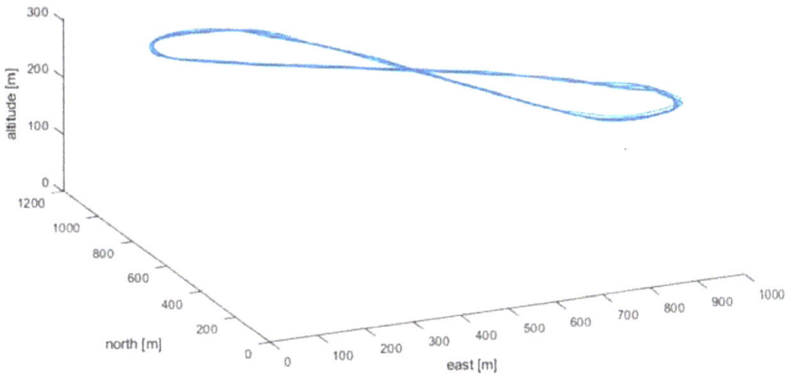

Figure 7. Section of the flight path from a single measurement flight.

Table 3. Mean standard deviation of velocity and altitude.

Velocity (m/s)		Altitude (m)	
Commanded	**Standard Deviation**	**Commanded**	**Standard Deviation**
20	0.23		0.09
21	0.23		0.11
22	0.21		0.12
24	0.31	300	0.12
26	0.25		0.13
28	0.22		0.11
30	0.22		0.08
32	0.20		0.08

Analysis of Static Horizontal Flight

In order for the evaluation to be considered valid using the mean value method, the "section of measurement" must be carried out at a velocity and an altitude as constant as possible. The evaluation of the individual legs shows that both velocity and altitude are measured with only a very small standard deviation (Table 3) depending on the flight speed.

As expected, high fluctuations were detected in the recorded propeller speed, which is also reflected in the measured thrust values. This is a result of controlling the airspeed by using the thrust. The resulting control action can be seen in the resulting fluctuations in measured thrust values (Figure 8). They do not mainly represent a scattering in the aerodynamic drag, but moreover the necessary control action to keep the airspeed constant. This is particularly the case during the two turns.

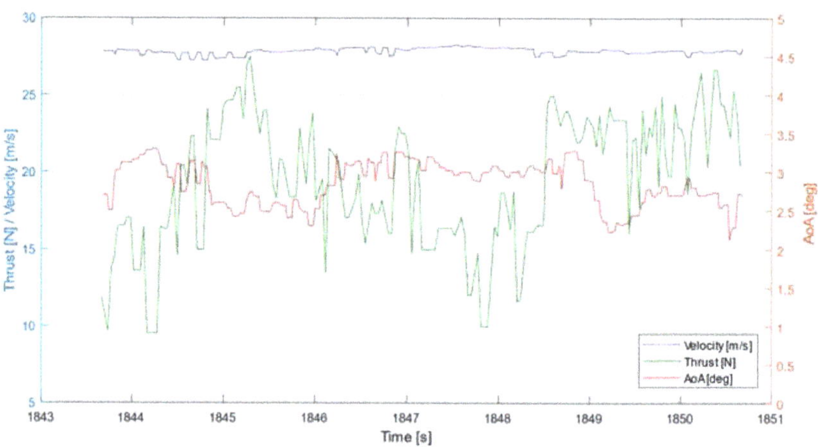

Figure 8. Example: Basic measurements on one single leg (with TAS 28 m/s).

The standard deviations caused by this must be taken into account accordingly in the evaluation and interpretation of the measured values. Nevertheless, the measured values on average give a good picture of the expected results and provide an excellent indication of the performance and aerodynamic quality of the aircraft.

For a first estimation and interpretation of the flight results, the values for thrust, angle of attack, lift and power are approximated as a function of 2nd order in dependence of the velocity as expected from the theoretical consideration. The assumption is based on the approximate relationship, that the lift increases linearly in relation to the AoA and the drag increases quadratic to the lift. Even if this procedure involves a degree of uncertainty in the interpretation of the results, it allows for an initial analysis of the aircraft characteristics, to provide input data for the aircraft design process.

6. Results

This section presents the results of the test flight presented in the previous section. Considering that the UAS should be used as a tool in aircraft design, the analysis focuses on the most interesting aspects for a design engineer. Figures 9 and 10 show the coefficients of lift and drag obtained from the measurement flights and the approximated resulting polar curve. As expected, due to the rpm variation described above, a significantly increased variation of the identified drag coefficient can be seen in Figure 10. Nevertheless, the values can be used for first design calculations.

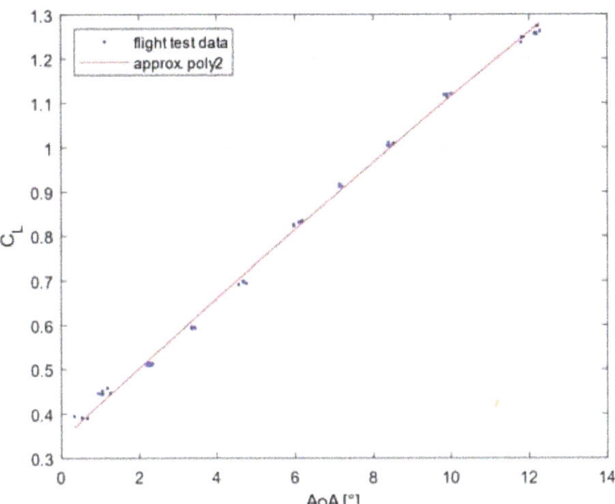

Figure 9. Lift polar.

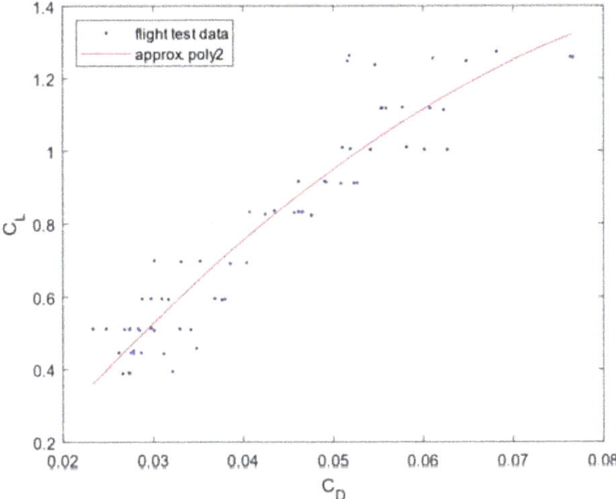

Figure 10. Drag polar.

Even if only a certain regime of the drag polar and the thrust to velocity curve is available, design points of interest (Table 4) can be estimated with the available identification. The points of maximum range and maximum endurance can directly be identified.

Table 4. Estimated points of max. range and max. endurance.

	Max. Range	**Max. Endurance**
Velocity	20.96 m/s	20.15 m/s
Glide Ratio	17.96	17.88
C_L	0.91	0.98
C_D	0.0507	0.0548
Thrust	19.06 N	19.04 N

The point of maximum range is given by the maximum of the glide ratio (C_L/C_D) and can be determined directly from Figures 10 and 11. The point of maximum endurance can be determined as point of minimum required thrust (min drag) in Figure 12.

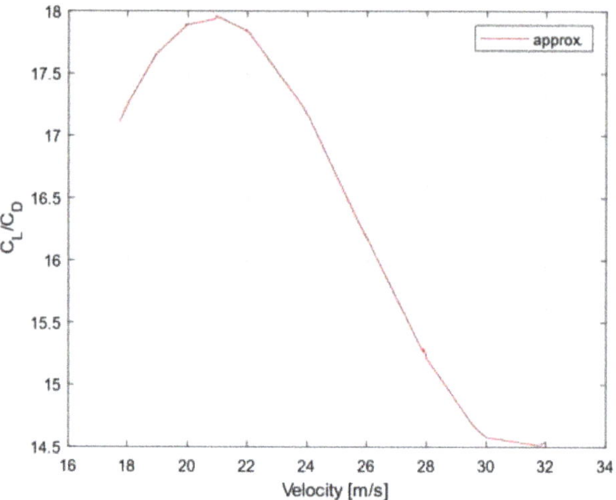

Figure 11. Glide ratio (C_L/C_D) to velocity.

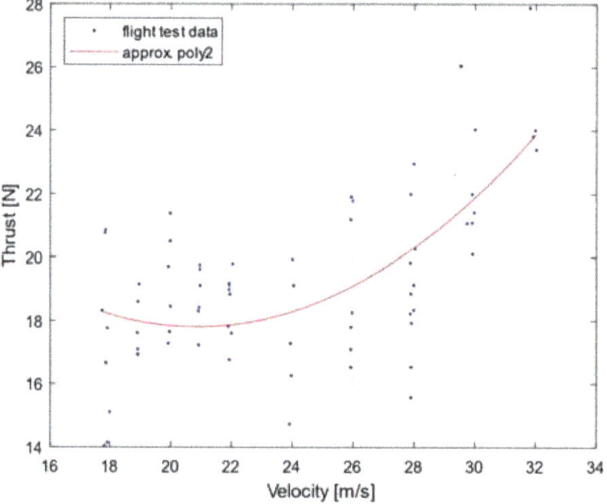

Figure 12. Thrust to velocity.

With the knowledge of the electrical power and the power output of the propeller in terms of thrust, the overall efficiency of the test platform can be determined (Figure 13).

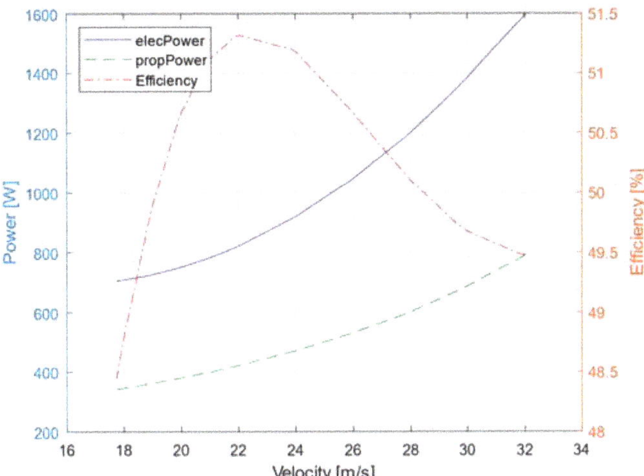

Figure 13. Performance and Efficiency of the free-flight test platform for different velocities.

7. Discussion

The measurements obtained from the flight tests provided the expected results with regard to the thrust curves and the resulting polar. As a first comparison between flight test results and a simulation of the wing airfoil, the gradient of lift coefficient versus the angle of attack was assessed. A comparison with the lift of the airfoil simulated in XFoil [13] shows a very good agreement with reality (Figure 14). The slight deviation of the slope of the two curves results from the influence of the lift distribution on the real wing caused by the boundary vortices.

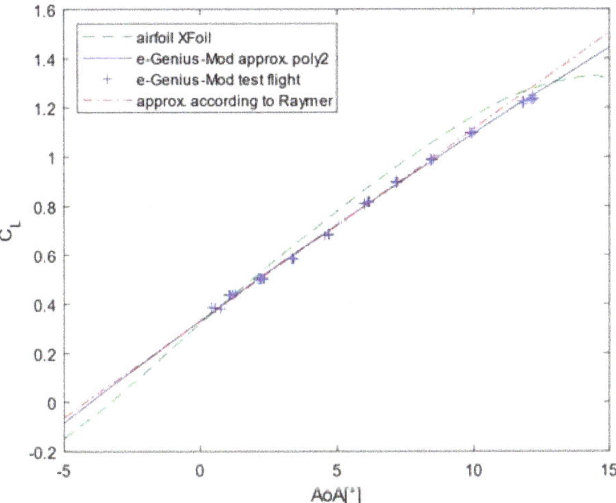

Figure 14. Comparison of XFoil airfoil simulation, semi empirical method and flight test data.

The amount of the deviation depends on the aspect ratio (AR) of the wing and can be calculated approximately [14] as follows with b describing the wing span and S the reference wing area:

$$\frac{dC_L}{dAoA} = \frac{AR}{(2+AR)} * 2\pi \text{ with } AR = \frac{b^2}{S} \tag{5}$$

The angle of attack for zero lift is found as −3.8° from interpolation of the measured values from real flight, which gives a good agreement with the value expected from the simulation. The corresponding installation position of the air data probe in relation to the wing's angle of attack is thus correctly recorded.

An additional validation method was performed, calculating the wing's lift curve slope with a semi-empirical formula [15], applicable for subsonic design.

$$C_{L_{AoA}} = \frac{2\pi AR}{2+\sqrt{4+\frac{AR^2\beta^2}{\eta^2}\left(1+\frac{(\tan AR_{max_t})^2}{\beta^2}\right)}}\left(\frac{S_{exp}}{S}\right)F \quad (6)$$

Within formula (6) η describes the so-called efficiency of the airfoil and is approximately 0.95 according to [15]. β^2 considers the flight velocity respectively the Mach number (M).

$$\beta^2 = 1 - M^2 \quad (7)$$

The wing geometry and the influence of the wing are considered in Equation (7) with AR_{max_t}, S_{exp} and F: AR_{max_t} is the aspect ratio of the wing section with the thickest airfoil chord location. The wing area exposed by the fuselage is considered with S_{exp} and the lift generated by the wing is approximated with F estimated via the diameter d of the fuselage tube under the wing. According to Raymer [15], these can be calculated as follows.

$$F = 1.07\left(1+\frac{d}{b}\right)^2 \quad (8)$$

To consider also the effects of the winglet the effective aspect ratio of the wing is estimated according to [14] taking into account the height of the winglets h in relation to the wing span.

$$AR_{eff} = AR\left(1+1.9\frac{h}{b}\right) \quad (9)$$

The polar of the calculated lift curve slope ($C_{L_{AoA}} = 0.0764$) is plotted in Figure 14 and the zero crossing is shifted by the estimated = 0.33. In a direct comparison of the part of the lift polar which can be considered as nearly linear between the semi-empirical and the measured lift curve slope with $C_{L_{AoA\ measured}} = 0.0787\, for\ (0^\circ < AOA < 5^\circ)$ a good match is achieved with only 3.01% deviation of the curve slope.

In the simplified approach used in this paper, the possible error in the results must also be considered. A relatively small error can be assumed for the lift coefficient. This error is made up of the accuracy of the air data probe (measurement error flow angle: <1.0°, velocities: <1.0 m/s) and a possible installation error which is corrected in the post processing. The drag measurement must be viewed with greater caution. The accuracy of the measuring system has been demonstrated under laboratory conditions, the variation of the measured thrust values due to the fluctuating speed controller (see Section 5) leads to a high standard deviation (mean 26.56%).

Nevertheless, the thrust measuring system generated usable data, which, however, are subject to fluctuations due to controlling thrust values. This must be considered in the evaluation. For further flight tests, control authority for the airspeed control should be reduced in order to ensure that the measured thrust values are smoother and there is less scattering. The proper functioning of the measuring system was verified in laboratory tests before and after the flight tests.

The measurement of the thrust finally gives information about the total drag, which is necessary to establish the drag polar and aerodynamic parameters of interest. From the illustrations in the "results" section the corresponding statements regarding the aerodynamic performance of the system can be derived. The values are within the predicted range. As expected, the scaling of the aircraft brings the optima for best range and flight duration very close together. By measuring the thrust during flight while simultaneously observing the electrical power consumption, a direct statement can also be made about the overall efficiency of the drive train. It can be seen from Figure 13 that the optimum efficiency is found in the range of the speed at which the system is operated in the glide ratio of best range. This is an important step to assess the impact of future design changes to the e-Genius Mod.

8. Conclusions

The measurement flights with the unmanned e-Genius-Mod test platform have shown that on-board thrust measurement provide a direct method to characterize the aircraft. The expectations to draw direct conclusions about the aerodynamic performance could be met. With the presented novel measurement system for direct, in-flight thrust measurements, the unmanned platform represents a useful tool in aircraft design as a "flying wind tunnel". It allows an identification of the flight performance and allows to evaluate the effects of new technologies and configurations. This is of particular interest for new configurations that use distributed electric propulsion to increase efficiency and to enable a future cleaner and greener aviation.

To improve the quality of the flight test results in the future and to allow for drag estimation from non-steady flight conditions, the aircraft dynamics will be included in the drag estimation through acceleration measurements from an inertial measurement unit. Using the thrust measurement system, the characterization of the aircraft could be completed to such an extent that further flight tests with modified settings and distributed propulsion systems can follow. Due to the performance measurements on the baseline configuration it will be possible in the future to draw direct comparisons to flight tests with modified configurations. The measurement system used has been verified and can now also be used for further configurations utilizing distributed propulsion. For this, the sensor unit will be miniaturized to be able to measure the thrust of each propulsion unit in flight. The additional measuring technology installed to record the aerodynamically relevant input variables perfectly complements the thrust measurement of the main propulsor. Therefore, the basic configuration of the e-Genius-Mod is now available for more extensive flight tests and offers ideal conditions for scaled flight tests of all kinds.

Author Contributions: Conceptualization, D.P.B.; methodology, D.P.B., J.D., and O.P.; investigation, D.P.B., J.D., and O.P.; writing—original draft preparation, D.P.B., and J.D.; writing—review and editing, D.P.B., J.D., O.P., S.N., and A.S.; supervision, W.F., and A.S.; project administration, D.P.B.; funding acquisition, D.P.B., S.N., W.F., and A.S. All authors have read and agreed to the published version of the manuscript.

Funding: This research was supported by Federal Ministry for Economic Affairs and Energy on the basis of a decision by the German Bundestag. Project: ELFLEAN—Electric wing tip propulsion system for the development of energy-efficient and noise reduced airplanes (20E1706).

Institutional Review Board Statement: Not applicable.

Informed Consent Statement: Not applicable.

Data Availability Statement: Data not yet available public.

Conflicts of Interest: The authors declare no conflict of interest.

References

1. Wilson, T.; Kirk, J.; Hobday, J.; Castrichini, A. Small scale flying demonstration of semi aeroelastic hinged wing tips. In Proceedings of the International Forum on Aeroelasticity and Structural Dynamics 2019, Savannah, GA, USA, 9–13 June 2019.
2. Süelözgen, Ö.; Wüstenhagen, M. Operational modal analysis for simulated flight flutter test of an unconventional aircraft. In Proceedings of the International Forum on Aeroelasticity and Structural Dynamics 2019, Savannah, GA, USA, 9–13 June 2019.
3. Kuehme, D.; Alley, N.R.; Phillips, C.; Cogan, C. Flight Test Evaluation and System Identification of the Area-I Prototype-Technology-Evaluation Research Aircraft (PTERA). In Proceedings of the AIAA Flight Testing Conference, Atlanta, GA, USA, 16–20 June 2014. [CrossRef]

4. Cogan, B.; Alley, N.; Hange, C.; Nguyen, N.; Spivey, D. *Flight Validation of Cruise Efficient, Low Noise, Extreme Short Takeoff and Landing (CESTOL) and Circulation Control (CC) for Drag Reduction Enabling Technologies*; Presentation, NASA Aeronautics Research Mission Directorate, Seedling Technical Seminar; NASA Aeronautics Research Mission Directorate: Washington, DC, USA, 2014.
5. Ortiz, P.; Alley, N. *Spanwise Adaptive Wing—PTERA Flight Test*; AIAA Aviation Forum: Atlanta, GA, USA, 2018; NASA Document ID 20180004640.
6. Bergmann, D.P.; Denzel, J.; Baden, A.; Kugler, L.; Strohmayer, A. Innovative Scaled Test Platform e-Genius-Mod—Scaling Methods and Systems Design. *Aerospace* **2019**, *6*, 20. [CrossRef]
7. Geiß, S.; Notter, S.; Strohmayer, A.; Fichter, W. Optimized Operation Strategies for Serial Hybrid-Electric Aircraft. In Proceedings of the AIAA Conference: Aviation Technology, Integration, and Operations Conference, Atlanta, GA, USA, 25–29 June 2018. [CrossRef]
8. Edwards, D. *Performance Testing of RNR's SBXC Using a Piccolo Autopilot, Technical Report*. 2008. Available online: http://soaring.goosetechnologies.com (accessed on 1 November 2020).
9. Norris, J.; Bauer, A.B. Zero-Thrust Glide Testing for Drag and Propulsive Efficiency of Propeller Aircraft. *J. Aircr.* **1993**, *30*, 505–511. [CrossRef]
10. Bronz, M.; Hattenberger, G. Aerodynamic Characterization of an Off-the-shelf Aircraft via Flight Test and Numerical Simulation. In Proceedings of the AIAA Flight Testing Conference, Washington, DC, USA, 13–17 June 2016. [CrossRef]
11. Stephan, J.; Pfeifle, O.; Notter, S.; Pinchetti, F.; Fichter, W. Precise Tracking of Extended Three-Dimensional Dubins Paths for Fixed-Wing Aircraft. *J. Guid. Control. Dyn.* **2020**, *43*, 2399–2405. [CrossRef]
12. Pfeifle, O.; Fichter, W.; Bergmann, D.; Denzel, J.; Strohmayer, A.; Schollenberger, M.; Lutz, T. Precision performance measurements of fixed-wing aircraft with wing tip propellers. In Proceedings of the AIAA Aviation 2019 Forum, Dallas, TX, USA, 17–21 June 2019. [CrossRef]
13. Baden, A. Entwurf eines Freiflugmodells des E-Motorseglers e-Genius unter Berücksichtigung der Aerodynamischen Vergleichbarkeit. Bachelor's Thesis, University of Stuttgart, Institute of Aircraft Design, Stuttgart, Germany, 2016.
14. Schlichting, H.; Truckenbrodt, E. *Aerodynamik des Flugzeuges*; Springer: Berlin/Heidelberg, Germany, 2001; Volume 2. [CrossRef]
15. Raymer, D.P. *Aircraft Design: A Conceptual Approach*, 2nd ed.; AIAA: Washington, DC, USA, 1989; ISBN 0-930403-51-7.

MDPI

Article

Modeling and Control of a Modular Iron Bird

Luciano Blasi [1], Mauro Borrelli [2], Egidio D'Amato [3,*], Luigi Emanuel di Grazia [1], Massimiliano Mattei [4] and Immacolata Notaro [1]

1 Department of Engineering, University of Campania Luigi Vanvitelli, 81031 Aversa, Italy; luciano.blasi@unicampania.it (L.B.); luigiemanuel.digrazia@unicampania.it (L.E.d.G.); immacolata.notaro@unicampania.it (I.N.)
2 Protom Group, 81043 Naples, Italy; mauro.borrelli@protom.it
3 Department of Science and Technology, University of Naples Parthenope, 81043 Naples, Italy
4 Department of Electrical Engineering and Information Technologies, University of Naples Federico II, 80125 Naples, Italy; massimiliano.mattei@unina.it
* Correspondence: egidio.damato@uniparthenope.it

Abstract: This paper describes the control architecture and the control laws of a new concept of Modular Iron Bird aimed at reproducing flight loads to test mobile aerodynamic control surface actuators for small and medium size aircraft and Unmanned Aerial Vehicles. The iron bird control system must guarantee the actuation of counteracting forces. On one side, a hydraulic actuator simulates the hinge moments acting on the mobile surface due to aerodynamic and inertial effects during flight; on the other side, the actuator to be tested applies an active hinge moment to control the angular position of the same surface. Reference aerodynamic and inertial loads are generated by a flight simulation module to reproduce more realistic conditions arising during operations. The design of the control action is based on a dynamic model of the hydraulic plant used to generate loads. This system is controlled using a Proportional Integral Derivative control algorithm tuned with an optimization algorithm taking into account the closed loop dynamics of the actuator under testing, uncertainties and disturbances in the controlled plant. Numerical simulations are presented to show the effectiveness of the proposed architecture and control laws.

Keywords: iron bird; hydraulic system; flight simulator; force control; PID control

Citation: Blasi, L.; Borrelli, M.; D'Amato, E.; di Grazia, L.E.; Mattei, M.; Notaro, I. Modeling and Control of a Modular Iron Bird. *Aerospace* **2021**, *8*, 39. https://doi.org/10.3390/aerospace8020039

Academic Editor: Andreas Strohmayer

Received: 15 December 2020
Accepted: 29 January 2021
Published: 2 February 2021

1. Introduction

An iron bird is often defined as an "aircraft which does not fly", used to validate the design and verify performance and stability [1] of aircraft components and systems. It is a mechanical representation of aircraft systems, including actuators, arranged on a frame instead of being inside the fuselage or the wings, completely visible, to test the integration of components as the integration of the actuation systems for aerodynamic surfaces and landing gears into the airframe, and their links to the power supplies and the flight control system.

Since the beginning of aviation, giants steps have been made on actuator technology. Currently, fly-by-wire aircraft use hydraulically supplied actuators in order to control mobile surfaces, but the birth of the All (or More) Electric Aircraft philosophy is changing this paradigm towards increasing the use of electrically supplied actuation systems that is usually called power-by-wire (PBW) actuation.

The advantages in using electrical versus hydraulic power on aircraft are well described in [2–5]. Good power density at the level of power network, more efficiency, more options and ease of command, dynamic reconfiguration of power paths are some of them. However, the technology maturity level of PBW is still low in terms of returns of operational experience. Moreover, some technological issues as poor local exchange of heat generated by energy losses have to be studied to increase their diffusion.

Electro-Mechanical Actuators (EMAs) represent a very good alternative also for small commercial aircraft and UAVs, because of the lower cost and the general lower maintenance effort needed with respect to Electro-Hydraulic Actuators (EHAs).

One of the earliest studies on EMA is [6], where the researches completed by Boeing and Rockwell in the eighties are summarized. Boeing concluded that their baseline aircraft would remain roughly the same mass when switched to EMAs, Rockwell predicted an increase in mass, but in both cases, when EMAs were introduced, the mass of the aircraft secondary power systems decreased.

In [7], a review of EMAs is presented, starting from a brief history on the use of this type of actuators on aircraft. It also explains the main ways to test an EMA, from tests in a room temperature situation, to tests in a thermal vacuum environment, to tests on an iron bird. Generally speaking, the iron bird testing is intended to validate also the linkages of the EMAs with other systems, e.g., the power system and the Flight Control System (FCS).

Iron birds are useful in both industrial and scientific worlds. In the literature, although several authors use iron birds to prove the effectiveness of their solution about flight control algorithms, without focusing on its design, their works represent useful examples and the state of the art about this topic. In [8], Airbus explains how the electrical flight control system used on its aircraft is validated through iron bird testing. In [9], the power plant system of a tilt rotor UAV was verified by means of iron bird ground tests. In [10], a modular iron bird was designed to test new concepts in the flight control area, however, the modules are used for a specific aircraft. In [11], a new methodology for Prognostic Health Management systems of electro-mechanical flight control validation using an iron bird is proposed. The MIB (Modular Iron Bird), which is currently in the construction phase at PROTOM, developed in collaboration with the University of Campania "L.Vanvitelli", is a new concept of geometrically simplified and modular iron bird to perform verification tests of equipment developed for FCSs of small/medium aircraft belonging to the general aviation and unmanned aerial vehicles. It is mainly designed to test the control and actuation system moving the aerodynamic control surfaces and their integration in the flight control system. On the other hand, the test objective is quite extensive, ranging from the analysis of the actuator dynamic response, stability, and accuracy, to the efficiency and power capability, to the response to fault.

Usually, iron birds are custom ground-based test device used for prototyping and integrating aircraft systems (e.g., actuators) during the development of new aircraft designs. Systems are installed into the iron bird so their functions can be tested both individually and in correlation with other systems. They are expensive and often not reusable, being economically inaccessible for many small companies. However, the growth of unmanned aircraft market and the lowering of costs in the design of small aircraft and their actuators, makes attractive the use of a network of affordable facilities able to test systems with limited costs. The MIB architecture tries to give an answer to this increasing need.

The paper is organized as follows. Section 2 deals with the MIB control system architecture description. The mathematical model of the plant to design and validate the proposed control laws is described in Section 3, whereas in Section 4 the force control law is presented together with the algorithm used to tune the control gains. Finally Section 5 illustrates some numerical simulation results.

2. The MIB Testing Facility

The basic elements composing MIB are:

- the Data Acquisition and Control System (DACS). This is deployed on different hardware components, namely:
  - the Real-Time Process Controller (RTPC) implementing the so called Process Control Algorithms for the tracking of desired loads on the aerodynamic surfaces and the actuator unit under test reference positions;

    - the host PC implementing the User Interfaces (UI) for test preparation, on-line monitoring of the test variables, data archiving, post-processing functions and off-line visualization of the test results;
- a given number of Test Stands (TS) (see Figure 1), actually five, which are composed of:
    - a loading hydraulic actuator linked to a rigid mobile surface;
    - a set of pressure, position and load sensors, needed for control purposes and to acquire and archive test data;
    - the EMA Unit Under Testing (UUT) which is mechanically loaded by the hydraulic actuator;
- the Hydraulic Power Unit (HPU) to supply hydraulic pressure to the actuators on all the TSs;
- the Electrical Power Systems (EPSs).

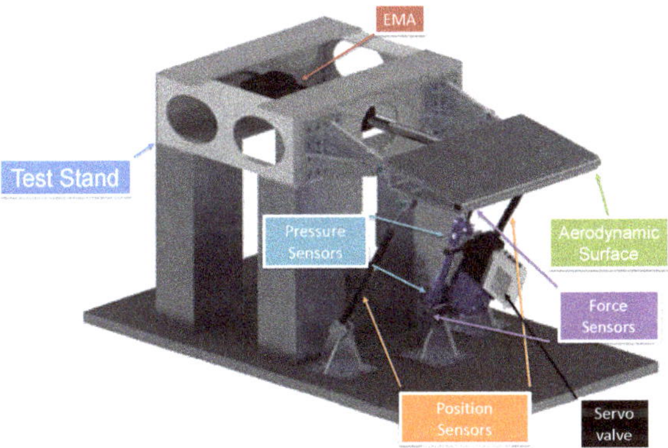

Figure 1. MIB (Modular Iron Bird) Test Stand overview.

The modularity of the proposed Iron Bird is based on the presence of several TS that can adapt to different actuators and layouts; actuators can be tested in parallel or one at a time. Also the hydraulic actuator simulating the aerodynamic and inertial loads can be chosen in function of the loads requested by the test. At the moment, a single rod actuator is adopted to mount a double load sensor, but in future a different actuator can be mounted. Moreover the flexibility offered by the automation system allows to plan different tests including those simulating the aircraft flight and related maneuvers.

2.1. The Process and Automation Control System

As shown in Figure 2, the MIB Process Control System implements the following subsystems:

- the test set up and automation functions (Green blocks),
- the real time simulator of the flight dynamics including flight controllers (Blue blocks),
- the actuators and sensor for the force control (Red blocks),
- the EMA actuator UUT (Dark Red block),
- the estimation of the surface position, the estimation of the force acting on the aerodynamic surface, and both the position and force controllers (Yellow blocks).

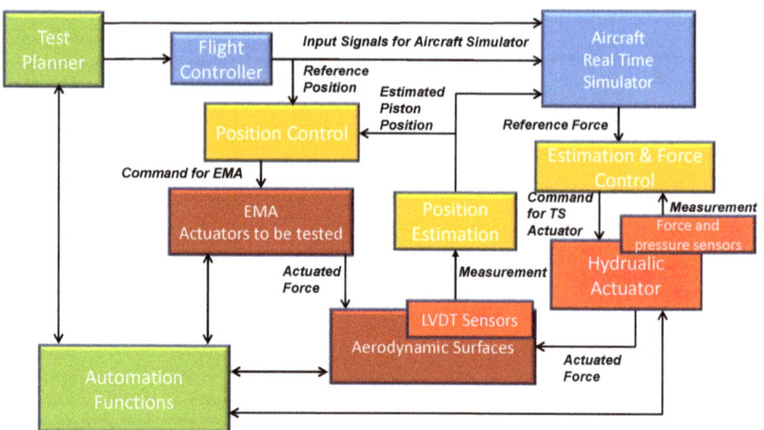

Figure 2. MIB Automation and Control System Architecture.

Red, dark red and yellow blocks are replicated for each TS active.

The MIB Test Planner (TP) manages the test setup procedure implementing the UI, and it is able to generate the input signals for the real time simulator and the reference signals to be passed to the RTPC.

The Automation Function block implements the functions to monitor the overall system health. These functions are implemented with a time step of 0.1 s.

The MIB Aircraft Real Time Simulator (ARTS) is a flight simulator, implementing the equations of the rigid body aircraft dynamics with moving surfaces, including the equations to compute hinge moments on mobile surfaces because these are needed to calculate the flight loads on EMA actuators. The use of simulators in the aircraft design and testing phase is well recognized by the scientific community [12].

The ARTS is also connected with a simulator of the flight control laws. Its inputs are the mobile surface positions, measured by sensors if the related stand is active, otherwise the desired position from the flight control laws is used. In this case the reference commands to the flight controller are produced by the TP with a time step of 0.1 s. The ARTS outputs are the loads to be actuated on the mobile surface, and hence to be passed to the Force Control block as reference signals. Gravity effects due to the aerodynamic surface installation are taken into account and compensated before generating reference loads.

Although this can be easily replaced, the standard flight control law module has a classical structure for fixed wing aircraft, and is based on three nested loops (see Figure 3, namely: a Stability Augmentation System (SAS), a Control Augmentation System (CAS), and an outer Autopilot. These loops can be activated by the User.

SAS control action implements a pitch, roll and yaw damper. CAS implements both a pitch and roll rate control. Finally the autopilot can control altitude, true air speed, and heading.

Due to the use of Matlab/Simulink simulation blocks, both the rigid six degrees of freedom (DoF) aircraft model with moving surfaces and the flight control law implemented can be readily replaced with a more detailed model including flexibility or other phenomena (e.g., ground effects during landing and take-off), and more complex control laws.

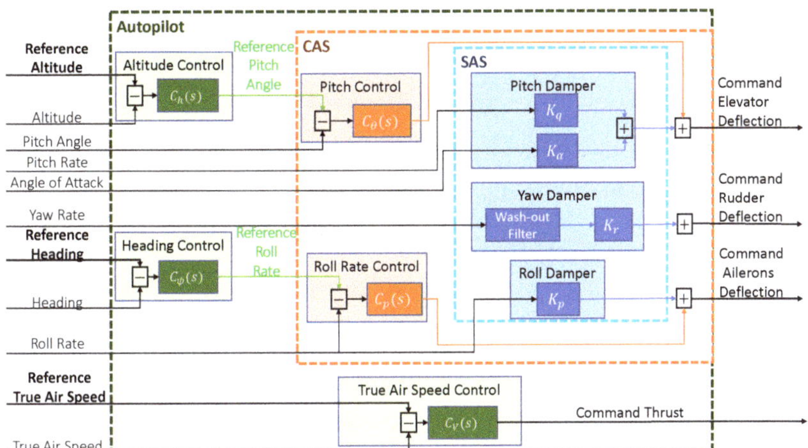

Figure 3. Schematic of the Flight Control Laws.

The Force Estimation and Control block implements the algorithm to estimate and control the load on the mobile surface and hence on UUT. The input to this block are the reference force to be actuated, the measurements from load cells and pressure sensors, and the position of the aerodynamic surface. The output is the command to the servo-valve controlling the flow-rates to the hydraulic actuator. Two force measurements are available at the top and the bottom of the hydraulic cylinder. On the basis of these measurements, pressure measurements into the cylinder chambers, and a position measurements to correct for geometrical effects, a more reliable estimation is obtained via Kalman Filtering.

Similarly, the Surface Position Estimator block can estimate the position of the mobile surfaces on the basis of the position measurement coming from two LVDTs and a speed measurement produced by a digital encoder. LVDTs are the position sensors shown in Figure 1 from which is possible to calculate the angular position of the mobile surfaces. This estimation block is divided from the Position Control block for flexibility purposes. In fact it may happen that the UUT is already equipped with a position control. In this case the MIB Position Control is disabled but a position estimation is still needed to drive the aircraft simulation.

2.2. *The Test Planner*

The Test Planner is a software module for the test preparation. It implements a GUI (Graphical User Interface) to set test parameters and runs on the host PC, in order to supply test parameters to the RTPC. During the test, variables are monitored with a refresh rate of 0.1 s.

Different kind of tests can be performed on MIB depending on the control modes foreseen for the EMA and for the hydraulic system, and on the use of the ARTS. EMA can be controlled in two different ways:

- EMA Direct Control: the EMA is directly controlled, bypassing the Position Control;
- EMA Closed Loop Position Control: the position controller is activated to compute the EMA control signal.

Hydraulic actuators can be controlled in two different ways:

- Hydraulic system Direct Control: the Force Controller is disabled in order to directly control the valve input signal. This test type can be useful in the first phases of the plant operation, in order to identify parameters of the hydraulic system dynamic model;
- Hydraulic system Closed Loop Control: the Force Controller is enabled and computes the servo-valve command on the basis of reference forces.

Finally, the ARTS can be operated in two different ways:

- Pre-programmed mode: the ARTS simply forwards waveforms designed through the TP GUI (both reference forces to the hydraulic system, and reference positions to the EMA controller).
- Self-consistent mode: the ARTS, on the basis of a reference manoeuvre defined by the TP, and the measured aerodynamic surface positions, calculates in real time the reference signals to the EMA position controllers and to the force controller.

The TP also provides initial conditions for the flight simulation if the ARTS is enabled. This is obtained via a suite of tools implemented in the TP module, including trim and linearization.

3. Control Oriented Plant Mathematical Modeling

Hydraulic systems are typically adopted in iron birds to generate loads for their compactness and performance. However, they also present some modeling and controllability issues, due to uncertainties and hard nonlinear dynamic behaviours. Non-linearities are mainly due to flow rates through servo-valves. Uncertainties and disturbances can derive from pressure fluctuation across the pumps or from fluid physical characteristics, such as the bulk modulus, but also from the deformation of the chambers and piston walls, and other effects related volume uncertainties.

Several modeling approaches can be found in the literature. In [1], modeling is based on Newton equation for UUT piston and for load actuator movement. In [13,14], the model is obtained directly in the Laplace domain for the load actuator. In [15,16], the mathematical model is based on Newton equation for the electro-hydraulic piston movements, and a mass balance is used for servo-valve modeling, controlled with input flow rate. In [17,18], the Newton equation is used for piston displacement: Coulomb friction model is adopted taking into account some model uncertainties. Friction, that is a relevant aspect in this field, is also modeled in [19–21]. Relevant works can be found in [22–26].

The objective of the following preliminary modeling is not to precisely describe the physical phenomenon behind the hydraulic system, but to formulate a sufficiently rich, though simple, parametric model, to be tuned with experimental data collected on the plant, on which a first validation of the controller structure and tuning of the controller gains can be done.

An important source of uncertainty in the modeling of the overall system is the actuator UUT which will change from test to test although, for the validation of the plant control system, a specific and known actuator will be used.

With the above premises, the following main modeling simplifications are made:

- the HPU (a Rexroth Cytropack system) comprehensive of an accumulator and regulation system of the pressure inside the accumulator is modeled as a whole with a simple uncertain dynamics;
- pipes dynamics are neglected. In fact, although they have been modeled, it has been verified in simulation that the dynamics are fast enough with respect to the characteristic times of interest, whereas their friction can be concentrated in the neighbouring elements;
- wall deformation are neglected as their uncertain effect on closed loop control can be somehow evaluated with a larger uncertainties on the air trapped in the hydraulic fluid;
- hard non-linearities as backlashes and dead zones and saturation are not taken into account in the controller design phase but simulated in the control performance evaluation.

One single TS is assumed to describe the mathematical modeling in more details. The hydraulic load generation system, schematically shown in Figure 4, is composed of four main elements: an open reservoir, an HPU, an electro-hydraulic servo-valve, and a cylinder connected to the movable surface on which the load is applied. The first two components are common to all the TSs.

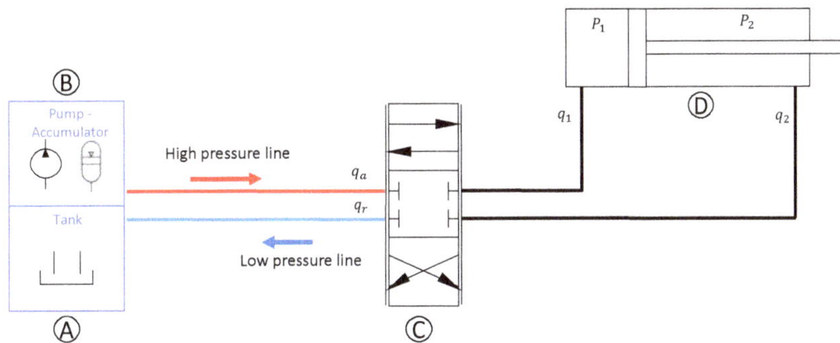

Figure 4. Schematic of the Hydraulic Load Generation System.

The HPU pump converts mechanical into hydraulic power, injecting fluid in the supply line. An accumulator is included in the power unit. This is a pressure storage reservoir in which a relatively small quantity of fluid is held under a pressure which is regulated by the system to a given value. The accumulator allows to react quickly to a transient flow rate demand, to smooth out possible pulsations, and avoid significant pressure losses. On the other hand the pump guarantees the mass flow rate needed by the hydraulic cylinder to actuate significant forces in short time.

The servo-valve modulates the flow rate into the two cylinder chambers moving the piston that generates the force on the UUT.

Other standard hydraulic and electric components, needed for a correct and safe operation of the plant, are not modeled and described, as not strictly related to process control validation and design. First of all, compressibility of the oil in the hydraulic circuit has to be taken into account because of the high pressure. In fact the fluid density ρ_r depends, assuming the definition of isothermal bulk modulus, on the fluid pressure P_r:

$$\rho_r = \rho_0 \cdot exp((P_r - P_0)/\beta) \tag{1}$$

where ρ_0 is the density at the atmospheric pressure P_0, and β is the fluid bulk modulus.

Due to the presence of the trapped air in the plant, the bulk modulus cannot be assumed constant, and consequently it is modeled as a nonlinear function of the hydraulic pressure:

$$\beta = \beta_0 \frac{1 + \alpha(\frac{P_0}{P})^{1/\gamma}}{1 + \alpha \frac{P_0^{1/\gamma}}{\gamma P^{\frac{\gamma+1}{\gamma}}} \beta_0} \tag{2}$$

where β_0 is the fluid bulk modulus at P_0, α is the trapped air volume ratio at P_0, P is the fluid pressure, γ is the air specific heat ratio.

Hydraulic systems need a finite volume of recirculating liquid to work. A reservoir permits to accumulate the fluid getting rid of the trapped air. The reservoir has a breather to maintain the pressure constant at the atmospheric value. The dynamic equations relating volume, pressure and flow rates are the following:

$$\dot{V}_r = q_{r,in} - q_{r,out} \tag{3}$$

$$P_r = \rho_r g h \tag{4}$$

where V_r is the liquid volume in the reservoir, $q_{r,in}$ and $q_{r,out}$ are the input and output volumetric flow rates, respectively, P_r is the pressure at the bottom of the reservoir, g is the gravity acceleration, $h = V_r/A_r$ is the height of fluid column, with V_r and A_r volume and cross section area of the cylindrical reservoir, respectively.

The hydraulic pump is a device used to convert mechanical power into hydraulic power. Modern HPUs can adapt fluid flow rate to keep a fairly constant pressure in an accumulator downstream the pump.

This accumulator is used to store a limited amount of fluid to smooth out pressure oscillations and to make the downstream components dynamic response less sensitive to the upstream conditions.

The whole HPU is modeled with the following uncertain first order system:

$$\dot{P}_p = K_H(q_p - q_v) \tag{5}$$

$$q_p = k_p(P_{p,ref} - P_p) + k_I \int (P_{p,ref} - P_p)d\tau \tag{6}$$

where P_p is the pressure in the accumulator. The maximum pressure delivered by the pump is 200 bar. q_p and q_v are the flow rate provided by the hydraulic pump, and the flow rate across the servo-valve, K_H is a constant obtained from the linearization of the system behavior around its operating conditions and is assumed to be uncertain. $P_{p,ref}$ is the reference pressure in the accumulator controlled by the HPU local pressure control system.

The force control is obtained with a servo-valve converting electrical command signals into a spool valve command to control the flow rate. Under simplifying assumptions and the absence of leakages, the servo-valve dynamics can be approximated as a first order dynamics:

$$\dot{X}_{sv} = -\frac{1}{\tau_{sv}}X_{sv} + \frac{u_{sv}}{\tau_{sv}} \tag{7}$$

$$X_{sv,min} <= X_{sv} <= X_{sv,max} \tag{8}$$

where X_{sv} is the servo-valve spool displacement, τ_{sv} is the time constant, and u_{sv} is the control action computed on the basis of error between desired force acting on the cylinder and the estimated one, $X_{sv,min}$ and $X_{sv,max}$ are the limits on the servo-valve opening variable. The flow rate through the servo-valve is

$$q_{sv} = C_d w X_{sv}\sqrt{\frac{2(P_u - P_d)}{\rho_{sv}}} \tag{9}$$

where C_d is the servo-valve discharge coefficient, $w = A_{sv}/X_{sv,max}$ is the servo-valve area gradient, A_{sv} is the servo-valve port area, ρ_{sv} is the fluid density, P_u is the pressure upstream the servo-valve, and P_d is the pressure downstream the servo-valve.

Assuming that P_i ($i = 1,2$) is the pressure in chamber i with $i = 1,2$, P_u and P_d are identified on the basis of the servo-valve position. If $X_{sv} \geq 0$, chamber #1 is connected to the fluid supply line, then $P_u = P_{hp}$, $P_d = P_1$; if $X_{sv} < 0$, chamber #1 is connected with the return line, then $P_u = P_1$, P_d is equal to the pressure of the return line that can be assumed as the pressure P_r in the reservoir. $\rho_{sv,s}$ and $\rho_{sv,r}$ being the servo-valve densities in the supply and return lines, respectively, the flow rates in the hydraulic actuator chambers are the following:

$$q_{sv,1} = \begin{cases} C_d w X_{sv}\sqrt{2(P_{hp} - P_1)/\rho_{sv,s}} & X_{sv} \geq 0 \\ C_d w X_{sv}\sqrt{2(P_1 - P_r)/\rho_{sv,r}} & X_{sv} < 0 \end{cases} \tag{10}$$

$$q_{sv,2} = \begin{cases} C_d w X_{sv}\sqrt{2(P_2 - P_r)/\rho_{sv,r}} & X_{sv} \geq 0 \\ C_d w X_{sv}\sqrt{2(P_{hp} - P_2)/\rho_{sv,s}} & X_{sv} < 0 \end{cases} \tag{11}$$

Hysteretic effects due to static friction, and deadband, as reported by the servo-valve manufacturer are also included in the dynamic modeling used to validate the control law.

In the hydraulic actuator, the piston is forced by the pressure difference between two chambers. Volume variation in chamber i is $\dot{V}_i = (-1)^i \dot{X} A_i$, $i = 1,2$. The chamber volume variation is related to pressure variations as follows:

$$\dot{P}_i = (q_i - \dot{V}_i)\frac{\beta}{V_i} \qquad i = 1,2 \tag{12}$$

$$\ddot{X}_p = ((P_1A_1 - P_2A_2) - b_p\dot{X}_p - k_pX_p - m_{ac}g\cos(X_p/l_{ac}) + R)/(m_p + J_{ac}/l_p^2 + J_{EMA}/l_p^2) \tag{13}$$

where X_p is the piston position with respect to the center of the cylinder, A_i the i-th chamber cross section, P_i the chamber pressure, q_i the i-th chamber input volumetric flow rate, b_p the damping coefficient, k_p the elastic coefficient, m_{ac} is the movable surface mass, l_{ac} the distance between the surface center of mass and the hinge axis, m_p the piston mass, l_p the distance between the piston and the hinge axis, J_{ac} and J_{EMA} the inertia moments of the surface and the EMA, respectively, with respect to the hinge axis, and R the force generated by EMA. The mechanical linkages are supposed rigid, however backlashes are introduced in the numerical simulations to evaluate their uncertain effects of control laws.

3.1. EMA Actuator under Testing

Special attention must be paid to the modeling of the EMA actuator UUT moving the mobile aerodynamic surface. In fact, being the object of the test, its model depends on the particular test itself. For the control system architecture validation and testing a specific EMA is used. This is modeled with a first order dynamic system

$$\dot{R} = -\frac{1}{\tau_{EMA}}R + \frac{u_{EMA}}{\tau_{EMA}} \tag{14}$$

where R is the generated force, τ_{EMA} is the EMA time constant, u_{EMA} is the control action generated by the position controller. In addition a rate limiter and a saturation is added for the force controller performance assessment.

In case the UUT comes as a black box to be tested, a procedure to identify the EMA dynamic model including non-linearities, based on the use of Neural Networks and Non-linear AutoRegressive eXogenous (NARX) dynamic models [27–30], has been formulated and validated with numerical simulations.

4. Force Control Algorithm

An interesting problem to be taken into account for the Process Controller, is that force control has to counteract the reaction of a position controller implemented on the UUT actuator side.

For this reason, the Force Control Algorithm must satisfy very tight requirements on the closed loop speed of response and the capability to counteract disturbances induced by the position control of the UUT.

In fact, both the hydraulic and EMA actuators are mechanically linked to the movable surface and each controller becomes a source of disturbance for the other. Looking at force control, the piston speed induced by position control, causes a variation of the volume of the two cylinder chambers, namely $V_i, i = 1,2$. Therefore, since the pressure rate in each chamber $\dot{P}_i, i = 1,2$ depends on the volume derivative $\dot{V}_i$ according to (12), if this effect is not properly compensated by dumping the necessary amount of fluid from one chamber to the other, strong overshoots of the controlled force occurs.

In the scientific literature, this problem has been dealt with in [1], where a Proportional-Integral (PI) controller with feed-forward compensation is adopted to counteract UUT speed disturbances on force control. This control approach, called *traditional*, suffers from synchronization accuracy issues. In [31] structured guidelines for the synthesis of dynamic force simulators that are required for the testing of high speed aerospace actuators are developed. Realistic and proven solutions at both test bench hardware and control design levels are provided. In [13], asynchronous controller is used to to regulate the loading actuator operating velocity synchronously with UUT actuator so as its actuator motion disturbance can be decoupled. In [32], the traditional control is achieved with an adaptive control approach based on the relationship between speed and current in electro-mechanical actuators.

In [15], it is shown that the closed loop poles of the UUT, controlled in position, are zeroes of the force driven open loop transfer function for load actuation. This leads to a performance limitation when a PID control is used for force control. A Quantitative Feedback Theory (QFT) based control technique is proposed to overcome this problem. QFT is used also in [16,33,34], a force controller is designed with a loop shaping technique.

In [35], several techniques are analysed: feedback force control with force direct measurement using a PI action, combinations of state feedback control schemes with state observers and velocity feedforward compensation actions. In [36], a PID controller is used and disturbances due to UUT speed are compensated with a signal proportional to a model based estimation of UUT acceleration.

A PID controller is proposed in [37], where three tuning methodologies are presented: optimal time tuning PID, optimal frequency tuning, and multi-objective PID. Also in [38], a PID is used for both load actuation and for rejecting UUT velocity disturbances. This controller is also used in [39] for servo-valve pressure control.

In [40], a feedforward controller and a feedforward inverse control with disturbance observer are used for load actuation. The former is used to reject disturbances due to UUT movements, the latter is used for the other disturbances. In [17,41] adaptive reference controls are presented. In [42], a fuzzy logic MIMO controller is proposed. Fuzzy logic is also used in [19,43] to tune a PID, in order to increase robustness. Fuzzy logic is also used in [44]. In [45], a Minimal Control Synthesis with integral action is used with adaptive gains, while in [20], a back-stepping based controller is adopted. Other works are based on H_∞ control [46], nonlinear adaptive optimal control strategies [47], sliding mode in [48] and an adaptive decoupling synchronous controller in [49].

In [50], a Model Predictive Controller is proposed by the authors dealing with loads generated by fast moving EMAs. The proposed controller uses a simplified model of hydraulic plant, achieved by neglecting the faster dynamics of pump, reservoir, pipes and accumulator. However, the computational burden of the predictive approach does not yet allow an online implementation with low-cost hardware solution.

The present work is focused on a preliminary classical PID control, tuned on the mathematical model of the plant, in view of a model calibration, controller parameters re-tuning, and first closed loop tests.

In fact force control is based on a Proportional-Integral-Derivative (PID) scheme, which computes the input to the servo-valve in order to act on the input/output flow rate in the two chambers of the hydraulic actuator to control pressure. The closed loop actuator control system has to be tuned to have a dynamic response which is significantly faster than the EMA position control time response. In this way the design of force control can be reasonably decoupled from the position controller dynamics.

Therefore, by taking into account the frequency separation between the two controllers, the force control gains have been optimized to guarantee a certain degree of robustness and given performance.

In practice the following quantities are defined:

- a set of uncertainties, defined in terms of variations with respect to nominal value of the following parameters:
  - Bulk modulus β_0
  - air trapped volume ratio α,
  - Servo-valve time constant τ_{sv}.
  - Discharge coefficient and servo-valve area product $C_d A_{sv}$,
  - EMA time constant τ_{EMA}.

A family of N_u plant models, implementing the above uncertainties, is defined:

$$\dot{x}_j = f_j(x_j, u) \tag{15}$$

$$y_j = h_j(x_j, u) \tag{16}$$

($j = 1, \ldots, N_u$), where x_j, y_j, and u are the state vector, the controlled output (force on the aerodynamic surface) and control input (servo-valve command), respectively;

- a set of N_S operating scenarios to evaluate the robust performance of the controller over the N_u models by means of the following cost function:

$$J(e_j^k(\cdot), u(\cdot), x_{0j}, t_0, t_f, w), \quad k = 1, \ldots, N_S \tag{17}$$

with $y_{ref}^k - y_j^k$ the output tracking error, $y_{ref,k}$ being the force reference signal, x_{0j} the initial state at time t_0, t_f a finite time for the cost function evaluation, and w a suitable vector of weights;

- a parametric structure for the force controller:

$$u(t) = u_{CL}(p, e_{[t_0,t]}(\cdot), t) \tag{18}$$

with $e = y_{ref} - y$. In our case, the vector p represents the gains of a PID control action to be optimized.

Therefore, the following optimization problem was solved:

$$\min_{p} \max_{\substack{j=1,\ldots,N_u \\ k=1,\ldots,N_S}} J(e_j^k, u_{CL}, x_{0j}, t_0, t_f, w) \tag{19}$$

subject to

$$\dot{x}_j = f_j(x_j, u(p, e_j^k(t), t)) \quad , x_j(t_0) = x_{0j} \tag{20a}$$

$$y_j = h_j(x_j, u(p, e_j^k(t), t)) \tag{20b}$$

$$\underline{u} \leq u(t) \leq \overline{u} \tag{20c}$$

$$\underline{x} \leq x_j(t) \leq \overline{x} \tag{20d}$$

The above optimization Problem, defined by (19) and (20), was solved in the discrete time using a Genetic Algorithm [51,52].

5. Numerical Result

Table 1 reports the main plant parameters used for numerical simulations. A single TS control implementing ailerons tests is considered.

The servo-valve is an adirectional control valve, direct operated, with integrated digital axis controller (IAC Multi Ethernet), Rexroth type 4WRPDH. The servo-valve opening variable in plots is a normalized servo-valve opening.

To optimize the controller gains, the following scenarios were considered:

- Scenarios without the Aircraft simulator:
  - force multi step (with increasing amplitude 100, 500 and 1000 N), constant pump pressure (15 MPa);
  - force multi step with constant pump pressure (80% of the nominal 15 MPa);
  - force multi step with constant pump pressure (120% of the nominal 15 MPa);
  - force multi step with sinusoidal pump pressure (15 MPa+ 0.15 sin($\omega * t$) MPa).
- Scenarios with the Aircraft simulator:
  - Aileron doublet (2.5 deg deflection for 9 s followed by −2.5 deg deflection for 9 s);
  - Aileron step (2.5 deg deflection).

In addition, the following model uncertainties were taken into account:

- $\pm 10\%$ of the nominal bulk modulus at atmospheric pressure;
- $\pm 10\%$ of the nominal air trapped volume ratio;
- $\pm 10\%$ of the servo-valve time constant;
- $\pm 10\%$ of the $C_d A_{sv}$ parameter;
- $\pm 10\%$ of EMA time constant.

Table 1. Plant parameters.

Parameter	Value
Bulk nominal modulus at atmospheric pressure, β_0 [GPa]	1.5
Pressure controlled by pump in the accumulator, P_{hp} [MPa]	15
Density at atmospheric pressure, ρ_0 [kg/m^3]	989
Servo-valve discharge coefficient times Servo-valve hole area, $C_d A_{sv}$ [cm^2]	0.03
Servo-valve equivalent time constant, τ_{sv} [s]	0.01
Chamber cross section, A [cm^2]	8.00
Dead volume of the chamber	5%
Piston mass, m [kg]	1.00
Fraction of trapped air at atmospheric pressure, α	0.01

To test the performance of the force controller, a campaign of simulations was performed. In each simulation, the above uncertainties were considered. In addition the following disturbances or additional phenomena were considered: hysterical effects of the servo-valve and mechanical links, servo-valve dead zone, piston end-stroke.

To evaluate the controller performance, responses to small step inputs (100 N for force and 0.1 cm for position) are considered in the presence of perturbed models. In Table 2, mean values of control performance indexes are shown. In particular, the mean quadratic error (MQE), the overshoot and the settling time (ST) are reported for the worst case. For the settling time the worst case is assumed to be the maximum for force control and the minimum for position control.

Table 2. Controllers performance: Mean Quadratic Error (MQE), Overshoot, Settling Time (ST).

Performance Parameters	Force	Position
Max MQE	200 N^2	4.6×10^{-4} cm^2
Max Overshoot	50%	30%
Perturbed ST	0.24 s	0.72 s

In the proposed Simulation #1, pre-programmed reference force and position signals were considered, as shown in Figures 5 and 6, respectively.

In Figure 7 the normalized servo-valve position time history is shown. Figure 8 shows the pressure provided by the pump including the sinusoidal disturbance. Lastly, in Figures 9 and 10 both cylinder chambers pressure are shown. It is worth noticing that the controlled force results is able to reject the disturbance represented by the piston movement. This is an important characteristic to be sure to evaluate the EMA in a well-emulated scenario.

Simulation #2, carried out with the same reference signals as Simulation #1, demonstrates the robustness of the controllers to measurement white gaussian noise on position and force measurements. The same quantities shown for Simulation #1 are shown in Figures 11–16. The controlled force is able to reject the disturbance represented by the piston movement also in presence of model uncertainties and measurement noise.

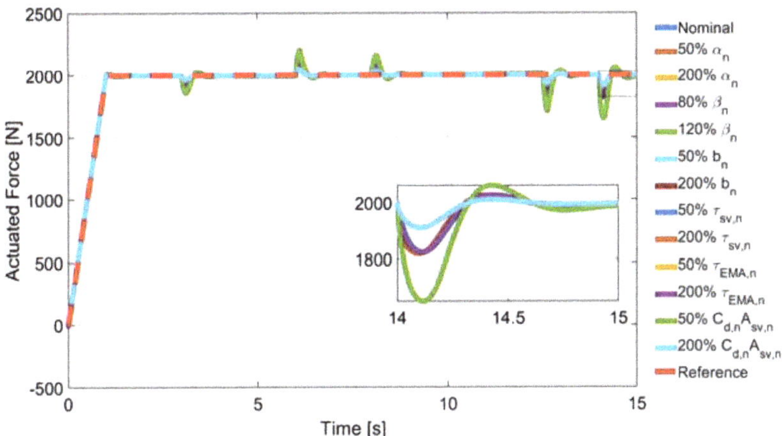

Figure 5. Simulation #1: Actuated force compared to the reference force.

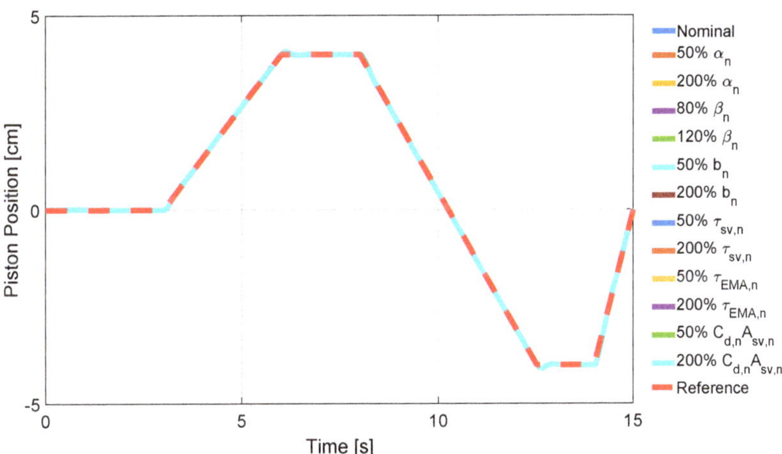

Figure 6. Simulation #1: Piston Position compared to the reference position.

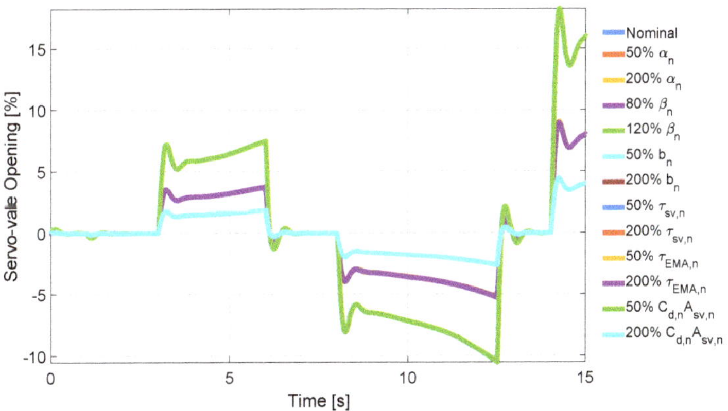

Figure 7. Simulation #1: Servo-valve opening command.

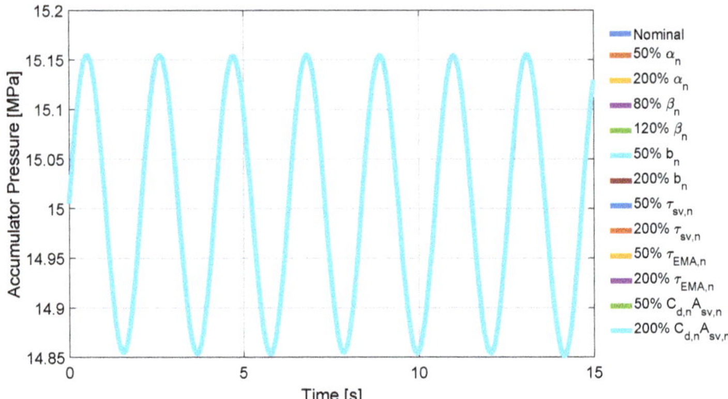

Figure 8. Simulation #1: Accumulator Pressure.

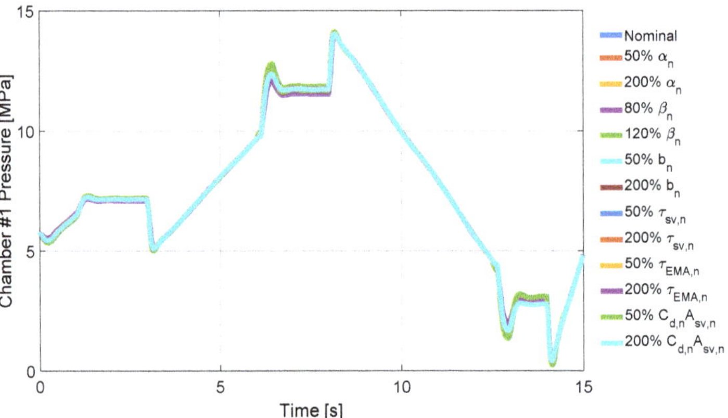

Figure 9. Simulation #1: Chamber #1 Pressure.

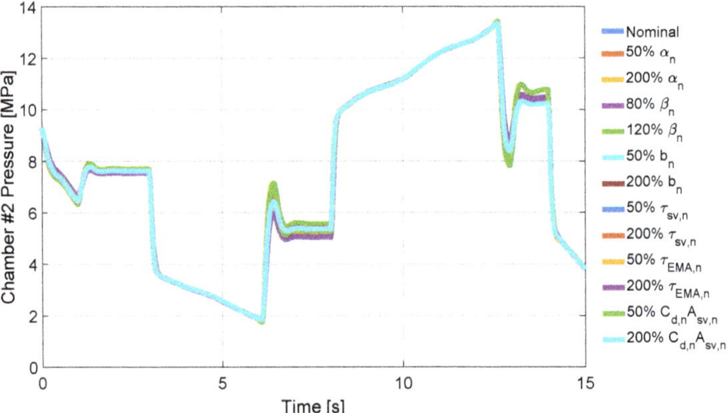

Figure 10. Simulation #1: Chamber #2 Pressure.

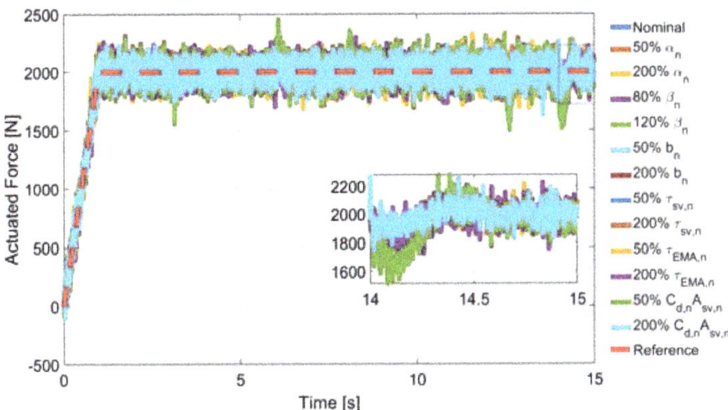

Figure 11. Simulation #2: Force actuated by the Hydraulic cylinder compared to the reference signal.

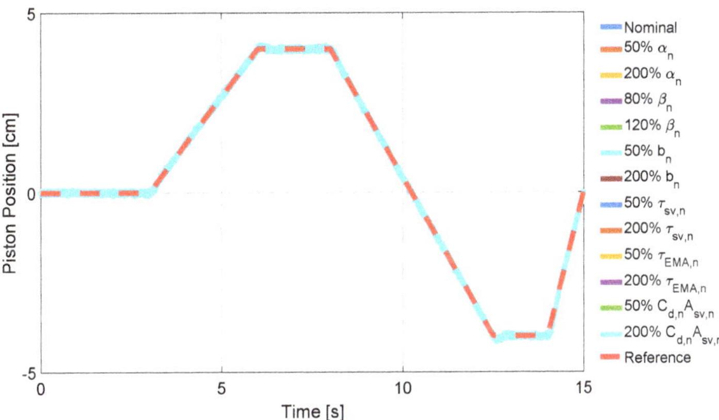

Figure 12. Simulation #2: Piston Position compared to the reference signal.

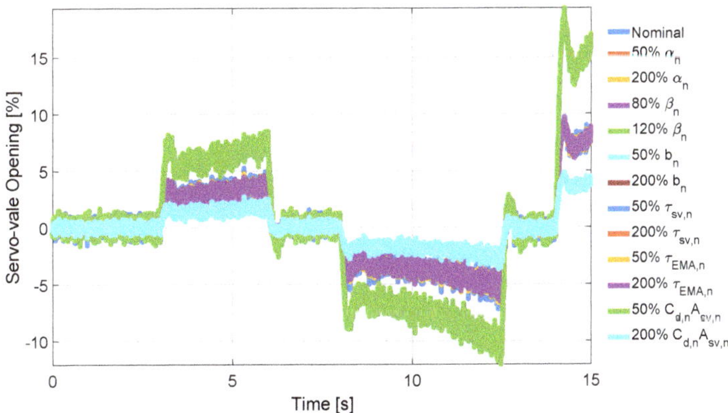

Figure 13. Simulation #2: Servo-valve opening command.

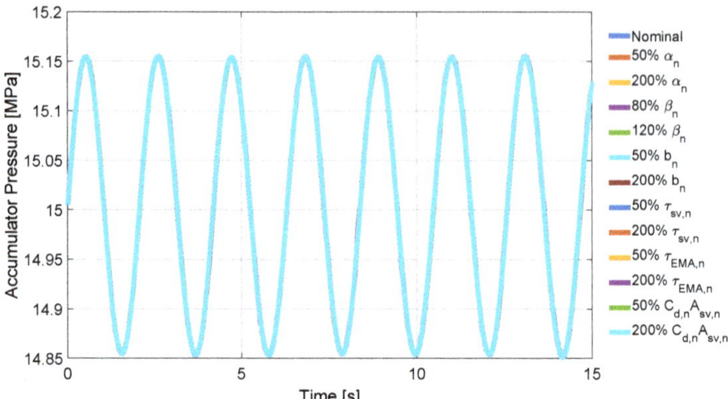

Figure 14. Simulation #2: Accumulator Pressure.

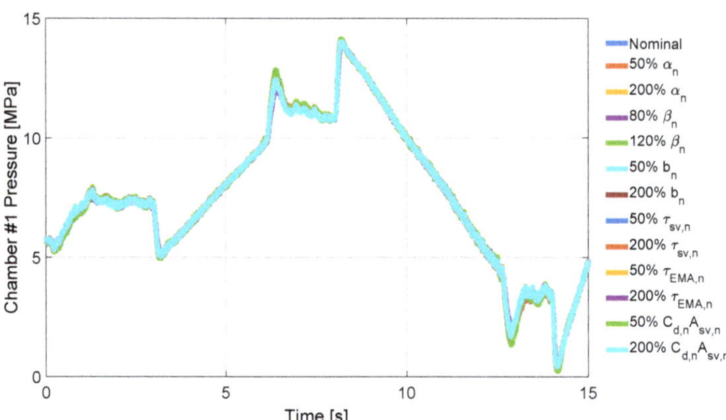

Figure 15. Simulation #2: Chamber #1 Pressure.

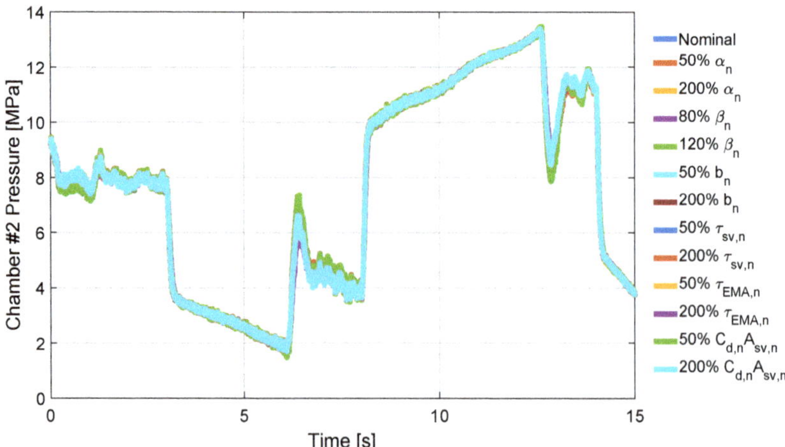

Figure 16. Simulation #2: Chamber #2 Pressure.

5.1. Numerical Results with the Flight Simulator in the Loop

An innovative feature of MIB is the capability of reproducing loads provided by a real time flight simulator including control laws.

Several tests involving the ARTS were carried out. In particular the following conditions, implying an aileron deflections, were simulated:

- without FCS:
  - a 0.044 rad step on ailerons deflection;
  - a ±0.044 rad doublet on ailerons deflection;
- with SAS active:
  - a 0.044 rad step on ailerons deflection;
  - a ±0.044 rad doublet on ailerons deflection;
- with CAS active:
  - an impulse (1.05 rad/s) on roll rate command;
  - a doublet (±1.05 rad/s) on roll rate command;
- with Autopilot active:
  - a step (0.088 rad) on the heading angle reference;
  - a ramp (0.26 rad/s) on heading angle reference;

The Cessna 172 was assumed as reference aircraft, whose main parameters are reported in Figure 17.

In the following Simulation #3, ARTS was used in "self-consistent mode" without flight control laws, and driven by a doublet input signal to the ailerons shown in Figure 18. The resulting force is shown in Figure 19. The reference force and position signals compared to the actual one in the presence of uncertainties are shown in Figures 20 and 21.

Figures 22 and 23 give an idea of the simulated maneuver, which is a sort of turn. Roll angle and roll angular speed are shown. It can be noted that reference signals produced by the ARTS are like smoothed by the aircraft dynamics. Therefore, force controller provides better results with respect to sharp reference signals.

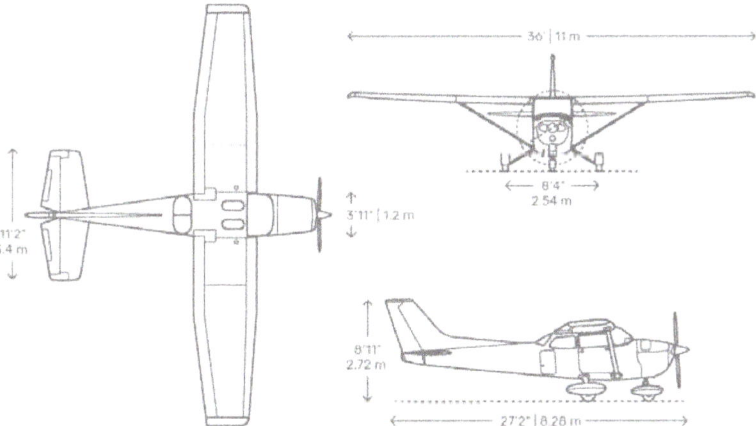

Figure 17. Simulated Aircraft-Cessna 172.

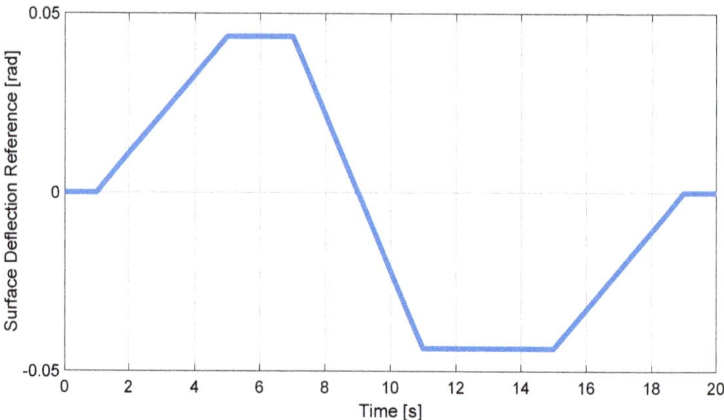

Figure 18. Simulation #3: Surface Deflection Reference.

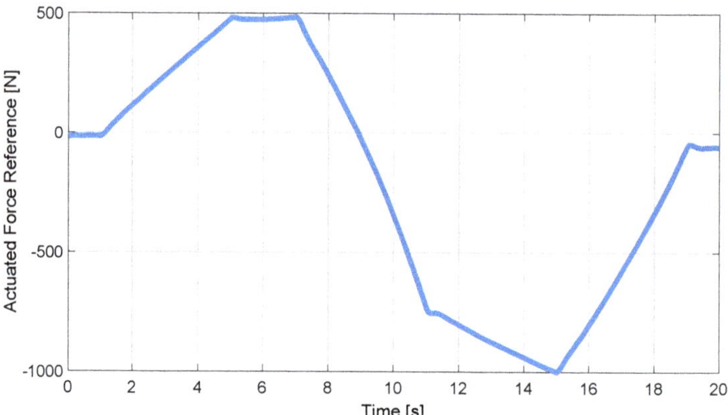

Figure 19. Simulation #3: Actuated Force Reference.

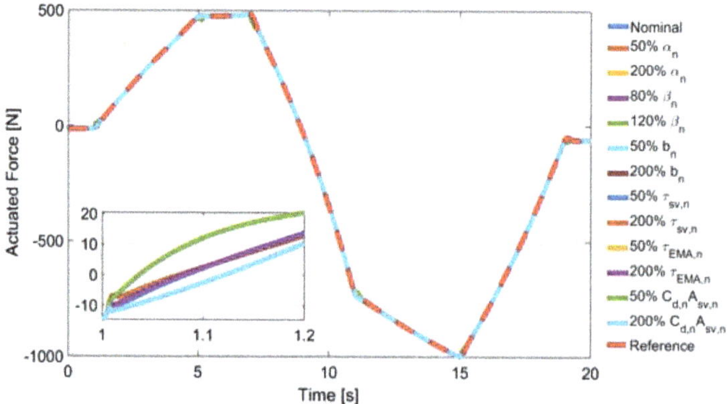

Figure 20. Simulation #3: Actuated Force compared to the reference signal.

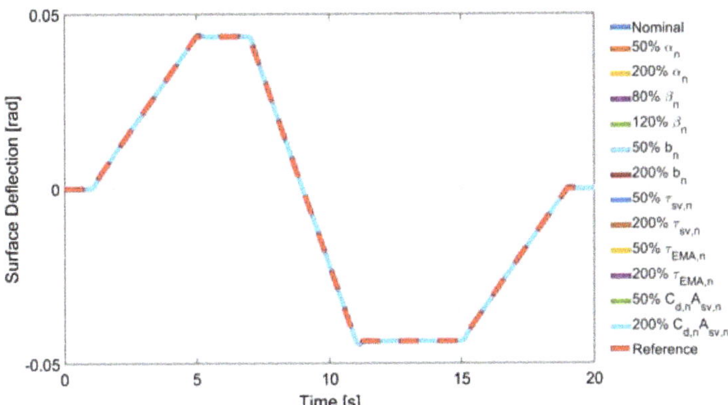

Figure 21. Simulation #3: Surface Deflection compared to the reference signal.

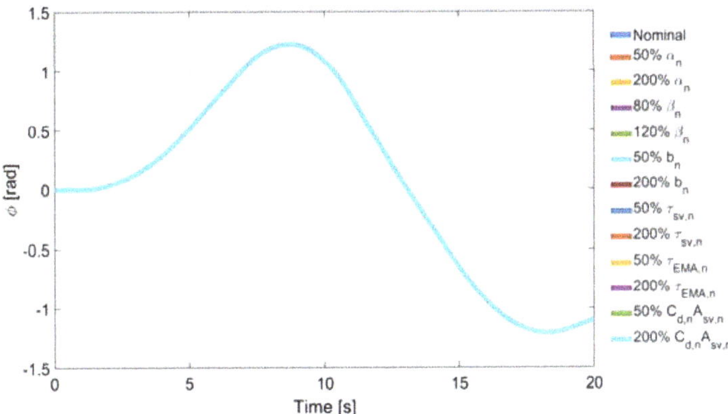

Figure 22. Simulation #3: Roll Angle from the Aircraft Simulator.

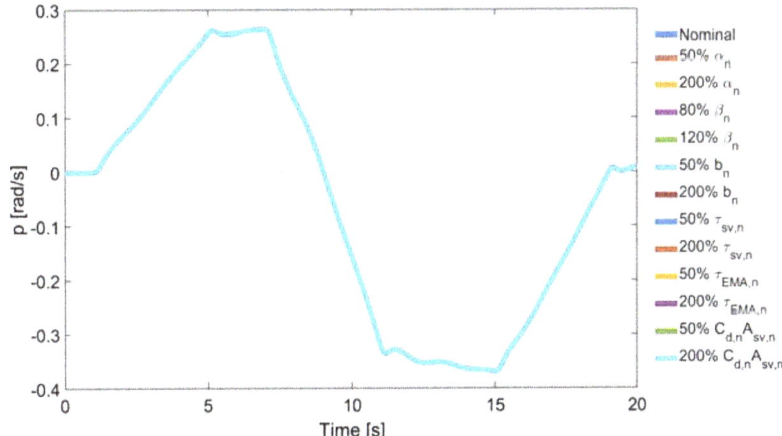

Figure 23. Simulation #3: Roll Rate from the Aircraft Simulator.

6. Conclusions

This paper describes the control architecture of the Modular Iron Bird which makes use of several test stands to increase flexibility and reduce costs for small commercial aircraft or unmanned aerial vehicles. This iron bird can reproduce preprogrammed loads to test single or multiple actuators, or realistic "in flight" load conditions thanks to the presence of a real time aircraft simulation module.

The design and tuning of the control law needs particular attention because of the concurrent action of both a force controller to guarantee the testing loads and a position controller to control the aerodynamic surface position. This required the use of high quality components, and suggested the use of a dynamic model of the plant during the design phase, to make a robust tuning of the force controller driving the hydraulic actuator.

In the plant design phase, a classical control approach has been adopted. Indeed, the force control algorithm is based on a PID, whose gains have been optimized to ensure robust performance in the presence of parametric uncertainties, and a closed loop response faster ($\sim$0.01 s response time) than the EMA position controller ($\sim$0.1 s response time). Results proved that the proposed solution works well in the presence of uncertainties and noise and hard nonlinearities neglected in the design of the control law such as hysteresis and dead zones. In particular, both the dynamic and static precision, is quite insensitive to the uncertainties and to the different scenarios considered to include possible tests on different kinds of aircraft.

Author Contributions: Conceptualization, methodology, formal analysis, investigation, writing, E.D., L.E.d.G., I.N., L.B., M.M. Data curation, visualization, software development, E.D., L.E.d.G., I.N. Funding acquisition, supervision M.B., M.M., L.B. All authors have read and agreed to the published version of the manuscript.

Funding: This research was partially funded by REGIONE CAMPANIA, M.I.B. (Modular Iron Bird) project, CUP B43D18000130007, FESR Campania 2014-2020.

Institutional Review Board Statement: Not applicable.

Informed Consent Statement: Not applicable.

Data Availability Statement: Data available on request due to restrictions. The data presented in this study are available on request from the corresponding author. The data are not publicly available due to NDA between Protom Group and university.

Conflicts of Interest: The authors declare no other conflict of interest.

References

1. Prasad, M.; Gangadharan, K. Aileron endurance test rig design based on high fidelity mathematical modeling. *CEAS Aeronaut. J.* **2017**, *8*, 653–671. [CrossRef]
2. Maré, J.C. *Aerospace Actuators 2: Signal-By-Wire and Power-By-Wire*; John Wiley & Sons: Hoboken, NJ, USA, 2017.
3. Wheeler, P.; Bozhko, S. The More Electric Aircraft: Technology and challenges. *IEEE Electrif. Mag.* **2014**, *2*, 6–12. [CrossRef]
4. Jean-Charles, M.; Jian, F. Review on signal-by-wire and power-by-wire actuation for more electric aircraft. *Chin. J. Aeronaut.* **2017**, *30*, 857–870.
5. Wilcox, D.C. Formulation of the kw turbulence model revisited. *AIAA J.* **2008**, *46*, 2823–2838. [CrossRef]
6. Rubertus, D.P.; Hunter, L.D. Electromechanical actuation technology for the all-electric aircraft. *IEEE Trans. Aerosp. Electr. Syst.* **1984**, *AES-20*, 243–249. [CrossRef]
7. Qiao, G.; Liu, G.; Shi, Z.; Wang, Y.; Ma, S.; Lim, T.C. A review of electromechanical actuators for More/All Electric aircraft systems. *Proc. Inst. Mech. Eng. Part C J. Mech. Eng. Sci.* **2018**, *232*, 4128–4151. [CrossRef]
8. Brière, D.; Traverse, P. AIRBUS A320/A330/A340 electrical flight controls-A family of fault-tolerant systems. In Proceedings of the FTCS-23 The Twenty-Third International Symposium on Fault-Tolerant Computing, Toulouse, France, 22–24 June 1993; pp. 616–623.
9. Hwang, S.; Choi, S. Ironbird ground test for tilt rotor unmanned aerial vehicle. *Int. J. Aeronaut. Space Sci.* **2010**, *11*, 313–318. [CrossRef]
10. Spangenberg, H.; Vechtel, D. Failure detection, identification and reconfiguration: applications for a modular iron bird. In Proceedings of the AIAA Modeling and Simulation Technologies Conference and Exhibit, Hilton Head, SC, USA, 20–23 August 2007; p. 6466.

11. De Martin, A.; Jacazio, G.; Sorli, M. Design of a PHM system for electro-mechanical flight controls: a roadmap from preliminary analyses to iron-bird validation. In Proceedings of the MATEC Web of Conferences, EDP Sciences, Sibiu, Romania, 5–7 June 2019; Volume 304, p. 04018.
12. Topczewski, S.; Narkiewicz, J.; Bibik, P. Helicopter Control During Landing on a Moving Confined Platform. *IEEE Access* **2020**, *8*, 107315–107325. [CrossRef]
13. Wang, C.; Jiao, Z.; Quan, L. Adaptive velocity synchronization compound control of electro-hydraulic load simulator. *Aerosp. Sci. Technol.* **2015**, *42*, 309–321. [CrossRef]
14. Wang, C.; Jiao, Z.; Quan, L. Nonlinear robust dual-loop control for electro-hydraulic load simulator. *ISA Trans.* **2015**, *59*, 280–289. [CrossRef]
15. Karpenko, M.; Sepehri, N. Electrohydraulic force control design of a hardware-in-the-loop load emulator using a nonlinear QFT technique. *Control Eng. Pract.* **2012**, *20*, 598–609. [CrossRef]
16. Nam, Y. Dynamic Characteristic Analysis and Force Loop Design for the Aerodynamic Load Simulator. *KSME Int. J.* **2000**, *14*, 1358–1364. [CrossRef]
17. Yao, B.; Chiu, G.T.; Reedy, J.T. Nonlinear adaptive robust control of one-dof electro-hydraulic servo systems. In *ASME International Mechanical Engineering Congress and Exposition (IMECE'97)*; FPST: Fallon, NV, USA, 1997; Volume 4, pp. 191–197.
18. Yao, B.; Bu, F.; Chiu, G.T. Non-linear adaptive robust control of electro-hydraulic systems driven by double-rod actuators. *Int. J. Control* **2001**, *74*, 761–775. [CrossRef]
19. Ullah, N.; Wang, S.; Aslam, J. Adative robust control of electrical load simulator based on fuzzy logic compensation. In Proceedings of the 2011 International Conference on Fluid Power and Mechatronics, Beijing, China, 17–20 August 2011; pp. 861–867.
20. Wang, C.; Jiao, Z.; Wu, S.; Shang, Y. A practical nonlinear robust control approach of electro-hydraulic load simulator. *Chin. J. Aeronaut.* **2014**, *27*, 735–744. [CrossRef]
21. Kim, W.; Shin, D.; Won, D.; Chung, C.C. Disturbance-observer-based position tracking controller in the presence of biased sinusoidal disturbance for electrohydraulic actuators. *IEEE Trans. Control Syst. Technol.* **2013**, *21*, 2290–2298. [CrossRef]
22. Shang, Y.; Liu, X.; Jiao, Z.; Wu, S. An integrated load sensing valve-controlled actuator based on power-by-wire for aircraft structural test. *Aerosp. Sci. Technol.* **2018**, *77*, 117–128. [CrossRef]
23. Wang, L.; Mare, J.C. A force equalization controller for active/active redundant actuation system involving servo-hydraulic and electro-mechanical technologies. *Proc. Inst. Mech. Eng. Part G J. Aerosp. Eng.* **2014**, *228*, 1768–1787. [CrossRef]
24. Manring, N.D.; Fales, R.C. *Hydraulic Control Systems*; John Wiley & Sons: Hoboken, NJ, USA, 2019.
25. Jelali, M.; Kroll, A. *Hydraulic Servo-Systems: Modelling, Identification and Control*; Springer Science & Business Media: Berlin, Germany, 2012.
26. Ali, H.H.; Fales, R.C. A review of flow control methods. *Int. J. Dyn. Control* **2021**, 1–8. [CrossRef]
27. Weerasooriya, S.; El-Sharkawi, M.A. Identification and control of a DC motor using back-propagation neural networks. *IEEE Trans. Energy Convers.* **1991**, *6*, 663–669. [CrossRef]
28. Saab, S.S.; Kaed-Bey, R.A. Parameter identification of a DC motor: an experimental approach. In Proceedings of the ICECS 2001 8th IEEE International Conference on Electronics, Circuits and Systems (Cat. No. 01EX483), Malta, Malta, 2–5 September 2001; Volume 2, pp. 981–984. [CrossRef]
29. Rubaai, A.; Kotaru, R. Online identification and control of a DC motor using learning adaptation of neural networks. *IEEE Trans. Ind. Appl.* **2000**, *36*, 935–942. [CrossRef]
30. Yassin, I.M.; Taib, M.N.; Rahim, N.A.; Salleh, M.K.M.; Abidin, H.Z. Particle Swarm Optimization for NARX structure selection—Application on DC motor model. In Proceedings of the 2010 IEEE Symposium on Industrial Electronics and Applications (ISIEA), Penang, Malaysia, 3–5 October 2010; pp. 456–462. [CrossRef]
31. Mare, J.C. Dynamic loading systems for ground testing of high speed aerospace actuators. *Int. J. Aircr. Eng. Aerosp. Technol.* **2006**, *78*, 275–282. [CrossRef]
32. Jiao, Z.X.; Gao, J.X.; Qing, H.; Wang, S.P. The velocity synchronizing control on the electro-hydraulic load simulator. *Chin. J. Aeronaut.* **2004**, *17*, 39–46. [CrossRef]
33. Niksefat, N.; Sepehri, N. Designing robust force control of hydraulic actuators despite system and environmental uncertainties. *IEEE Control Syst. Mag.* **2001**, *21*, 66–77.
34. Di Rito, G.; Denti, E.; Galatolo, R. Robust force control in a hydraulic workbench for flight actuators. In Proceedings of the 2006 IEEE Conference on Computer Aided Control System Design, 2006 IEEE International Conference on Control Applications, 2006 IEEE International Symposium on Intelligent Control, Munich, Germany, 4–6 October 2006; pp. 807–813.
35. Conrad, F.; Jensen, C. Design of hydraulic force control systems with state estimate feedback. *IFAC Proc. Vol.* **1987**, *20*, 307–312. [CrossRef]
36. Bertucci, A.; Mornacchi, A.; Jacazio, G.; Sorli, M. A force control test rig for the dynamic characterization of helicopter primary flight control systems. *Procedia Eng.* **2015**, *106*, 71–82. [CrossRef]
37. Liu, G.; Daley, S. Optimal-tuning PID control for industrial systems. *Control Eng. Pract.* **2001**, *9*, 1185–1194. [CrossRef]
38. Jacazio, G.; Balossini, G. Real-time loading actuator control for an advanced aerospace test rig. *Proc. Inst. Mech. Eng. Part I J. Syst. Control Eng.* **2007**, *221*, 199–210. [CrossRef]
39. Pan, H.L.; Yan, J. Implement of electro-Hydraulic servo control of aero variable stroke plunger pump. In *Advanced Materials Research*; Trans Tech Publication: Stafa-Zurich, Switzerland, 2012; Volume 443, pp. 313–318.

40. Zhu, Z.; Tang, Y.; Shen, G. Experimental investigation of a compound force tracking control strategy for electro-hydraulic hybrid testing system with suppression of vibration disturbances. *Proc. Inst. Mech. Eng. Part C J. Mech. Eng. Sci.* **2017**, *231*, 1033–1056. [CrossRef]
41. Bu, F.; Yao, B. Nonlinear adaptive robust control of hydraulic actuators regulated by proportional directional control valves with deadband and nonlinear flow gains. In Proceedings of the 2000 American Control Conference (ACC), Chicago, IL, USA, 28–30 June 2000; Volume 6, pp. 4129–4133.
42. Pratumsuwan, P.; Junchangpood, A. Force and position control in the electro-hydraulic system by using a MIMO fuzzy controller. In Proceedings of the 2013 IEEE 8th Conference on Industrial Electronics and Applications (ICIEA), Melbourne, VIC, Australia, 19–21 June 2013; pp. 1462–1467.
43. Wang, X.; Wang, S.; Wang, X. Electrical load simulator based on velocity-loop compensation and improved fuzzy-PID. In Proceedings of the 2009 IEEE International Symposium on Industrial Electronics, Seoul, Korea, 5–8 July 2009; pp. 238–243.
44. Li, X.; Zhu, Z.C.; Rui, G.C.; Cheng, D.; Shen, G.; Tang, Y. Force loading tracking control of an electro-hydraulic actuator based on a nonlinear adaptive fuzzy backstepping control scheme. *Symmetry* **2018**, *10*, 155. [CrossRef]
45. Gizatullin, A.; Edge, K. Adaptive control for a multi-axis hydraulic test rig. *Proc. Inst. Mech. Eng. Part I J. Syst. Control Eng.* **2007**, *221*, 183–198. [CrossRef]
46. Piché, R.; Pohjolainen, S.; Virvalo, T. Design of robust controllers for position servos using H-infinity theory. *Proc. Inst. Mech. Eng. Part I J. Syst. Control Eng.* **1991**, *205*, 299–306. [CrossRef]
47. Yao, J.; Jiao, Z.; Shang , Y.; Huang, C. Adaptive nonlinear optimal compensation control for electro-hydraulic load simulator. *Chin. J. Aeronaut.* **2010**, *23*, 720–733.
48. Kallu, K.D.; Wang, J.; Abbasi, S.J.; Lee, M.C. Estimated reaction force-based bilateral control between 3dof master and hydraulic slave manipulators for dismantlement. *Electronics* **2018**, *7*, 256. [CrossRef]
49. Shi, C.; Wang, X.; Wang, S.; Wang, J.; Tomovic, M.M. Adaptive decoupling synchronous control of dissimilar redundant actuation system for large civil aircraft. *Aerosp. Sci. Technol.* **2015**, *47*, 114–124. [CrossRef]
50. Borrelli, M.; D'Amato, E.; di Grazia, L.E.; Mattei, M.; Notaro, I. MPC load control for aircraft actuator testing. In Proceedings of the 2020 7th International Conference on Control, Decision and Information Technologies (CoDIT), Prague, Czech Republic, 29 June–2 July 2020; Volume 1, pp. 587–592. [CrossRef]
51. Drabble, D.; Ponnapalli, P.V.S.; Thomson, M. G.A. Optimisation of PID Controllers—Optimal Fitness Functions. In *Developments in Soft Computing*; John, R., Birkenhead, R., Eds.; Physica-Verlag HD: Heidelberg, Germany, 2001; pp. 183–190.
52. Tran, H.K.; Son, H.H.; Duc, P.V.; Trang, T.T.; Nguyen, H.N. Improved Genetic Algorithm Tuning Controller Design for Autonomous Hovercraft. *Processes* **2020**, *8*, 66. [CrossRef]

 aerospace

MDPI

Article

Validation of a Simulation Tool for an Environmentally Friendly Aircraft Cargo Fire Protection System

Arnav Pathak, Victor Norrefeldt * and Marie Pschirer

Fraunhofer Institute for Building Physics IBP, 83626 Valley, Germany; arnav.pathak@ibp.fraunhofer.de (A.P.); marie.pschirer@ibp.fraunhofer.de (M.P.)

* Correspondence: victor.norrefeldt@ibp.fraunhofer.de

Abstract: One of the objectives of the CleanSky-2 project is to develop an Environmentally Friendly Fire Protection (EFFP) system to substitute halon for the aircraft cargo hold. For this, an aircraft demonstrator including the cargo hold was equipped with a nitrogen-based fire suppression system. The demonstrator is located in the Flight Test Facility (FTF) low-pressure vessel and can thus be subjected to realistic cruise pressure conditions and take-off and descent pressure profiles. As a design tool, a zonally refined simulation model to predict the local oxygen and nitrogen concentration distribution in the cargo hold has been developed using the Indoor Environment Simulation Suite (IESS). The model allows for fast transient simulations of the suppression system operation. This paper presents a model validation case of knockdown during cruising, followed by a holding phase and descent.

Keywords: cargo fire protection; simulation; fire suppression; testing

Citation: Pathak, A.; Norrefeldt, V.; Pschirer, M. Validation of a Simulation Tool for an Environmentally Friendly Aircraft Cargo Fire Protection System. *Aerospace* **2021**, *8*, 35. https://doi.org/10.3390/aerospace8020035

Academic Editor: Spiros Pantelakis

Received: 30 November 2020

Accepted: 27 January 2021

Published: 30 January 2021

1. Introduction

Homogeneous and efficient fire suppression agent distribution inside an aircraft cargo hold is the key for protecting the aircraft from a cargo hold fire and for achieving the fire protection goals. Aircraft cargo with fire suppression systems must provide fire protection for several hours. Thus, the system must be adequate to maintain a safe fire suppressive atmosphere inside the cargo hold for the specified diversion time.

The aircraft cargo fire suppression system has two major phases, a knockdown phase that diminishes fire either by cooling the fire or reducing the oxygen concentration as described in the Minimum Performance Standard (MPS) [1], followed by the holding phase to maintain a fire suppressive environment inside the cargo hold throughout the flight and landing phase. The main design parameters for the system sizing are the cargo hold volume, the agent target concentration, and the air tightness. For example, the cargo door seal is subjected to the gradient between the pressurized fuselage and ambient conditions at cruising and thus leaks air. This air is replaced with fresh air aspired through the Pressure Management System (PMS) to ensure the balance of airflows. A controlled supply of additional agent to compensate for the air ingress is thus necessary during the holding phase. Additionally, the PMS prevents pressure peaks during agent injection that would lead to opening of the rapid decompression panels by allowing air to leave the cargo hold.

Today, a large quantity of halon from bottles is released to perform the knockdown within two minutes, and then the fire suppressive environment is maintained within the cargo hold by constant metering of halon. However, due to high Ozone Depletion Potential (ODP) of halon, the aerospace industry is looking for halon-free systems. The use of nitrogen as a suppression agent, diluting the cargo hold oxygen concentration below the flammability point, is one such alternative under research within the CleanSky2 "Environmentally Friendly Fire Protection" (EFFP) project. Within this project, a simulation toolchain is developed that predicts the agent concentration distribution within the cargo hold, and a physical full-scale demonstration is implemented in the Flight Test Facility, a low-pressure chamber hosting the front section of a former in-service A310 including the underfloor area and forward cargo hold.

This paper presents transient validation results of a nitrogen knockdown followed by a holding phase and descent.

Several modelling activities have been performed for halon replacement technology with water mist, inert gases, and solid-propellant gas-generator suppression systems [2–5]. These models include CFD simulations and single or two-zone models. However, these models simulate agent distribution at constant pressure, single operating points or are non-locally refined. In cargo fire suppression systems, the transient phases of knock-down, holding, and descent have significant impacts on the agent distribution inside the cargo hold. The location of air ingress ports, such as the pressure management system (PMS, Section 2.2.1), or leakages in the cargo interiors, further influences the local agent distribution gradient in the cargo hold.

Single or two-zone models are quick to run; however, these models do not provide a local agent distribution inside the cargo hold. CFD models provide the details of agent distribution, but the computation cost is very high, especially when transient phases are considered. The IESS uses a hybrid simulation approach, where the high momentum flow regime close to the discharge nozzle has been pre-simulated by CFD, and the results of this near-field domain have been integrated in a zonal model of the cargo hold.

In this paper, a validation result for such a transient simulation is shown. For this, a nitrogen-based knockdown and holding system has been integrated in the Flight Test Facility aircraft mock-up. The cargo hold has been equipped with sensors to measure the local oxygen concentration. A realistic cabin pressure profile of take-off, cruising, and descent is implemented using a low-pressure vessel, which is able to generate an ambient pressure similar to cruising conditions (750 hPa, corresponding to an equivalent height of 8.000 ft.).

2. Materials and Methods

In this study, both numerical modelling and experimental testing are used. The following sections describe the simulation method applied with the Indoor Environment Simulation Suite (IESS, Section 2.1), as well as the experimental method (Section 2.2) in the Flight Test Facility (FTF).

2.1. Indoor Environment Simulation Suite (IESS)

The Indoor Environment Simulation Suite (IESS, Figure 1) provides indoor climatic simulations using the zonal approach. In contrast to CFD or multi-zone models, the zonal modelling approach subdivides the indoor space into typically 10^2 to 10^3 zones. In addition to this airflow modelling, the IESS provides interfaces for walls, sources and sinks, radiation, conduction, and species distribution. Through this, a transient, multiphysics simulation is enabled. A toolchain has been developed to ease the setup, customization, and post-processing of the models.

Figure 1 shows a typical modelling workflow. Starting from a CAD geometry of the space envelope, the IESS Model Generator exports a Modelica model with zonal subdivision. This integrated model contains the following major submodels:

- Zonally subdivided air volume model;
- Models of ventilation sinks and sources;
- Zonally subdivided wall facet models including conduction in the material, convective heat exchange with adjacent air volumes, and radiative heat exchange with other surfaces;
- Interfaces to integrate boundary conditions that are pre-simulated by CFD.

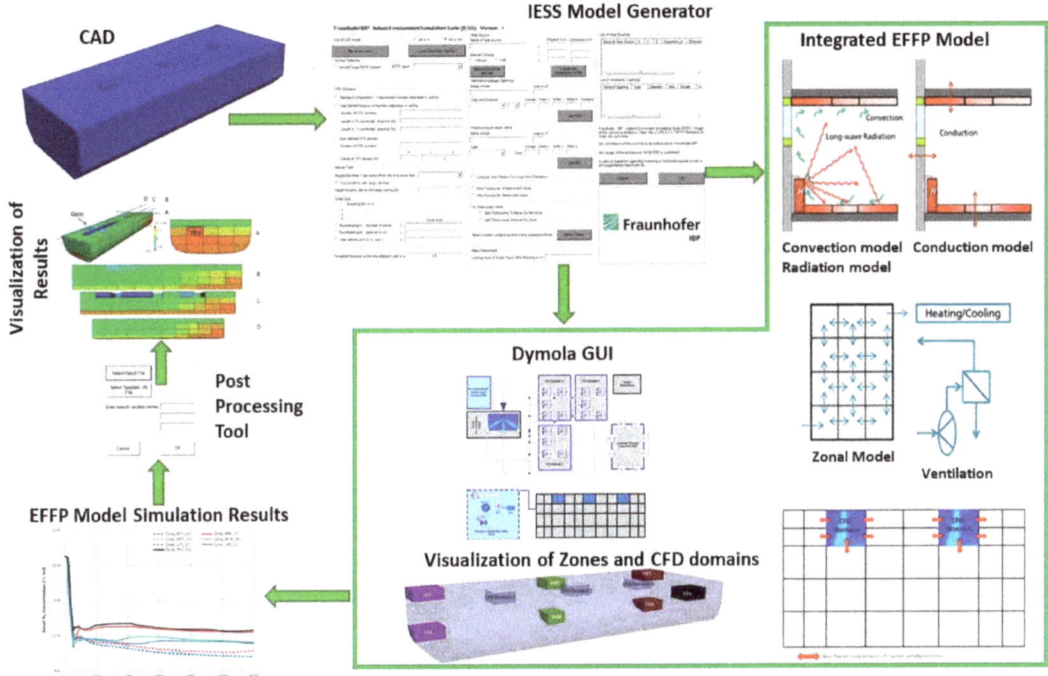

Figure 1. Indoor Environment Simulation Suite (IESS) workflow.

A rudimentary graphical user interface for setting the boundary conditions is produced along with a visualization of the zones' distribution. After the simulation, results are displayed as transient plots, and a post processing tool generates 3D visualizations of scalar quantities, such as temperature or concentrations in the zones. In the following sections, each of these bricks is described in further detail.

2.1.1. IESS Model Generator

The IESS Model Generator is a pre-processing tool developed in C++. It translates a CAD model exported into the .stl format to Modelica code that can be simulated in the Modelica simulation environment [6]. This tool automatically distributes the zonal grid, attributes the adjacencies of zones and walls and provides top-level parameters such as source intensities, exterior temperature, etc., for easy model customization.

2.1.2. Airflow Model (VEPZO)

The aim of a zonal model is to perform quick simulations of the airflow pattern and temperature distribution in an indoor space. Therefore, an air volume is subdivided into several discrete zones. Zonal models are a compromise between the more complex CFD calculations and the approximation of a perfectly mixed air volume as assumed by single node models. A new formulation of such models has been developed by Fraunhofer IBP, the VElocity Propagating Zonal model (VEPZO model). The VEPZO model uses the airflow velocity as a property of a zone and a viscous loss model in order to better match the physics of airflows. The two main components of the VEPZO model are a volume model and a flow model. The volume model considers mass conservation and conservation of enthalpy while the flow model calculates the mass flow rate from the pressure difference. The advantages of the VEPZO model are as follows: only a limited amount of information on the boundary conditions is necessary, in addition to the delivery of a local resolution of air temperatures, agent concentrations, and air flow within a domain as well as a moderate computational cost, which finally allows parametric variation for transient situations [7]. Zones are coupled to the CFD domain by exchanging mass flow at the boundary, to walls by convective heat exchange, and to other sources and sinks such as leakage flows, internal heat sources, etc.

The cargo hold model has been subdivided into 11 × 5 × 4 zones. Three CFD domains (cf. Section 2.1.6) provide inlet conditions for the agent injection. Leakage through the cargo door (cf. Section 2.2.1) is represented by a sink. The pressure management system (PMS, cf. Section 2.2.1) is implemented as a two-way flow model to ensure that the overall balance of airflow sums to zero and to avoid pressure peaks in the simulation.

Wall facets are distributed according to the zonal decomposition and exchange heat with the adjacent zones by convection.

2.1.3. Longwave Radiation Model Coupled with Zonal Model (RADZO)

Especially in applications with increased temperature differences between surfaces, the longwave radiation generates an important heat flow and thus impacts surface temperatures. The radiation model coupled with the zonal model (RADZO) computes view factors between finely refined surface mesh elements and allocates accurate view factors to the course zonal grid mesh. The RADZO model calculates longwave radiant heat exchange between all "n" surfaces [8]. The accurate estimation of surface temperatures leads to a more accurate prediction of air temperature stratification and thus airflow pattern and agent distribution.

2.1.4. Conduction through the Enclosure Model (ENCZO)

Enclosure models are based on a suite of thermal capacitances and thermal resistances. Different materials are implemented in the model by adjusting these characteristic parameters. The enclosure can be simulated as a single or multiple layer model. The Model Generator exports a template wall model in the initial code that is later customized to the actual wall layers by the modeler. In the application presented here, conduction through honeycomb panels is implemented.

2.1.5. Convection Model Coupled with the Zonal Model (CONZO)

Since the boundary layer flow is not resolved in the zonal model, the numerical determination of the convective heat transfer coefficient h_c is not possible. This depends on the orientation of the surfaces, the temperature difference relative to the air, and the prevailing flow conditions. The IESS-CONZO library therefore provides a collection of correlations for various situations and predicts the h_c value based on the local air exchange in a zone [9].

2.1.6. CFD Domain

The CFD domain was specifically developed for increasing the accuracy of the zonal modelling approach for the cargo fire suppression application. The inert agent typically is provided from pressurized bottles and is distributed through nozzles in the ceiling of the cargo hold. This results in a local high momentum flow pattern around the inlet nozzle. The scale of this flow is far smaller than the typical size of a zone, but it determines the flow field in the entire compartment. In order to maintain the high simulation speed possible with the zonal modelling approach, the concept of a CFD zone has been selected. In this approach, the nearfield around the injection nozzle is pre-simulated by CFD. Due to the small size of this domain, the CFD computation converges relatively quickly. The simulation result is used as a supply/sink boundary condition in the zonal model in order to transmit the effect of the nozzle to the large space of the cargo hold.

To numerically solve the high momentum flow and turbulent eddies in the vicinity of the agent discharge, the CFD simulation of the Airbus single aisle nozzle and the A350 nozzle was performed prior to setting up the cargo hold zonal model.

In order to verify the general approach, these CFD simulations were validated by tests on a small prototype single nozzle-cavity demonstrator (Figure 2, top). The nozzle was connected to a nitrogen cylinder, and flow rates and pressure were controlled. Anemometers were placed in the nozzle jet regions, and readings were compared to the CFD calculation (Figure 2, bottom).

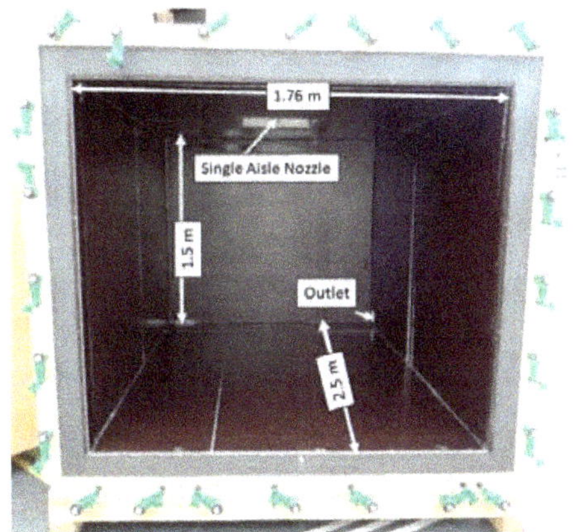

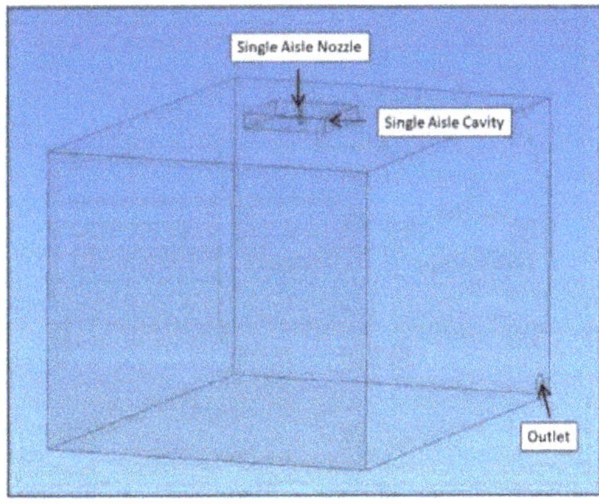

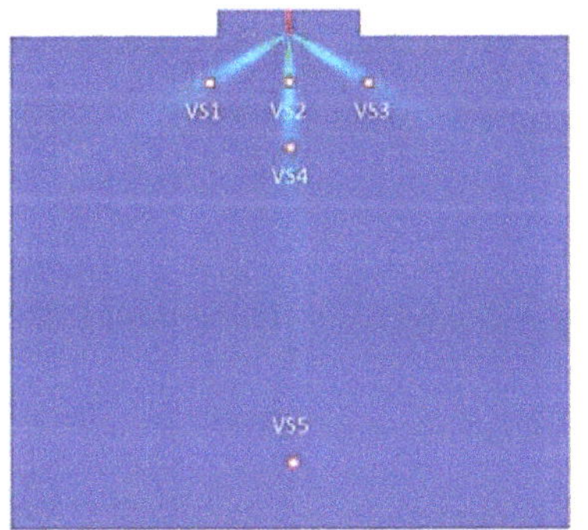

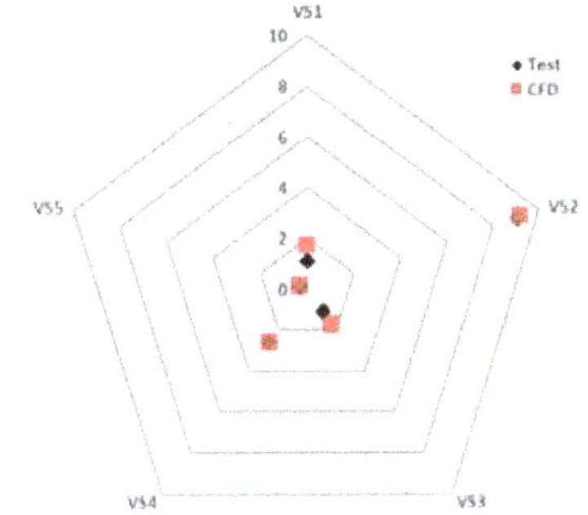

Figure 2. Prototype of a box demonstrator with single aisle cavity and nozzle (**top left**), CAD model of the prototype box demonstrator (**top right**), simulation result, and anemometer location (**bottom left**), validation result in m/s (**bottom right**).

The decline of the flow velocity below 2 m/s was used as a criterion to generate a box around the CFD domain (Figure 3). The mass flow across the surfaces of this box, the average temperature, and concentration are provided as tabulated boundary conditions for the zonal model of the cargo hold.

2.1.7. Integrated Cargo Hold Model

Figure 4 summarizes the workflow used to generate the model of the A310 forward cargo hold. From the geometrical definition of the cargo hold setup, the IESS Model Generator automatically creates the zonal Airflow, Radiation, Conduction, and Convection models. The pre-simulated near-field CFD model of the injection nozzle is integrated in three positions as boundary conditions for mass flow and supply temperature/concentrations. Each of the blocks represents a volume that is further connected to flow models, leakage models or convection models depending on the location in the cargo hold. The pressure management system (PMS) model is connected to the aft wall of the cargo hold.

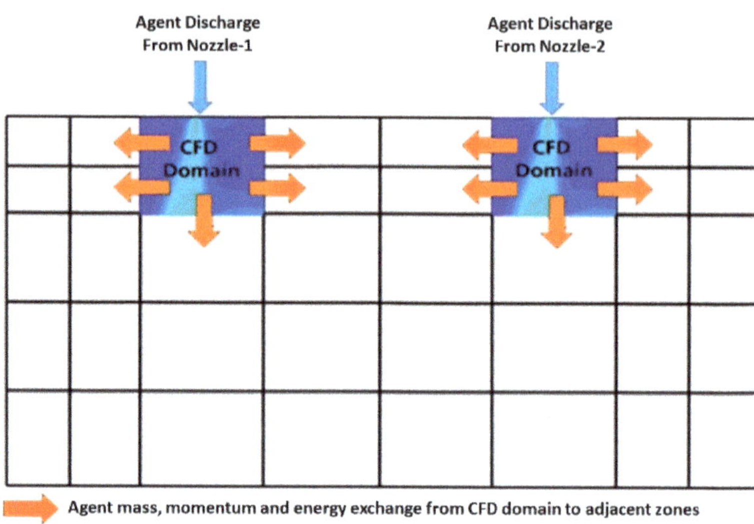

Figure 3. Schematic zonal cargo hold model coupled with two CFD domains.

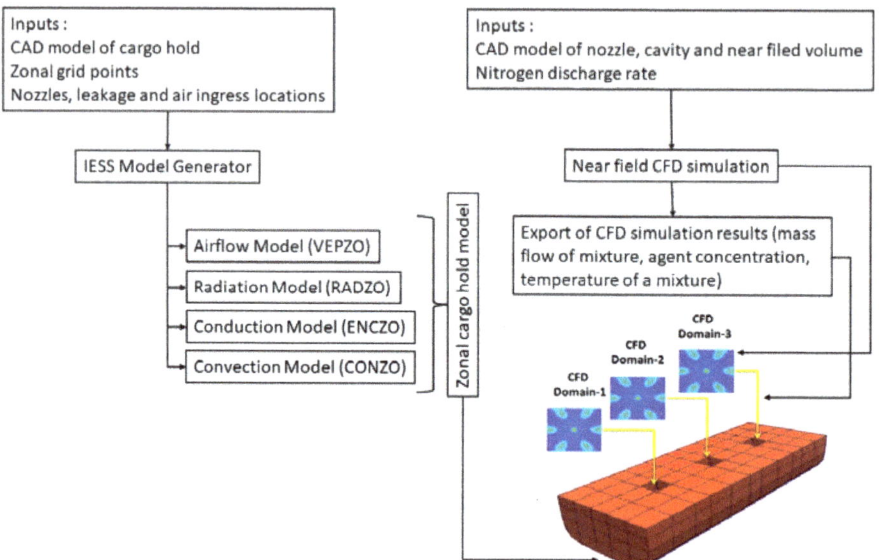

Figure 4. Workflow for cargo model generation.

This model has been used for parametric studies for different discharge profiles, different descent profiles, and different fire suppression strategies [10] and is now validated with experiments.

2.2. Test Setup

Experiments were conducted in the A310 mock-up of the Flight Test Facility (FTF) located at the Fraunhofer-Institute for Building Physics in Holzkirchen, Germany. A schematic view of the FTF is presented in Figure 5. The front part of a former in-service twin-aisle long range aircraft containing a cabin, crown, galley, cockpit, avionics bay, cargo, and bilge was placed in a low-pressure vessel. Through the variation of the pressure in the vessel, the cabin pressure evolution of a real flight can be simulated. The mock-up is equipped with a ventilation system to replicate the environmental control system (ECS). In order to generate a similar heat load in the cabin, thermal dummies were placed on the seats. Through this, a realistic airflow pattern in the cabin was ensured. Recirculation air was aspired from the triangular area, and exhaust air was ejected from the bilge.

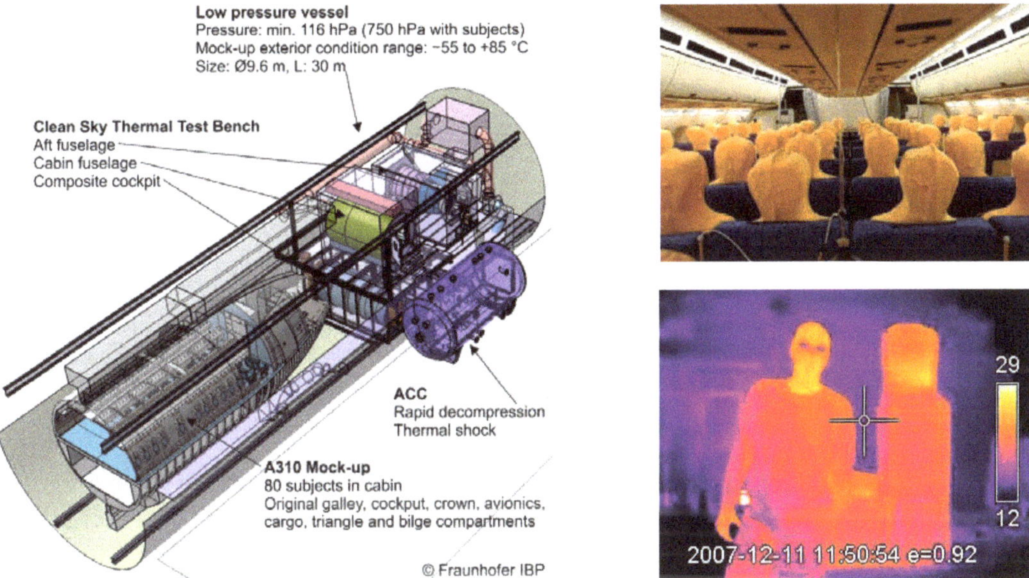

Figure 5. Overview of the Flight Test Facility, cabin equipped with thermal dummies, and IR picture of a human compared to a dummy.

2.2.1. Cargo Bay Refurbishment

For the environmentally friendly fire protection system tests, the cargo bay was refurbished (Figure 6, left). The main items of the refurbishment are as follows:

- Original lining and ceiling panels from spare parts and A350 production series (except the center line, where Plexiglas was used to keep the agent distribution line visible).
- Original panel sealing to meet airtightness requirements.
- Integration of high pressure piping and three injection nozzles with a protective cavity in the ceiling.
- Integration of the pressure management system (PMS) allowing the equalization of pressure between the cargo bay and adjacent bays.
- Cargo door leakage simulation according to the MPS standard [1] to generate the in-flight leakage across the door seal. In the test setup, this leakage was replicated by a piccolo tube distributed along the door seal. This tube is connected to a flow-controlled extraction fan.
- Oxygen concentration sensors are distributed in the cargo hold to assess the local oxygen and nitrogen concentrations.

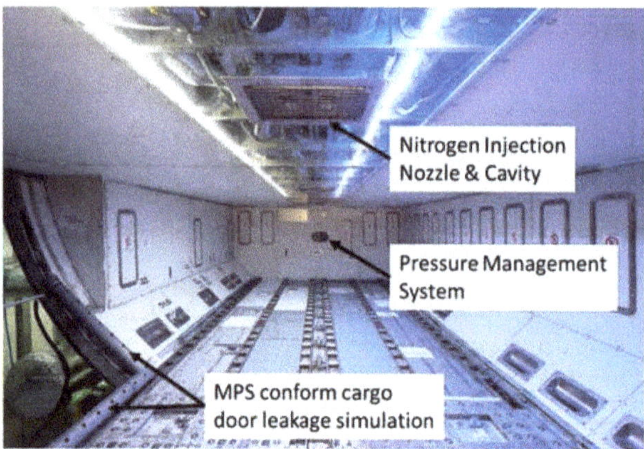

Figure 6. Refurbished cargo hold (**left**) and integrated OBIGGS (**right**).

The nitrogen needed to perform the inertion task was taken from industrial bottles placed outside the low pressure vessel. The bottles were on a scale to measure the consumed amount of nitrogen. Furthermore, an On-Board Inert Gas Generating System (OBIGGS, Figure 6, right) demonstrator, derived from the fuel tank inertion technology, has been integrated in the flight test facility. The OBIGGS consists of selective air separation membranes that separate hot pressurized air into a nitrogen rich fraction used for cargo bay inertion and an oxygen rich fraction dumped overboard.

2.2.2. Typical Test

At the beginning of the test, heat dummies in the cabin and the cabin ventilation system are turned on. Then, the pressure in the low pressure vessel is reduced from ambient pressure to 750 hPa (8000 ft.). The leakage flow simulation through the cargo door is activated. Due to this suction, air from the underfloor area enters the cargo bay through the PMS. When the pressure, airflow rates, and temperatures are stabilized, the test begins.

- Phase 1: Knockdown

A sufficiently large amount of nitrogen is supplied within 3 min to bring down the oxygen concentration in the cargo hold below the flammability point. Dinesh et al. [11] suggest that the remaining oxygen level should be 13% or lower. The exact amount of nitrogen depends on the cargo hold size and the target concentration.

- Phase 2: Holding

A metered flow of pure bottled nitrogen or nitrogen enriched air (NEA) provided by the OBIGGS is supplied. The required flow rate depends on the air tightness of the door seal as this air is replaced by the ingress of new, fresh air through the PMS into the cargo hold.

- Phase 3: Descent

The descent phase is critical in terms of oxygen concentration due to the repressurization from 750 hPa to ground pressure. This repressurization is performed by supplying ambient air through the PMS. Thus, a noticeable amount of fresh air enters the cargo hold and increases the oxygen concentration. There are two strategies to cope with this: either the holding system increases its flow accordingly or the oxygen concentration is kept sufficiently low prior to descent to meet the requirement at the end of the descent.

3. Test and Model Validation Results

In this section, the test and model validation results are presented. The presented test uses NEA from the OBIGGS during the holding phase and follows the strategy to provide a sufficiently low oxygen concentration in the cargo hold to meet the requirement of concentrations below 13% after descent.

Test Sequence and Result

For model validation, the test sequence set out in Table 1 was performed. Times reflect the elapsed minutes. It simulates a flight where the fire suppression system is activated during cruising and operates until landing at the airport.

Table 1. Performed test sequence.

Test Sequence	Start	End	Performed
Ground	0	5	normal operation of the cabin
Takeoff	5	23	reduction of pressure to 750 hPa
Normal cruise	23	37	normal operation of the cabin
Knockdown (KD)	37	41	supply of 41.5 kg of nitrogen from bottles to reach an initial oxygen concentration below 10%.
Holding without leakage simulation	41	51	supply of 9 L/s nitrogen-enriched air with residual O_2-concentration of 7–8%
Holding with leakage simulation	51	60	supply of 9 L/s nitrogen-enriched air (NEA) from OBIGGS with residual O_2-concentration of 7–8% Activation of cargo door leakage at 9 L/s
Descent	59	68	Re-pressurization to ambient pressure, leakage, and NEA flow maintained
Ground	68	70	NEA flow and leakage turned off

The pressure chart is shown in Figure 7 (top). To increase the pressure in the cargo bay from 750 hPa to 940 hPa, an air amount of approx. 11.7 m^3 is supplied during descent. Taking into account the time of 9 min to reach ground pressure, this corresponds to an average air ingress rate of 21.6 L/s, thus noticeably higher than the leakage rate though the door seal.

Figure 7, bottom shows the oxygen concentration in the front upper left and back lower right position. It is obvious that the knock-down leads to an instantaneous drop in the oxygen concentration in the cargo hold. During the holding phase, NEA is supplied with a residual oxygen concentration of 7–8%. The measurement indicates that the oxygen concentration is uniform in the cargo bay and no relevant gradient builds up with or without the activated door leakage system. The reason for this is that leakage and the holding supply rate are both balanced at 9 L/s, thus no major ingress of fresh air occurs during this phase. While the door leakage is not activated, the supply of NEA leads to an outflow of air through the PMS. This outflow is redirected to the door leakage when the system is started.

The re-pressurization leads to a higher gradient of oxygen concentration throughout the cargo hold. The pressure management system and thus the ingress of fresh air is located at the rear wall. Therefore, the local concentration becomes higher in this region during descent. Nevertheless, the requirement to remain below 13% is still met at this "weakest" point.

The experiment shows that the OBIGGS stably delivers NEA with 7–8% residual oxygen concentration throughout the mission profile.

The same mission profile, starting with knockdown and ending with landing, was applied to the simulation model. The zonal modelling approach accurately predicts the magnitude of the oxygen concentration (Figure 8, left). The deviation between simulated and measured oxygen concentrations remains below 1%.

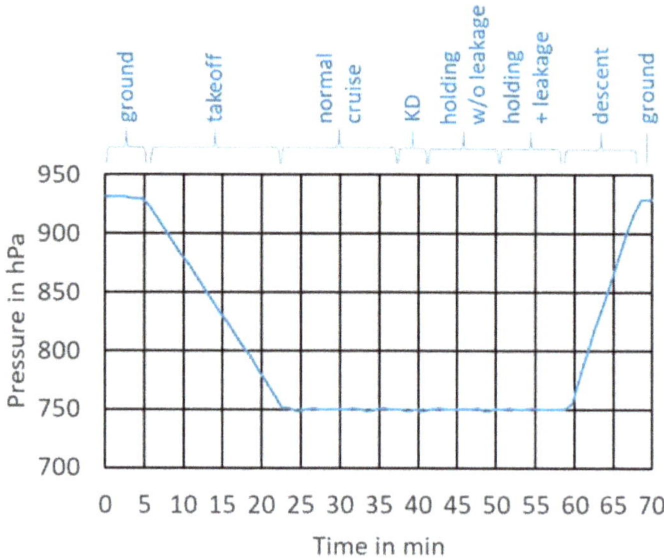

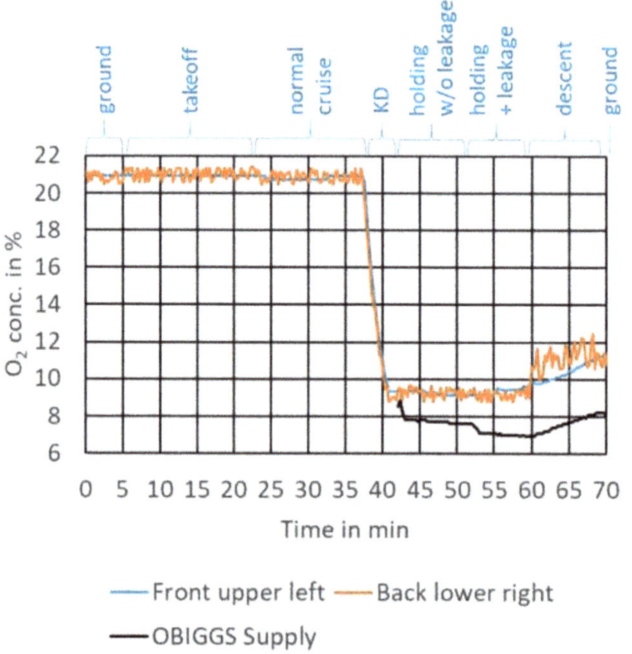

Figure 7. Top: Pressure boundary condition in the test, **Bottom**: Evolution of the cargo hold oxygen concentration in the extremities of the bay.

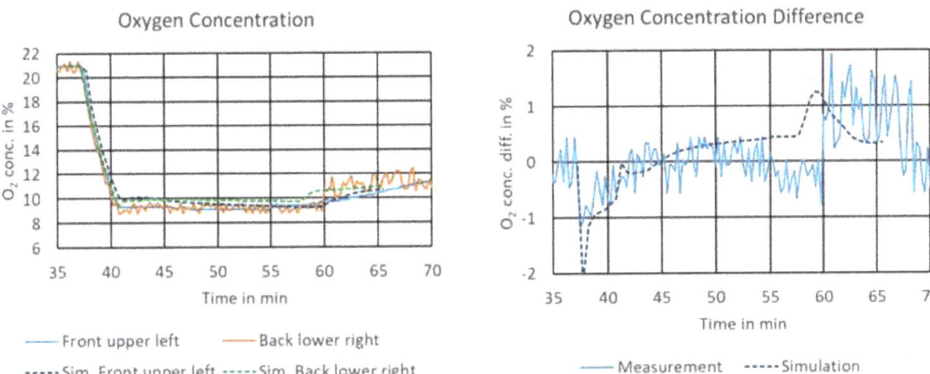

Figure 8. Left: Comparison of measured (bold) and simulated (dotted) oxygen concentrations. **Right**: Oxygen concentration difference between the extremities.

The measured and simulated gradients of the oxygen concentration are shown (Figure 8, right). For this, both sensor readings were subtracted from the concentrations in the zones representative of the sensors. The concentration difference was accurately predicted with a maximum deviation of the simulated and measured difference of less than 1%. The dynamic response of the simulation model was faster than the actual test data. One reason could be the time delay in the sensors to react to such changes or the integration scheme used by the Modelica/Dymola solver.

An exemplary plot of the oxygen distribution during descent, together with the location of the evaluated zones in the cargo hold is shown in Figure 9.

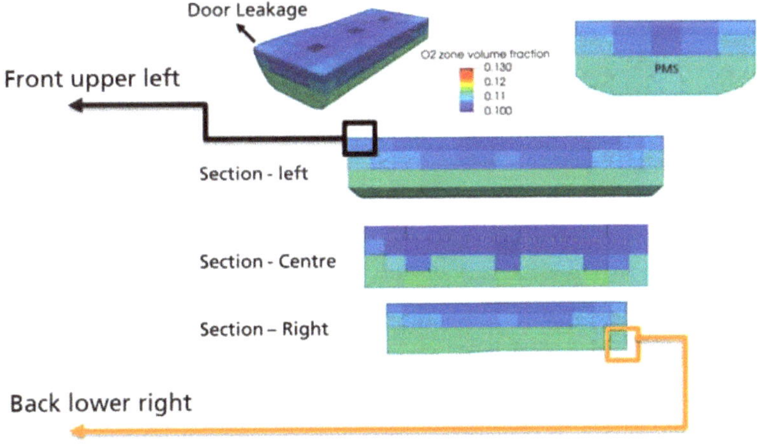

Figure 9. Simulated oxygen gradient during descent.

4. Conclusions

This paper presents a model validation case for the agent distribution of environmentally friendly fire protection systems. The validated model can now serve to predict critical load cases for the cargo hold, the effect of geometry change, e.g., when considering a different aircraft size, and is extendable for other agents. Through this approach, a transient design tool for the cargo fire protection mission has been developed.

From the model validation it is deduced that the IESS is capable of predicting the oxygen concentration in the cargo hold for a transient mission profile within 1% accuracy. For a model-based nitrogen fire suppression system design, this maximal detected deviation suggests aiming for a design setpoint of 12% oxygen concentration in order to reliably meet the requirement of a residual oxygen concentration of 13%.

The presented example considers an empty cargo hold. Future research will consider the effect of containers placed in the cargo hold leading to the obstruction of the airflow paths.

Author Contributions: Conceptualization, V.N. and A.P.; methodology, A.P.; software, A.P.; validation, A.P.; formal analysis, M.P.; investigation, M.P.; resources, V.N.; data curation, M.P.; writing—original draft preparation, V.N.; writing—review and editing, A.P.; visualization, V.N.; supervision, A.P.; project administration, V.N.; funding acquisition, V.N. All authors have read and agreed to the published version of the manuscript.

Funding: The work was conducted with financial support from the Clean Sky 2 program under Grant Agreement number: LPA-IADP CS2-LPA-GAM-2018-2019-01. The authors are responsible for the content of this publication.

Data Availability Statement: A file with data used for the plots in this paper is uploaded together with the paper.

Conflicts of Interest: The authors declare no conflict of interest.

Abbreviations

Acronym	Explanation
CAD	Computer-Aided Design
CFD	Computational Fluid Dynamics
ECS	Environmental Control System
EFFP	Environmentally Friendly Fire Protection
FTF	Flight Test Facility
IESS	Indoor Environment Simulation Suite
KD	Knockdown
MPS	Minimum Performance Standard [1]
NEA	Nitrogen-Enriched Air
OBIGGS	On-Board Inert Gas Generation System
ODP	Ozone Depletion Potential
PMS	Pressure Management System

References

1. Reinhardt, J.W. *Minimum Performance Standard for Aircraft Cargo Compartment Halon Replacement Fire Suppression Systems*; 2nd Update; US Department of Transportation Federal Aviation Administration: Springfield, VA, USA, 2005.
2. Hetrick, T.; Todd, M. Investigation of Hold Time Calculation Methodologies for Total Flooding Clean Extinguishing Agents. In Proceedings of the SUPDET Conference, Orlando, FL, USA, 5–8 March 2007.
3. Li, Y.F.; Chow, W. A Zone Model in Simulating Water Mist Suppression on Obstructed Fire. *Heat Transf. Eng.* **2006**, *27*, 99–115. [CrossRef]
4. Vaari, J. A transient one-zone computer model for total flooding water mist fire suppression in ventilated enclosures. *Fire Saf. J.* **2002**, *37*, 229–257. [CrossRef]
5. Yoon, S.S.; Kim, H.Y.; Hewson, J.C.; Suo-Anttila, J.M.; Glaze, D.J.; Desjardin, P.E. A Modeling Investigation of Suppressant Distribution from a Prototype Solid-Propellant Gas-Generator Suppression System into a Simulated Aircraft Cargo Bay. *Dry. Technol.* **2007**, *25*, 1011–1023. [CrossRef]
6. Pathak, A.; Norrefeldt, V.; Lemouedda, A.; Grün, G. The Modelica Thermal Model Generation Tool for Automated Creation of a Coupled Airflow, Radiation Model and Wall Model in Modelica. In Proceedings of the 10th International Modelica Conference, Lund, Sweden, 10–12 March 2014; Linköping University Electronic Press: Linköping, Sweden, 2014; pp. 115–124.
7. Norrefeldt, V.; Grün, G. VEPZO—Velocity Propagating Zonal Model for the prediction of airflow pattern and temperature distribution in enclosed spaces. In Proceedings of the 9th International Modelica Conference, Munich, Germany, 3–5 September 2012.
8. Pathak, A.; Norrefeldt, V.; Grün, G. Modelling of radiative heat transfer in Modelica with a mobile solar radiation model and a view factor model. In Proceedings of the 9th International Modelica Conference, Munich, Germany, 3–5 September 2012.

9. Norrefeldt, V.; Pathak, P.; Siede, M.; Lemouedda, A. Thermal Model and Thermal Model Generation Tool for Business Jet Applications. In Proceedings of the Greener Aviation Conference, Brussels, Belgium, 12–14 March 2014.
10. Pathak, A.; Norrefeldt, V.; Grün, G. Simulation tool for environment friendly aircraft cargo fire protection system evaluation. In Proceedings of the Aerospace Europe Conference, Bordeaux, France, 25–28 February 2020.
11. Dinesh, A.; Benson, C.M.; Holburn, P.; Sampath, S.; Xiong, Y. Performance evaluation of nitrogen for fire safety application in aircraft. *Reliab. Eng. Syst. Saf.* **2020**, *202*, 107044. [CrossRef]

Article

Performance and Emissions of a Microturbine and Turbofan Powered by Alternative Fuels†

Radoslaw Przysowa [1,*], Bartosz Gawron [1], Tomasz Białecki [1], Anna Łęgowik [1] and Jerzy Merkisz [2] and Remigiusz Jasiński [2]

[1] Instytut Techniczny Wojsk Lotniczych (ITWL), ul. Ksiecia Boleslawa 6, 01-494 Warsaw, Poland; bartosz.gawron@itwl.pl (B.G.); tomasz.bialecki@itwl.pl (T.B.); anna.legowik@itwl.pl (A.Ł.)

[2] Faculty of Civil and Transport Engineering, Poznan University of Technology, ul. Piotrowo 3, 60-965 Poznan, Poland; jerzy.merkisz@put.poznan.pl (J.M.); remigiusz.jasinski@put.poznan.pl (R.J.)

* Correspondence: radoslaw.przysowa@itwl.pl

† This paper is an extended version of our paper published in 10th EASN International Conference on Innovation in Aviation & Space to the Satisfaction of the European Citizens.

Abstract: Alternative fuels containing biocomponents produced in various technologies are introduced in aviation to reduce its carbon footprint but there is little data describing their impact on the performance and emissions of engines. The purpose of the work is to compare the performance and gas emissions produced from two different jet engines—the GTM-140 microturbine and the full-size DGEN380 turbofan, powered by blends of Jet A-1 and one of two biocomponents: (1) Alcohol-to-Jet (ATJ) and (2) Hydroprocessed Esters and Fatty Acids (HEFA) produced from used cooking oil (UCO) in various concentrations. The acquired data will be used to develop an engine emissivity model to predict gas emissions. Blends of the mineral fuel with synthetic components were prepared in various concentrations, and their physicochemical parameters were examined in the laboratory. Measurements of emissions from both engines were carried out in selected operating points using the Semtech DS gaseous analyzer and the EEPS spectrometer. The impact of tested blends on engine operating parameters is limited, and their use does not carry the risk of a significant decrease in aircraft performance or increase in fuel consumption. Increasing the content of biocomponents causes a noticeable rise in the emission of CO and slight increase for some other gasses (HC and NO_x), which should not, however, worsen the working conditions of the ground personnel. This implies that there are no contraindications against using tested blends for fuelling gas-turbine engines.

Keywords: turbofan; microturbine; sustainable aviation fuel; ATJ; HEFA; emissions; alternative fuel; biocomponent; combustion; fuel blend; drop-in fuel; synthesized kerosene

Citation: Przysowa, R.; Gawron, B.; Białecki, T.; Łęgowik, A.; Merkisz, J.; Jasiński, R. Performance and Emissions of a Microturbine and Turbofan Powered by Alternative Fuels. *Aerospace* **2021**, *8*, 25. https://doi.org/10.3390/aerospace8020025

Academic Editor: Spiros Pantelakis
Received: 15 December 2020
Accepted: 15 January 2021
Published: 21 January 2021

1. Introduction

In order to reduce CO_2 emissions, as well as make use of inedible raw materials from renewable sources, alternative fuels containing biocomponents produced in various technologies are introduced in aviation [1–3]. However, there is much less experience in using biofuels to propel aircraft than in using mineral fuels. Currently, ASTM D 7566 standard allows for seven synthetic fuel production technologies to be used in aircraft turbine engines (Table 1), including Alcohol-to-Jet (ATJ) and Hydroprocessed Esters and Fatty Acids (HEFA). ATJ and HEFA belong to well-established sustainable aviation fuels (SAF) which do not require modifications in engines, aircraft or ground infrastructure (drop-in fuels) if used within the permissible mixing ratio.

Air-quality measurements clearly show that a single take-off or landing operation noticeably increases the concentration of toxic compounds in the air [4]. The widespread use of gas-turbine engines contributes to the deterioration of air quality in and around airports, and the negative impact of emitted gases and particles on health is of concern within ground personnel and neighbouring communities. Although emissions of all large engines are controlled during aircraft certification, the high intensity of air operations causes an accumulation of pollutants. Therefore, it is necessary to widely monitor air quality [5], especially in the vicinity of airports. After many years, the introduction of new rules for the certification of aircraft engines by ICAO (International Civil Aviation Organization) began, which is associated with the use of a new measurement methodology, which has been proposed in publications over the past years [6]. Recent research efforts [7] are aimed at measuring gaseous and particulate emissions under real conditions and better understanding their toxicity and impact on health [8–10].

Alternative fuels are expected not to increase emissions and provide a comparable or better engine performance. Blends with the ratio of biocomponents lower than the maximum allowable one are used mainly due to the higher cost of the alternative fuel or for fear of their impact on engine durability or performance. Therefore, when an alternative fuel is introduced into the fleet, the ratio of the biocomponent is gradually increased, observing its impact on the parameters and health of the engines.

New types of alternative fuels are synthesised in small quantities, insufficient to power large engines. This is why microturbines are often used to test new blends [11–14] despite the fact that their structure differs considerably from one of the commercial engine.

More and more often, scientific activity is focused on the development of models of emissions of toxic compounds from aviation [3] and their validation in real flight conditions and laboratory tests. To describe various fuels and blends, it is necessary to define their thermophysical parameters. In zero-dimensional models of engines, which are developed in GSP [15] or GasTurb [16,17], a simplified analytical description of combustion and tabulated values of temperature for fuel blends are often used. In the modelling of combustors, thermodynamic equations are used to describe combustion and heat transport [18,19]. When designing new combustors with reduced emissions, complex numerical CFD models are used [20,21]. An engine emissions data bank [22] and machine learning [23] models using statistical methods [24] or artificial intelligence are also used to predict emissions.

Table 1. Certified processes for Sustainable aviation fuel [25].

Abbreviation	Conversion Process	Possible Feedstocks	Ratio
FT-SPK	Fischer-Tropsch hydroprocessed synthesized paraffinic kerosene	Coal, Natural Gas Biomass	$\leq$50%
HEFA-SPK	Synthesized paraffinic kerosene produced from hydroprocessed esters and fatty acids	Bio-Oils, Animal Fat Recycled Oils	$\leq$50%
SIP	Synthesized kerosene isoparaffins produced from hydroprocessed fermented sugars	Biomass used for sugar production	$\leq$10%
SPK/A	Synthesized kerosene with aromatics derived by alkylation of light aromatics	Coal, Natural Gas Biomass	$\leq$50%
ATJ-SPK	Alcohol-to-jet synthetic paraffinic kerosene	Biomass from ethanol or isobutanol production	$\leq$50%
CHJ	Catalytic Hydrothermolysis Jet from processed fatty acid esters and fatty acids	Triglycerides such as soybean oil, jatropha oil	$\leq$50%
HC-HEFA	Synthesized paraffinic kerosene from hydroprocessed hydrocarbons, esters and fatty acids	Algae	$\leq$10%

In this work, both a microturbine and full-size engine are used to generalise some results that can only be obtained with a microturbine. The purpose of the study is to compare the performance and gas emissions produced from two different jet engines—the GTM-140 microturbine and the DGEN380 geared turbofan. The acquired data will be used to develop an engine emissivity model to predict emissions of exhaust gas compounds.

2. Methods

Blends of the mineral fuel with synthetic components were prepared in various concentrations and their physicochemical parameters were examined in the laboratory. The engines were powered by blends of Jet A-1 and one of two biocomponents: (1) ATJ and (2) HEFA produced from used cooking oil (UCO). Measurements of gas emissions from the GTM-140 microturbine were carried out in selected operating points using the Semtech DS gaseous analyzer and the EEPS spectrometer. Similar emission measurements were made for the DGEN380 engine in a test cell. Measurements were averaged for each operating point, visualised and compared. Results were analyzed in the scope of physicochemical parameters of fuel blends, engine operating parameters and gas emissions (Figure 1).

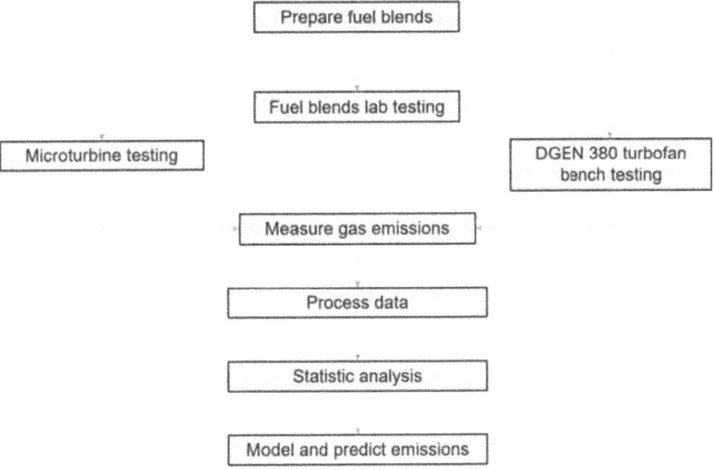

Figure 1. Testing methodology.

2.1. Fuel Lab Testing

In the fuel lab , physicochemical properties of prepared blends were tested such as: density at 15 °C, viscosity at −20 °C and −40 °C, calorific value, the aromatics and naphthalenes content, flash point, crystallisation temperature, non-smoking smoke point and distillation. The selected properties are important for the combustion process and engine operation and are also defined in the ASTM D1655-18a and ASTM D7566-18 documents.

2.2. GTM-140 Microturbine

The GTM-140 microturbine (Figure 2) consists of a single-stage radial compressor, driven by a single-stage axial turbine, and an annular combustion chamber with a set of evaporators. It operates in the range of 33,000–120,000 rpm and can produce take-off thrust up to 140 N. The tested variant of the microturbine played the role of a combustor rig and was devoid of a converging nozzle. For this reason, the generated thrust was limited to 70 N, and consequently SFC values were high (above 200 kg/kN/h). However, in this work, the absolute values of the parameters were less important than their relative changes in response to the increased biocomponent ratio.

Figure 2. GTM-140 microturbine.

The test bench is equipped with a portable GA60 gas analyzer (Figure 3) which was used in previous emission research [26–28]. Exhaust gases are sampled from the engine nozzle using a probe and delivered through a heated exhaust hose to the instrument. The analyzer is equipped with electrochemical sensors to measure O_2, CO, NO, NO_2, SO_2 gases and two NDIR (nondispersive infrared) sensors for infrared measurement (CO_2 and C_xH_y). The exhaust gas components are measured with the resolution of 0.1 ppm and uncertainty of 5%. The analyzer also enables the measurement of exhaust gas temperature through a thermocouple built inside the probe. Operating parameters and emissions are presented and stored by the data acquisition system developed in LabVIEW [29] Additionally, the Semtech DS gaseous analyzer (Figure 4) and Engine Exhaust Particle Sizer (EEPS 3090, Figure 5) operated by Poznan University of Technology were used with a respective exhaust sampling system (Figure 6). Exhaust gases were introduced to the Semtech DS analyzer through a probe maintaining the temperature of 191 °C and the exhaust sample was directed to the flame-ionizing detector (FID) where HC concentration was measured. Then the sample was cooled down to temperature of 4 °C and the concentration measurement of NO_x (NDUV analyzer), CO, CO_2 (NDIR analyzer) was performed. In the following sections, raw emission results expressed in ppm or percents are presented.

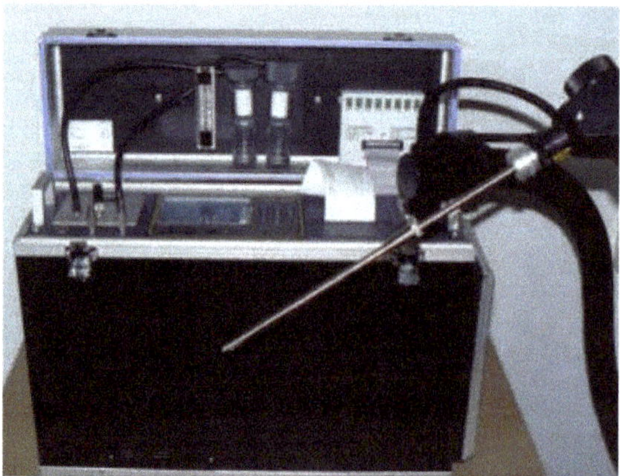

Figure 3. GA-60 emissions analyser.

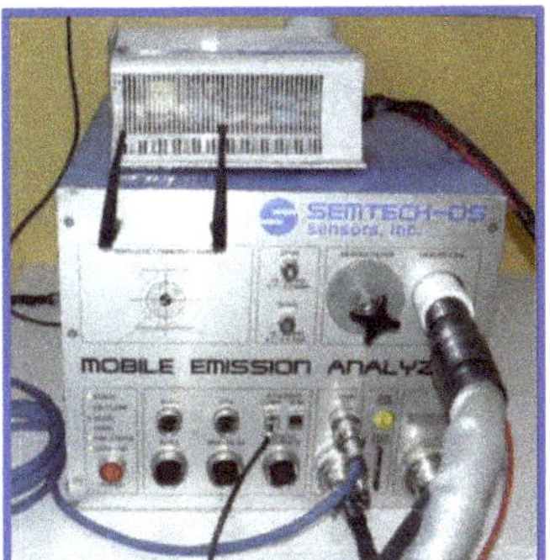

Figure 4. SEMTECH DS gaseous emission analyzer.

Figure 5. Engine Exhaust Particle Sizer (EEPS) 3090.

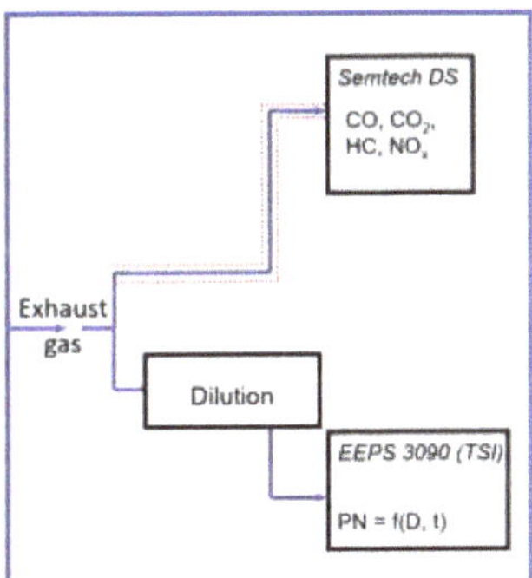

Figure 6. Exhaust gas sampling.

2.3. DGEN 380 Turbofan

DGEN 380 is a high bypass ratio (7.6) geared turbofan, producing 255 daN of thrust (Figure 7). It has a layout similar to modern commercial engines, low fuel consumption and is well instrumented. The data acquisition system of the WESTT test cell (Figure 8) enables data acquisition and analysis of several engine performance parameter such as thrust, fuel consumption, temperature and pressure. The turbofan was developed for ultralight aircraft but has not been certified yet, so it is not covered by the ICAO emissions databank. In this work, the Semtech DS analyzer was used to measure CO, CO_2, HC and NO_x emissions. These are probably the first published results of gaseous emissions for this engine.

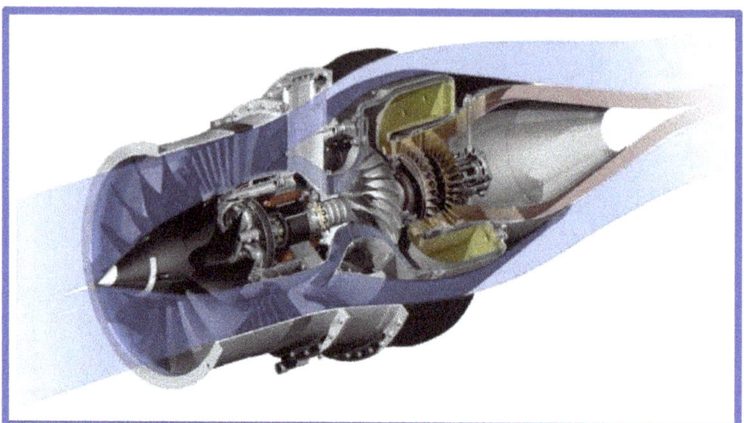

Figure 7. DGEN 380 geared turbofan.

Figure 8. Test cell.

During tests of the DGEN380 turbofan with biocomponents, a finite amount of fuel in the tank was used to maintain a constant blend ratio throughout the test. This means that the tank was not constantly refilled with fuel as was the case with testing on the Jet A-1 fuel. With the tank almost empty, the flow resistance may have increased, but no lower fuel pressure was observed, nor was the fuel pump running at higher RPM. Nevertheless, it turned out that the fuel flow was slightly lower (by 2%) and as a result, it the nominal thrust was not achieved, despite the PLA set at 100% (Section 3.4).

2.4. Engine Testing

Engine emissions significantly depend on its operating mode (Figure 9), so they are tested in selected operating points that belong to the Landing and Take-Off cycle (LTO), according to the ICAO procedure [30]. Engine test steps correspond to the following operating modes: takeoff, climb, cruise, approach and taxi/ground idle. The tests performed on both types of engines consisted of a series of operating points, where the speed was increased in subsequent steps (Figure 10). Performance parameters and emissions were measured and averaged over 30 s for the microturbine and 60 s for the turbojet. The parameters are presented versus thrust which better represents LTO operating modes.

Tests with ATJ blends were performed one day under stable ambient conditions while tests with HEFA were performed one year later under similar conditions. The fuel was not switched during each test-run, so both startups and shutdowns were carried out on the same fuel as the main test. No problem was observed when starting or shutting down the microturbine on alternative fuels. In the case of the turbofan, restarting it after a shutdown was unsuccessful when it was fuelled with an ATJ blend. However, it was not caused by the fuel but by the low oil pressure due to air in the oil.

When comparing the individual blends, an identical mission profile was repeated, consisting of a series of operating points. Efforts were made to ensure that the rotational speeds were the same in subsequent tests, but it was difficult, especially for intermediate speeds. The engine operation range was selected using PLA and FADEC and it was not possible to fine-tune speed or thrust. There are slight but significant speed differences between the test-runs, which is especially important when analysing changes in engine performance.

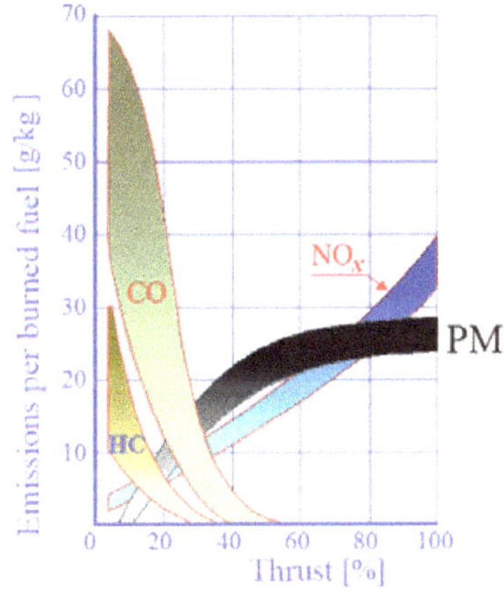

Figure 9. Emissions vs. thrust.

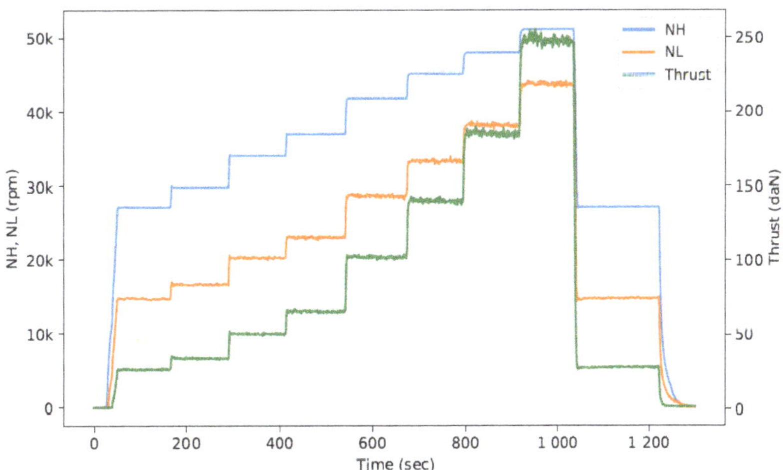

Figure 10. Engine test profile for DGEN 380.

As the certified fuels were used in this project and the running time was short, the engines were not inspected for solid deposits or hot corrosion [31] after the testing. This can be done with a borescope or by stripping down the engine [32,33]. However, the DGEN 380 turbofan was tested later with Jet A-1 fuel without any problems while the microturbine has been continuously operated on various alternative fuels and regularly inspected when replacing the rotor bearings. No significant damage or deposits were observed in the hot section.

3. Results

3.1. Fuel Lab Testing

Lab testing confirmed that the mineral fuel meets the requirements of the ASTM D1655 standard, while ATJ and HEFA biocomponents and their blends comply with ASTM D7566. Table 2 presents the selected physicochemical properties of all tested fuels. The results show that neat biocomponents and their blends are characterised by a lower density than Jet A-1. This has an impact on fuel mass flow. Moreover, biocomponents have a slight higher calorific value, which may result in higher values of the produced heat and a higher EGT. As it is well known, higher combustion temperature increases NO_x emissions. Moreover, neat bio-components are free from aromatics which generally have the least desirable combustion characteristics among kerosene's major components.

Table 2. Fuel lab testing results.

Fuel	Density at 15 °C kg/m³	Viscosity at −20 °C mm²/s	Calorific Value MJ/kg	Aromatics (*v*/*v*) %	Naphtha-Lenes (*v*/*v*) %	Flash Point °C	Freezing Point °C	Smoke Point mm
ASTM	775–840	Max 8.0	Min 42.8	Max 25	Max 3.0	Min 38	Max −40	Min 18
Jet A-1	798	3.40	43.2	16.7	0.58	49.5	−63.5	20
5% ATJ	796	3.45	43.3	15.7	0.55	49.0	−65.5	23
20% ATJ	790	3.57	43.4	13.0	0.46	49.0	−66.5	25
30% ATJ	786	3.66	43.4	11.3	0.40	49.0	−66.8	28
50% ATJ	776	3.65	43.6	8.8	0.27	44.5	−60.0	30
ATJ	759	4.78	44.0	0.0		47.5	−67.5	
Jet A-1	796	3.25			0.55	49	−62.6	23
5% HEFA	794	3.29			0.52	48	−62.8	27
20% HEFA	787	3.40			0.44	46	−59.6	28
30% HEFA	783	3.47			0.39	46	−56.0	
HEFA	752	4.09	44.2			45	−39.9	

3.2. Microturbine—ATJ

During engine tests, the operating parameters such as thrust (Figure 11) and SFC (Figure 12) remain stable, with insignificant differences between tests. Plots below (Figures 13–16) present the CO, CO_2, HC and NO_x emissions of engine fuelled by blends with increasing ratio of the biocomponent (Jet A-1, 50% blend of Jet A-1 with the ATJ component and neat ATJ component). Values are averaged for 30 s while error bars show the triple standard deviation from the mean.

The results show that the addition of the ATJ component caused an increase in CO emissions compared to the Jet A-1 aviation fuel in all operating points (Figure 13). The greater the proportion of ATJ in the mixture, the greater the increase in CO emissions. In terms of CO_2 emissions (Figure 14), ATJ component causes a slight decrease in CO_2 emissions in relation to the neat Jet A-1 fuel.

In the case of HC emissions (Figure 15), the change in emissions of this parameter with the addition of the ATJ component is not conclusive. At the first two rotational speeds, an increase in HC emissions was obtained after adding the ATJ component in relation to Jet A-1. At three successive operating points, the addition of the ATJ component resulted in a decrease in HC emissions for the Jet A-1 fuel.

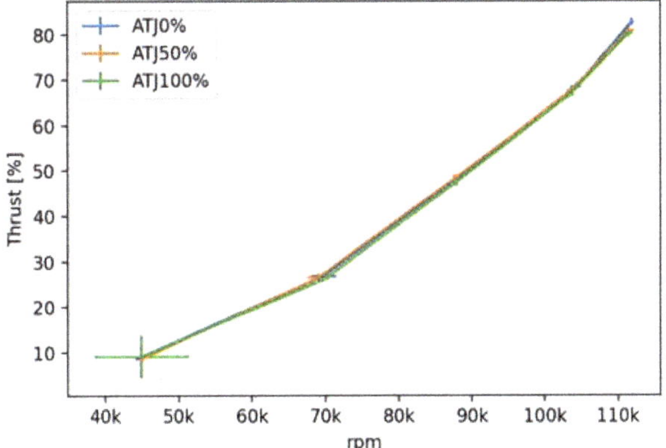

Figure 11. GTM-140/ATJ: Thrust.

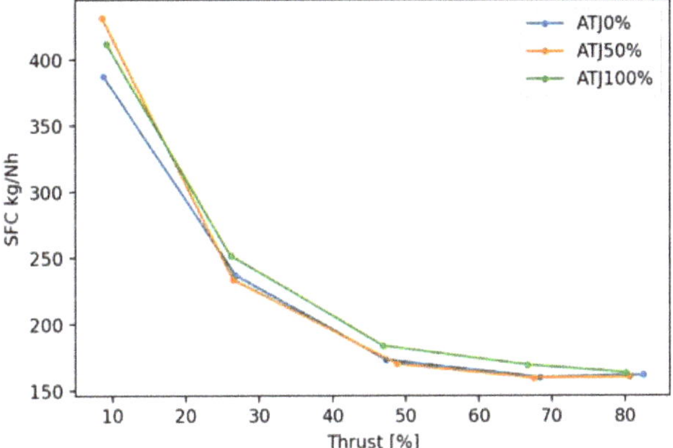

Figure 12. GTM-140/ATJ: SFC.

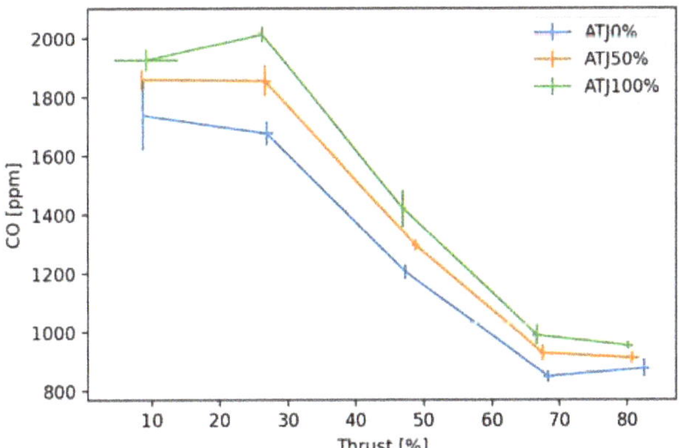

Figure 13. GTM-140/ATJ: CO emission.

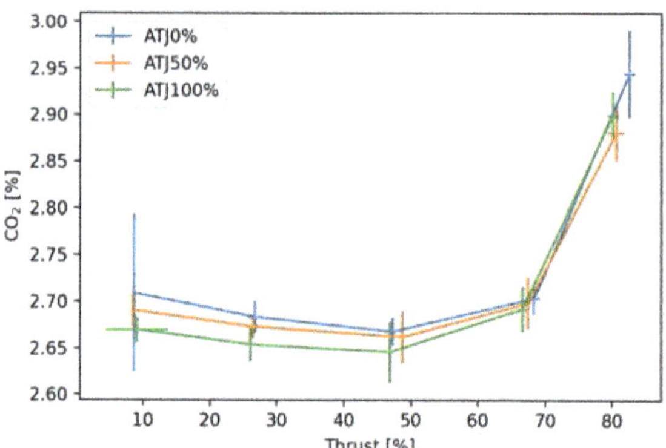

Figure 14. GTM-140/ATJ: CO_2 emission.

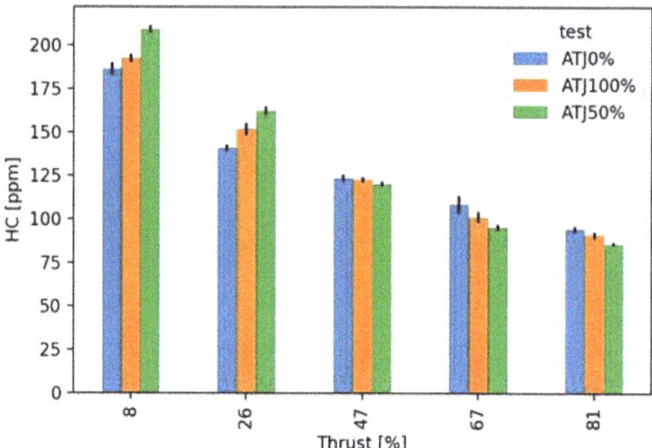

Figure 15. GTM-140/ATJ: HC emission.

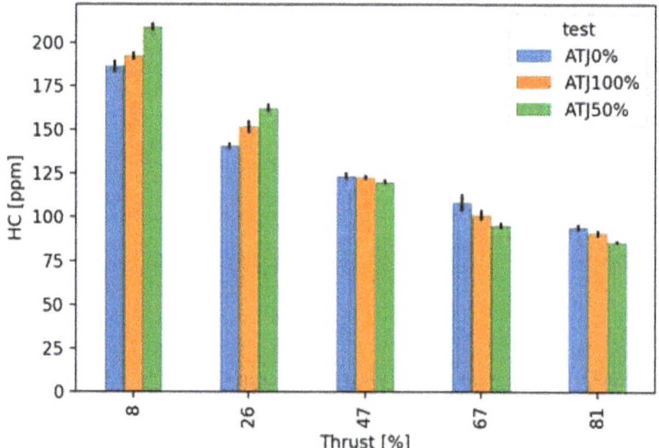

Figure 16. GTM-140/ATJ: NO_x emissions.

3.3. Microturbine—HEFA

Similarly, the test-runs with HEFA blends did not affect significantly engine performance (Figures 17 and 18). Adding the HEFA component to aviation fuel causes an increase in CO emissions (Figure 19) in all analysed operating points while CO_2 shows no clear trend (Figure 20). In terms of NO emissions, a similar direction of changes for this parameter was obtained, that is, the HEFA component generally increases NO emissions (Figure 21). The reverse trend was obtained for the emissions of hydrocarbons (HC, Figure 22). Adding HEFA component resulted in decrease in HC emissions in relation to pure Jet A-1 fuel.

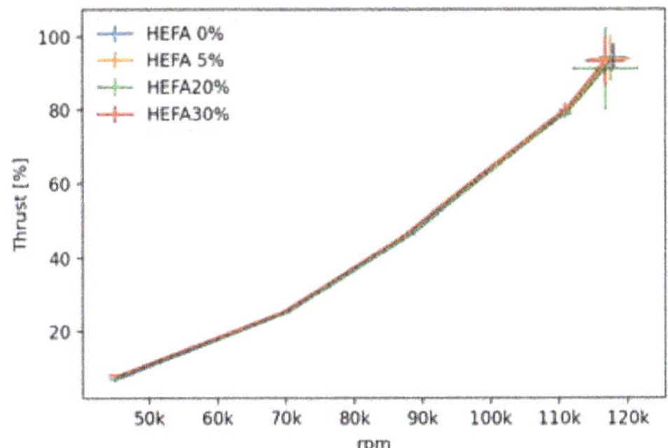

Figure 17. GTM-140/HEFA: Thrust.

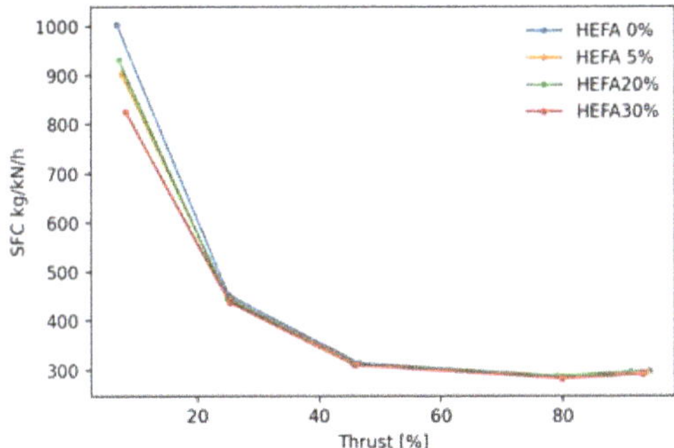

Figure 18. GTM-140/HEFA: SFC.

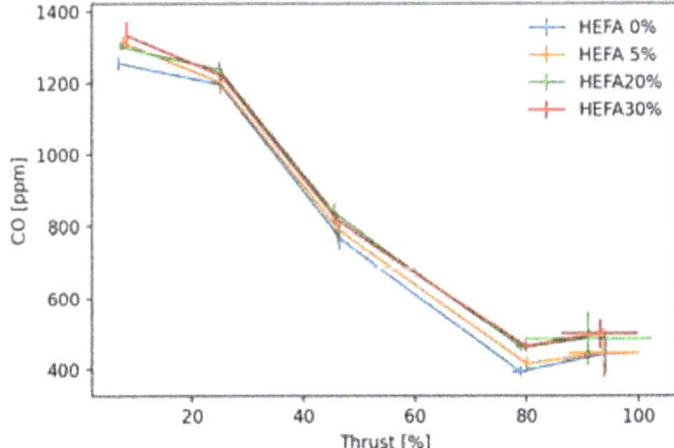

Figure 19. GTM-140/HEFA: CO emission.

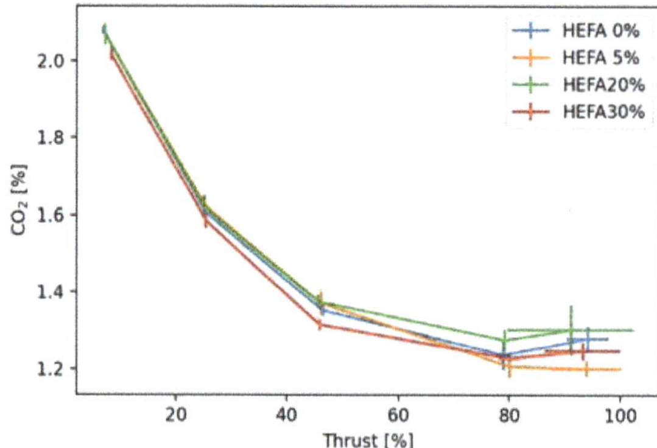

Figure 20. GTM-140/HEFA: CO_2 emission.

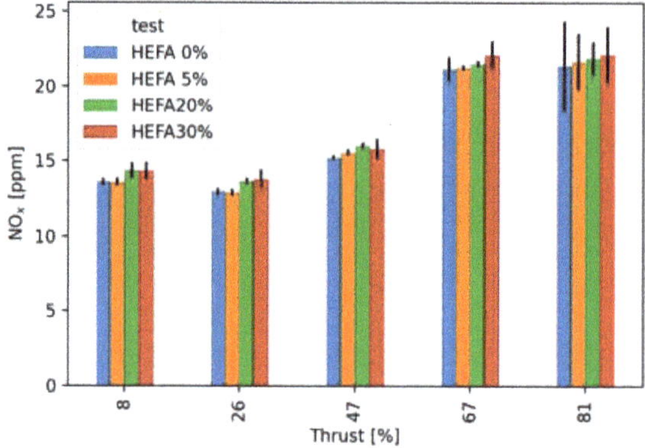

Figure 21. GTM-140/HEFA: NO_x.

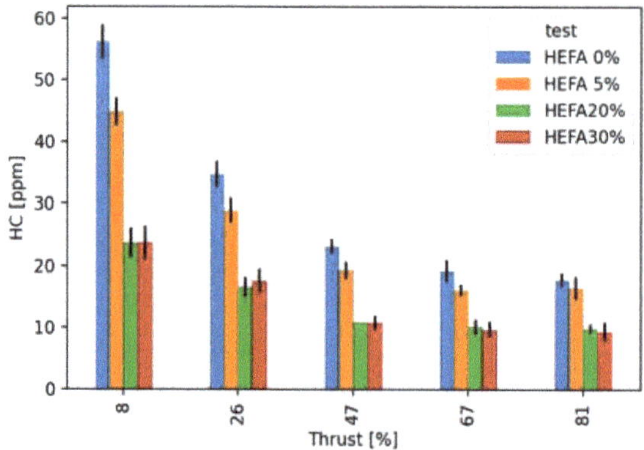

Figure 22. GTM-140/HEFA: HC emission.

3.4. *Turbofan*

Based on the collected performance data of the DGEN 380 turbofan, it can be concluded that the relative differences in thrust (Figures 23 and 24) and fuel consumption (Figures 25 and 26) between the tests are usually less than 1%. The biggest ones are for the take-off range, but they do not exceed 5% (Figures A1 and A2). They are more related to the testing and engine control methods than to the used fuel blend.

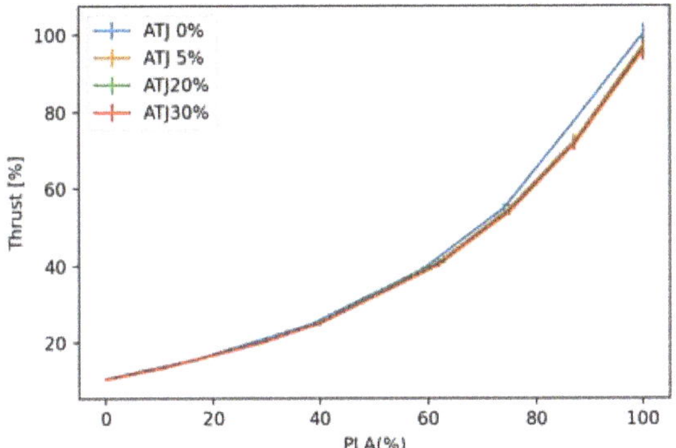

Figure 23. DGEN 380/ATJ: Thrust vs. PLA.

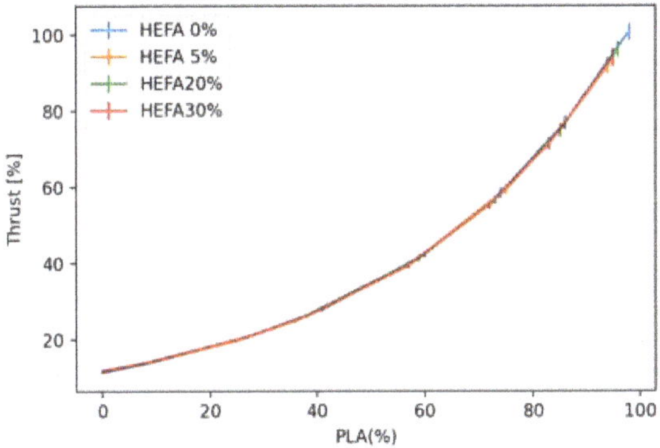

Figure 24. DGEN 380/HEFA: Thrust vs. PLA.

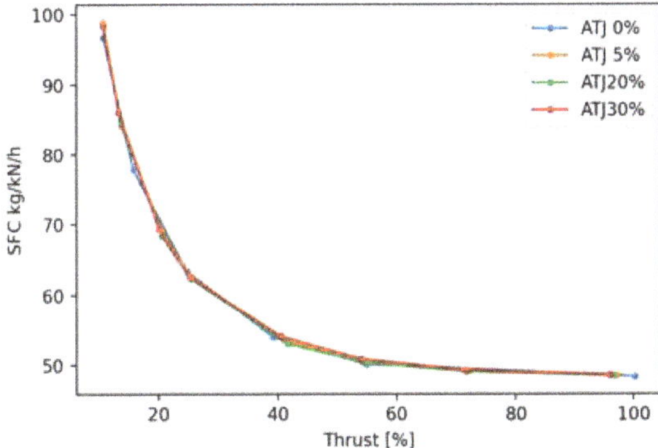

Figure 25. DGEN 380/ATJ: SFC.

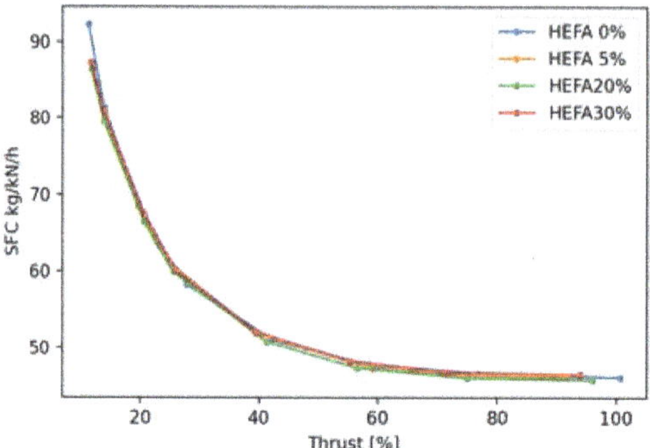

Figure 26. DGEN 380/HEFA: SFC.

When analysing the operating parameters, there is no clear trend of their change due to the use of blends with biocomponents. To associate the differences with some fuel characteristics, the model of engine and its control system is necessary. For example, when using ATJ, a slight increase in exhaust gas temperature (EGT) was observed with the increasing ratio of the biocomponent (Figure 27), but a similar trend was not observed for HEFA (Figure 28).

It is noteworthy that despite its name, PLA is proportional to rotational speed instead of thrust (Figures 23 and 24). This is the effect of FADEC operation which was not modified in this work.

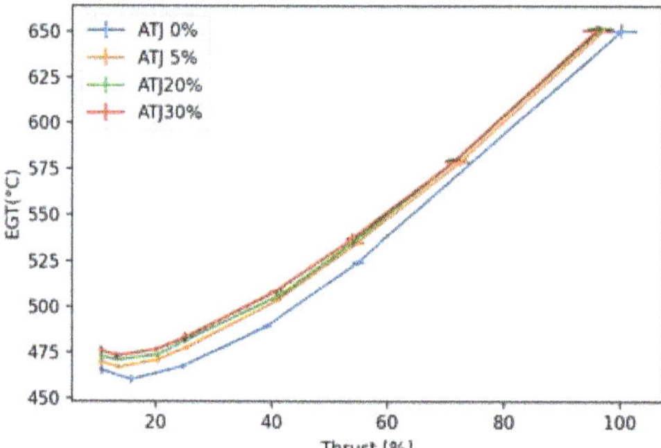

Figure 27. DGEN 380/ATJ: EGT.

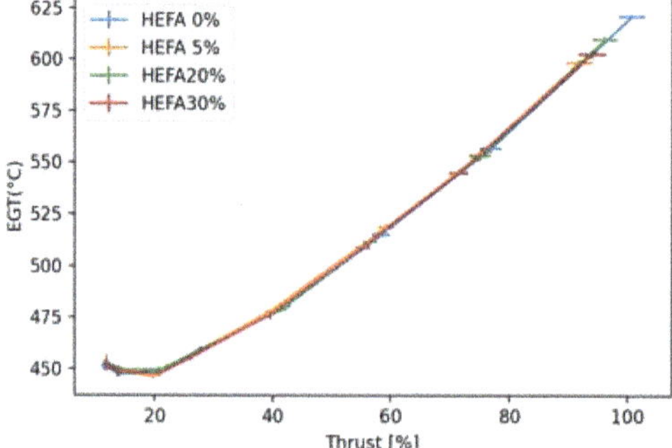

Figure 28. DGEN 380/HEFA: EGT.

Emission-wise, adding the ATJ component to the mineral fuel resulted in a clear increase in CO emissions (Figure 29) and a slight increase in CO_2 emissions (Figure 30) in all the analysed operating points of the DGEN 380 turbofan. The formation of CO is related to incomplete combustion of the fuel, so this process is more apparent when running on a DGEN 380 engine for blends with the ATJ component. In the case of the HEFA component, its addition resulted in a clear increase in CO emission (Figure 31), a slight increase in CO_2 emissions (Figure 32) in all analysed operating points.

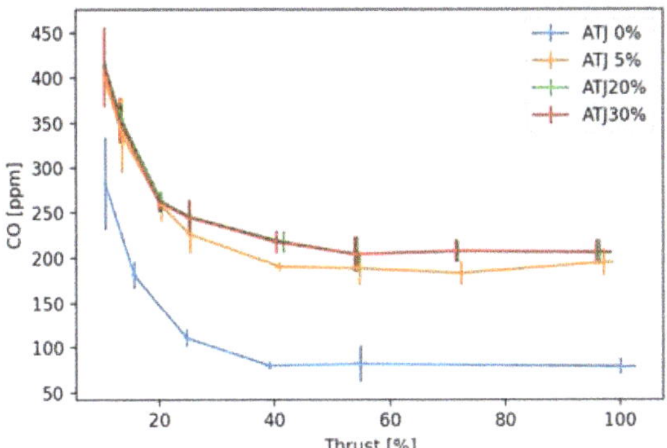

Figure 29. DGEN 380/ATJ: CO.

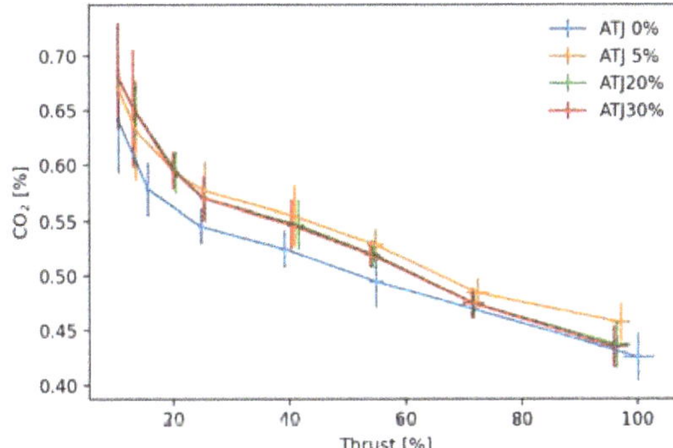

Figure 30. DGEN 380/ATJ: CO_2.

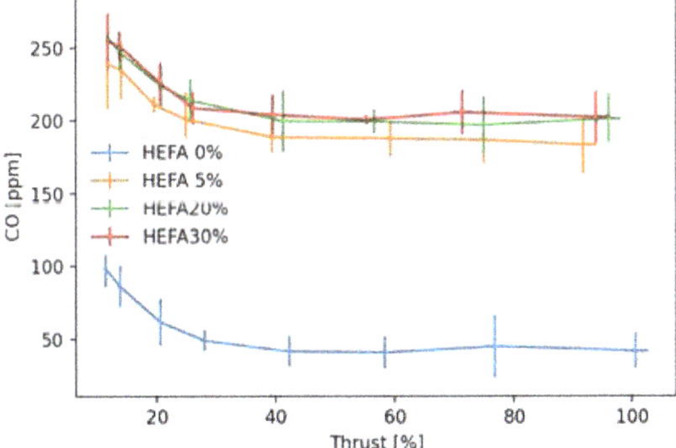

Figure 31. DGEN 380/HEFA: CO.

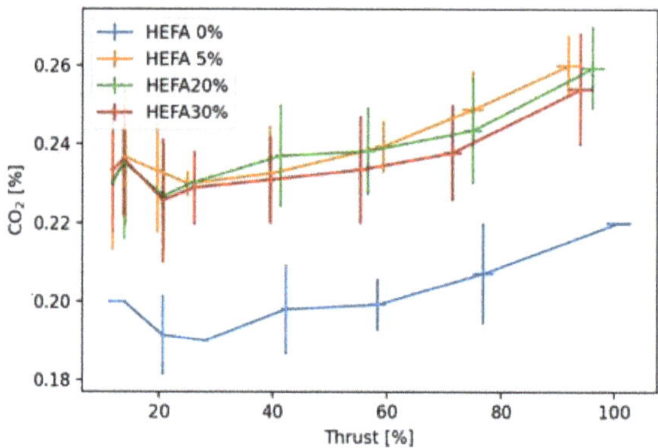

Figure 32. DGEN 380/HEFA: CO_2.

Interestingly, the absolute emission values are almost twice as high for ATJ as for HEFA. Relatively low numbers of the CO_2 emissions for both biofuels are due to the dilution of the exhaust gas. It corresponds to the distance between the measuring probe and the engine, which was chosen in relation to the diameter of the outlet nozzle. Placing the probe in close proximity to the engine would make the results more uncertain. In particular, directing the gas stream directly at the probe makes it impossible to take a sample, as the turbulence around the probe caused by high dynamic pressure disturb the measurements.

A significant increase in CO emissions observed for the DGEN380 engine for both biocomponents has not been completely explained. Different gas dilution for ATJ and HEFA tests is addressed by relating emissions to the CO_2 level which is equivalent to the known method of correcting emission to 15% oxygen [34]. The obtained relation of CO to CO_2 behaves well (Figures 33 and 34) and confirms that the observed CO increase is not a simple measurement error.

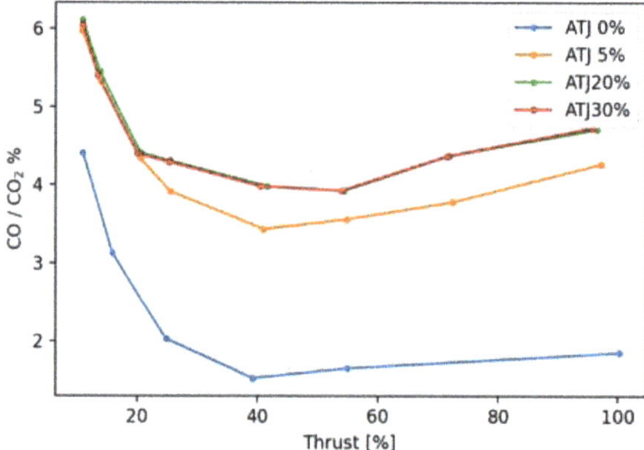

Figure 33. DGEN 380/ATJ: CO vs. CO_2.

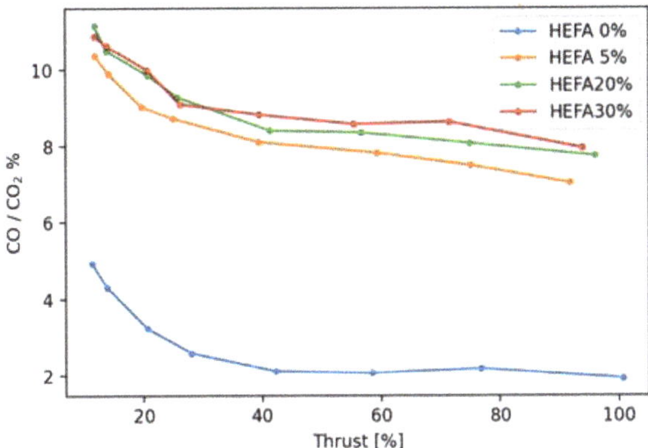

Figure 34. DGEN 380/HEFA: CO vs. CO_2.

In general, CO emissions are related to incomplete combustion and dominate at low speeds. In this case, the increase was approximately 150 ppm for any speed and for all blends including biocomponents. This engine has a reverse flow combustor which is not typical for turbofans but the CFD simulations confirmed the effective combustion of bioethanol and moderate emissions [35]. Custom design of the combustor may have an impact, but should not increase CO emission so much as a result of adding tiny amounts of biocomponent (5%).

4. Conclusions

For both engines, the analysis of engine performance parameters showed that the tested blends differ so little from the mineral fuel that their impact on the engine operating parameters is limited, and their use does not carry the risk of a significant decrease in aircraft performance or increase in fuel consumption.

The experimental emissions results for the DGEN 380 turbofan have never been published elsewhere. In relation to the previous tests of the microturbine carried out in ITWL, the emissions for the ATJ biocomponent and intermediate ratios of HEFA were studied for the first time. In addition, the Semtech analyzer of a class better resolution was used. Particulate emissions were also measured, but their analysis will be the subject of a separate publication.

It was found that increasing the content of biocomponents causes a noticeable increase in the emission of CO and some other gasses (HC and NO_x), which should not, however, worsen the working conditions of the ground personnel. Deeper understanding of the effects of fuel blends on engine performance and emissions requires complex engine models, describing, in particular, its combustor and control system.

The use of small engines, and especially microturbines, for alternative fuels and emissions testing is debatable. They are not scaled large engines, but their structure differ significantly, especially in terms of combustors. Heat cycle losses and SFC are much higher due to relatively large tip clearances. Although the overall emission trends are maintained, the attempt to scale the results from the microturbine to a larger engine was partially successful.

The acquired data will be used to develop an engine emissivity model to predict gas emissions. Statistical methods and the analytical combustion model seem to be suitable for linking the thermophysical parameters of the fuel with the operating parameters and emissions of engines of a basic structure.

Author Contributions: Conceptualization: B.G., T.B., A.Ł. and J.M.; Data curation: R.P., B.G. and R.J.; Investigation: R.P., B.G., A.Ł. and R.J.; Methodology, R.P., B.G., T.B., A.Ł. and R.J.; Project administration: R.P.; Software: R.P.; Supervision: R.P., T.B. and J.M.; Validation: B.G., T.B., J.M. and R.J.; Visualization: R.P. and A.Ł.; Writing—original draft: R.P. and B.G.; Writing—review & editing, R.P. All authors have read and agreed to the published version of the manuscript.

Funding: This research received no external funding.

Data Availability Statement: The data presented in this study are available on request from the corresponding author.

Acknowledgments: Our special thanks are extended to Aleksander Olejnik from the Military University of Technology in Warsaw for providing the DGEN 380 turbofan and Anna Mikołajczyk for performing engine tests. We also wish to thank Jacek Pielecha from the Poznań University of Technology for for his contribution to emission measurements and substantive support.

Conflicts of Interest: The authors declare no conflict of interest.

Abbreviations

The following abbreviations and symbols are used in this manuscript:

ATJ	Alcohol-to-Jet
CFD	computational fluid dynamics
CO	carbon monoxide
CO_2	carbon dioxide
EASN	European Aeronautics Science Network
EGT	exhaust gas temperature
FADEC	full authority digital engine control
FID	flame-ionizing detector
GSP	Gas turbine Simulation Program
HC	hydrocarbons
HEFA	Hydroprocessed Esters and Fatty Acids
ICAO	International Civil Aviation Organization
ITWL	The Air Force Institute of Technology in Warsaw
LTO	landing and take-off (cycle)
NDIR	non-dispersive infrared (sensor)
NDUV	non-dispersive ultraviolet (analyzer)
NO_x	nitrogen oxides
PLA	power lever angle
PM	particulate mater
rpm	revolutions per minute
SAF	sustainable aviation fuel
SFC	specific fuel consumption
SPK	synthetic paraffinic kerosene
UCO	used cooking oil
WESTT	Whole Engine Simulator Turbine Technology

Appendix A

Figures A1 and A2 show the SFC difference between blends and the Jet A-1 fuel.

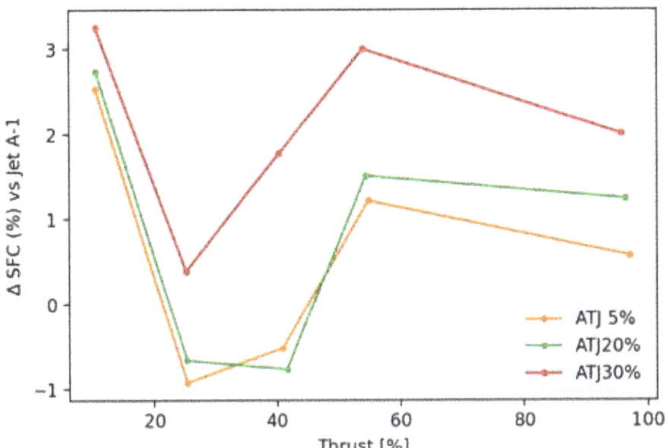

Figure A1. DGEN 380/ATJ: SFC changes.

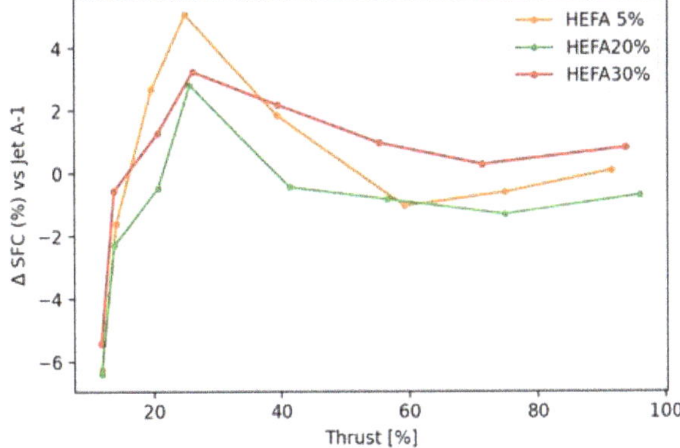

Figure A2. DGEN 380/HEFA: SFC changes.

References

1. Lieuwen, T.C.; Yang, V. *Gas Turbine Emissions*; Cambridge University Press: Cambridge, UK, 2013; Volume 38.
2. Merkisz, J.; Markowski, J.; Pielecha, J. *Selected Issues in Exhaust Emissions from Aviation Engines*; NOVA Science: New York, NY, USA, 2014; p. 209.
3. *Military User's Guide for the Certification of Aviation Platforms on Synthetic Jet Fuels TR-AVT-225*; NATO Science and Technology Organization: Neuilly-sur-Seine, France, 2018.
4. Jasinski, R. Mass and number analysis of particles emitted during aircraft landing. *E3S Web Conf.* **2018**, *44*, 00057. [CrossRef]
5. Chwaleba, A.; Olejnik, A.; Rapacki, T.; Tuśnio, N. Analysis of capability of air pollution monitoring from an unmanned aircraft. *Aviation* **2014**, *18*, 13–19. [CrossRef]
6. Jasiński, R.; Markowski, J.; Pielecha, J. Probe Positioning for the Exhaust Emissions Measurements. *Procedia Eng.* **2017**, *192*, 381–386. [CrossRef]
7. Archilla, V.; Hormigo, D.; Sánchez-García, M.; Raper, D. AVIATOR-Assessing aViation emission Impact on local Air quality at airports: Towards Regulation. In *Proceedings of the MATEC Web of Conferences*; EDP Sciences: Les Ulis, France, 2019; Volume 304, p. 02023. [CrossRef]
8. Gawron, B.; Białecki, T.; Janicka, A.; Górniak, A.; Zawiślak, M. An innovative method for exhaust gases toxicity evaluation in the miniature turbojet engine. *Aircr. Eng. Aerosp. Technol.* **2017**, *89*, 757–763. [CrossRef]
9. Janicka, A.; Zawiślak, M.; Zaczyńska, E.; Czarny, A.; Górniak, A.; Gawron, B.; Białecki, T. Exhausts toxicity investigation of turbojet engine, fed with conventional and biofuel, performed with aid of BAT-CELL method. *Toxicol. Lett.* **2017**, *280*, S202. [CrossRef]

10. Gawron, B.; Białecki, T.; Janicka, A.; Zawiślak, M.; Górniak, A. Exhaust toxicity evaluation in a gas turbine engine fueled by aviation fuel containing synthesized hydrocarbons. *Aircr. Eng. Aerosp. Technol.* **2020**, *92*, 60–66. [CrossRef]
11. Habib, Z.; Parthasarathy, R.; Gollahalli, S. Performance and emission characteristics of biofuel in a small-scale gas turbine engine. *Appl. Energy* **2010**, *87*, 1701–1709. [CrossRef]
12. Chiariello, F.; Allouis, C.; Reale, F.; Massoli, P. Gaseous and particulate emissions of a micro gas turbine fuelled by straight vegetable oil-kerosene blends. *Exp. Therm. Fluid Sci.* **2014**, *56*, 16–22. [CrossRef]
13. Badami, M.; Nuccio, P.; Pastrone, D.; Signoretto, A. Performance of a small-scale turbojet engine fed with traditional and alternative fuels. *Energy Convers. Manag.* **2014**, *82*, 219–228. [CrossRef]
14. Allouis, C.; Amoresano, A.; Capasso, R.; Langella, G.; Niola, V.; Quaremba, G. The impact of biofuel properties on emissions and performances of a micro gas turbine using combustion vibrations detection. *Fuel Process. Technol.* **2018**, *179*, 10–16. [CrossRef]
15. Kluiters, S.C.A.; Visser, W.P.J.; Rademaker, E.R. *A New Combustor and Emission Model for the Gas Turbine Simulation Program GSP*; Technical Report; NLR: Amsterdam, The Netherlands, 2014.
16. Gaspar, R.M.; Sousa, J.M. Impact of alternative fuels on the operational and environmental performance of a small turbofan engine. *Energy Convers. Manag.* **2016**, *130*, 81–90. [CrossRef]
17. GasTurb GmbH. *GasTurb 13 Design and Off-Design Performance of Gas Turbines*; GasTurb GmbH: Aachen, Germany, 2018.
18. Mador, R.; Roberts, R. A pollutant emissions prediction model for gas turbine combustors. In Proceedings of the 10th Propulsion Conference, Lake Tahoe, NV, USA, 31 October–2 November 1973; American Institute of Aeronautics and Astronautics: Reston, VA, USA, 1974. [CrossRef]
19. Rizk, N.; Mongia, H. Emissions predictions of different gas turbine combustors. In Proceedings of the 32nd Aerospace Sciences Meeting and Exhibit, Reno, NV, USA, 10–13 January 1994; American Institute of Aeronautics and Astronautics: Reston, VA, USA, 1994. [CrossRef]
20. Kulczycki, A.; Kaźmierczak, U. Method of preliminary evaluation of biocomponents influence on the process of biofuels combustion in aviation turbine engines. *J. KONES* **2017**, *24*, 83–90. [CrossRef]
21. Dolmatov, D.; Hajivand, M. On Low-Emission Annular Combustor Based on Designing of Liner Air Admission Holes. *NTU KhPI Bull. Power Heat Eng. Process. Equip.* **2018**, *13*, 83–94. [CrossRef]
22. EASA. *Introduction to the ICAO Engine Emissions Databank Background*; EASA: Köln, Germany, 2018.
23. Allaire, D.L.; Waitz, I.A.; Willcox, K.E. A comparison of two methods for predicting emissions from aircraft gas turbine combustors. *Proc. ASME Turbo Expo* **2007**, *2*, 899–908. [CrossRef]
24. Filippone, A.; Bojdo, N. Statistical model for gas turbine engines exhaust emissions. *Transp. Res. Part D Transp. Environ.* **2018**, *59*, 451–463. [CrossRef]
25. *ASTM D7566-20b, Standard Specification for Aviation Turbine Fuel Containing Synthesized Hydrocarbons*; Technical Report; ASTM International: West Conshohocken, PA, USA, 2020. [CrossRef]
26. Gawron, B.; Białecki, T. Measurement of exhaust gas emissions from miniature turbojet engine. *Combust. Engines* **2016**, *167*, 58–63. [CrossRef]
27. Gawron, B.; Białecki, T. Impact of a Jet A-1/HEFA blend on the performance and emission characteristics of a miniature turbojet engine. *Int. J. Environ. Sci. Technol.* **2018**, *15*, 1501–1508. [CrossRef]
28. Gawron, B.; Białecki, T.; Janicka, A.; Suchocki, T. Combustion and emissions characteristics of the turbine engine fueled with HeFA blends from different feedstocks. *Energies* **2020**, *13*, 1277. [CrossRef]
29. Gawron, B.; Białecki, T. The laboratory test rig with miniature jet engine to research aviation fuels combustion process. *J. Konbin* **2015**, *36*, 79–90. [CrossRef]
30. ICAO CAEP10; ICAO Steering Group. *Environmental Technical Manual, Volume II Procedures for the Emissions Certification of Aircraft Engines, Second Edition—2014 + Revisions—Feb 2016*; ICAO: Montreal, QC, Canada, 2016; Volume II.
31. Olzak, B.; Szymczak, J.; Szczepankowski, A. Gaseous erosion and corrosion of turbines. *J. Pol. CIMAC* **2007**, *2*, 199–204.
32. Szczepankowski, A.; Szymczak, J.; Przysowa, R. The Effect of a Dusty Environment upon Performance and Operating Parameters of Aircraft Gas Turbine Engines. In *STO-MP-AVT-272 Impact of Volcanic Ash Clouds on Military Operations*; Jackson, R., Ed.; NATO Science and Technology Organization: Neuilly-sur-Seine, France, 2017; pp. 1–13. [CrossRef]
33. Dzięgielewski, W.; Gawron, B. Badanie przydatności biokomponentów I generacji do paliw stosowanych do turbinowych silników lotniczych (Suitability of 1st generation biocomponents for fuelling aircraft gas-turbine engines). *Res. Work. Air Force Inst. Technol.* **2018**, *30*, 221–234. [CrossRef]
34. *Output-Based Regulations: A Handbook for Air Regulators*; Technical Report; U.S. Environmental Protection Agency Combined Heat and Power Partnership: Washington, DC, USA, 2004.
35. Pi, J.; Fu, J.; Jiang, S.; Kong, Q.G. Properties of the small-size turbofan engine burning biofuels. *Biofuels* **2017**, *8*, 579–583. [CrossRef]

Article

Proof of Concept Study for Fuselage Boundary Layer Ingesting Propulsion

Arne Seitz [1,*], Anaïs Luisa Habermann [1], Fabian Peter [1], Florian Troeltsch [1], Alejandro Castillo Pardo [2], Biagio Della Corte [3], Martijn van Sluis [3], Zdobyslaw Goraj [4], Mariusz Kowalski [4], Xin Zhao [5], Tomas Grönstedt [5], Julian Bijewitz [6] and Guido Wortmann [7]

1 Bauhaus Luftfahrt e.V., Willy-Messerschmitt-Str. 1, 82024 Taufkirchen, Germany; anais.habermann@bauhaus-luftfahrt.net (A.L.H.); fabian.peter@bauhaus-luftfahrt.net (F.P.); florian.troeltsch@bauhaus-luftfahrt.net (F.T.)
2 Whittle Laboratory, University of Cambridge, 1 JJ Thomson Av., Cambridge CB30DY, UK; ac2181@cam.ac.uk
3 Faculty of Aerospace Engineering, Delft University of Technology, 2629 Delft, The Netherlands; b.dellacorte@tudelft.nl (B.D.C.); m.vansluis@tudelft.nl (M.v.S.)
4 Faculty of Power and Aeronautical Engineering, Warsaw University of Technology, Pl. Politechniki 1, 00-661 Warsaw, Poland; goraj@meil.pw.edu.pl (Z.G.); mkowalski@meil.pw.edu.pl (M.K.)
5 Division of Fluid Dynamics, Department of Mechanics and Maritime Sciences, Chalmers University of Technology, 412 96 Gothenburg, Sweden; xin.zhao@chalmers.se (X.Z.); tomas.gronstedt@chalmers.se (T.G.)
6 Engineering Advanced Programs, MTU Aero Engines AG, 80995 Munich, Germany; julian.bijewitz@mtu.de
7 Rolls-Royce Electrical, Rolls Royce Deutschland Ltd., 91058 Erlangen, Germany; guido.wortmann@rolls-royce-electrical.com
* Correspondence: arne.seitz@bauhaus-luftfahrt.net

Citation: Seitz, A.; Habermann, A.L.; Peter, F.; Troeltsch, F.; Castillo Pardo, A.; Della Corte, B.; van Sluis, M.; Goraj, Z.; Kowalski, M.; Zhao, X.; et al. Proof of Concept Study for Fuselage Boundary Layer Ingesting Propulsion. *Aerospace* **2021**, *8*, 16. https://doi.org/10.3390/aerospace8010016

Received: 14 December 2020
Accepted: 9 January 2021
Published: 13 January 2021

Abstract: Key results from the EU H2020 project CENTRELINE are presented. The research activities undertaken to demonstrate the proof of concept (technology readiness level—TRL 3) for the so-called propulsive fuselage concept (PFC) for fuselage wake-filling propulsion integration are discussed. The technology application case in the wide-body market segment is motivated. The developed performance bookkeeping scheme for fuselage boundary layer ingestion (BLI) propulsion integration is reviewed. The results of the 2D aerodynamic shape optimization for the bare PFC configuration are presented. Key findings from the high-fidelity aero-numerical simulation and aerodynamic validation testing, i.e., the overall aircraft wind tunnel and the BLI fan rig test campaigns, are discussed. The design results for the architectural concept, systems integration and electric machinery pre-design for the fuselage fan turbo-electric power train are summarized. The design and performance implications on the main power plants are analyzed. Conceptual design solutions for the mechanical and aero-structural integration of the BLI propulsive device are introduced. Key heuristics deduced for PFC conceptual aircraft design are presented. Assessments of fuel burn, NOx emissions, and noise are presented for the PFC aircraft and benchmarked against advanced conventional technology for an entry-into-service in 2035. The PFC design mission fuel benefit based on 2D optimized PFC aero-shaping is 4.7%.

Keywords: boundary layer ingestion; propulsive fuselage; wake-filling; turbo-electric; proof-of-concept; wind tunnel; fan rig; multi-disciplinary aircraft design; collaborative research

1. Introduction

Novel propulsion systems and their synergistic integration with the airframe are expected to play a key role in achieving aviation's long-term sustainability targets [1,2]. Therefore, significant further improvements in propulsion system overall efficiency will be required [3]. This includes the need for both, an ultra-efficient power supply to the propulsive devices as well as an ultra-efficient production of the required thrust by the propulsive devices. In order to achieve maximum propulsive efficiencies, the realization of extremely low specific thrust configurations is required. Under conventional propulsion system integration paradigms, the associated large propulsor diameters create a complex array of issues at the vehicular level including geometric installation challenges, aircraft drag penalties due to increased nacelle wetted areas, as well as airframe structural weight penalties in case under-wing power plant installation becomes impossible.

A particularly promising approach to elude these drawbacks and to achieve significant further improvements in overall vehicular propulsive efficiency is known from the field of marine propulsion. Ship propellers are installed at the stern of the vessel in order to utilize the kinetic energy that is contained in the boundary layer flow around the vessel's body for the production of thrust. This principle of energy recuperation via boundary layer ingesting (BLI) propulsion is also applicable to airborne systems [4]: the kinetic energy in the boundary layer flow is induced by surface skin friction as the body moves relative to the fluid. The reactive force on the body is known as viscous or skin-friction drag. In order to maintain a steady motion, the total drag of the vehicle needs to be balanced by the thrust force delivered by the propulsor. In conventional aircraft propulsion installation, propulsive thrust is produced against still air. Looking from a stationary perspective, in this case, any produced thrust results in a jet excess momentum flow in opposite direction of the vehicular motion, i.e., kinetic energy lost in the wake of the vehicle. At the same time, the kinetic energy content of the boundary layer flow around the wetted body is lost in the vehicular wake, too. By ingesting the boundary layer flow at the aft of the vehicle's body, the required thrust force is produced against the fluid being in motion together with the body. For the given thrust force, the jet flow excess momentum, and thus, the jet kinetic energy loss in the wake is reduced. At the same time, the wake kinetic energy loss associated with the ingested share of the boundary layer flow is reduced or totally eliminated, a mechanism also referred to as wake-filling.

For large commercial aircraft, the share of viscous drag in cruise typically ranges between 60–70% of the total drag. Almost half of this share may be attributed to the fuselage body, making it the most interesting airframe component to be utilized for the purpose of wake-filling propulsion integration [5]. A most straightforward way to realize fuselage wake-filling is by full annular (360°) BLI through a single propulsor encircling the very aft-section of the fuselage—also referred to as propulsive fuselage Concept (PFC). This paper summarizes the key results and findings obtained from technology readiness level (TRL) 3 research and innovation activities for a PFC aircraft featuring a turbo-electrically powered BLI fuselage fan (FF) that were performed as part the recently completed European Commission (EC) funded project CENTRELINE.

1.1. Literature Survey of Fuselage BLI Propulsion

The utilization of BLI as a means to increase aircraft propulsive efficiency through wake-filling has been subject to theoretical treatise over several decades (e.g., Smith and Roberts [6] (1947), Goldschmied [7] (1954), Smith [4] (1993) and Drela [8] (2009)). A first patent based on the effect of BLI and wake-filling propulsion was filed by Betz and Ackeret in 1923 [9]. A first patent describing an explicit concept for fuselage wake-filling propulsion integration was filed in 1941 [10]. Initial experimental studies related to fuselage BLI and wake-filling were conducted for the boundary layer-controlled airship body concept proposed by Goldschmied in 1957 [11]. More recently, low-speed wind tunnel experiments were performed on generic streamlined body by ONERA [12] and TU Delft [13]. Experiments have also been performed at MIT for the D8 configuration [14]. A detailed analysis of the aerodynamics of a boundary layer ingesting fan was performed in a low-speed experimental fan rig at the University of Cambridge [15]. Examples of existing aircraft utilizing aft-fuselage propulsion integration, however not explicitly designed to maximize wake-filling, include the Douglas XB-42 (1944), the RFB Fantrainer (1978), the LearAvia LearFan 2100 (1981), and the Grob GF 200 (1991).

Over the last two decades, a variety of concepts and low-TRL studies featuring propulsive devices to exploit the effect of fuselage wake-filling by BLI have been published. Beside blended wing body designs with integrated BLI propulsion such as the Silent Aircraft Initiative "SAX-40" [16] and NASA's "N3-X" configuration [17], a number of tube-and-wing aircraft layouts equipped with fuselage BLI propulsors have been presented. Noted examples include NASA's "FuseFan" concept [18], the MIT "D8" concept [19], Bauhaus Luftfahrt "Claire Liner" [20] and "Propulsive Fuselage" [5] concepts, the EADS/AGI "VoltAir" [21], the Boeing "SUGAR Freeze" [22], and the NASA "STARC-ABL" [23,24]. The first multidisciplinary design study for large transport category aircraft featuring full annular fuselage BLI propulsion was performed as part of the EC funded research project DisPURSAL [25].

In parallel to the CENTRELINE project, an increasing level of research effort in the field of tightly-coupled wake-filling propulsion has been observed. The focus of recent research activities can be found in the aerodynamic optimization of affected aircraft components (e.g., [26–29]) and the design and optimization of the BLI propulsion system (e.g., [30–35]). With most studies still based on numerical simulation by computational fluid dynamics (CFD), also the development of dedicated experimental testing capabilities has been progressing significantly [36].

1.2. Outline of the CENTRELINE Project

Funded as part of the European Union's Horizon 2020 Framework Programme, the "ConcEpt validatioN sTudy foR fusElage wake-filLIng propulsioN integration", in short "CENTRELINE" (grant agreement no. 723242) was dedicated to perform the proof-of-concept and initial experimental validation for the PFC approach. Coordinated by Bauhaus Luftfahrt, the collaborative research during the 42-month project was conducted by a consortium of key stakeholders from European industry, research, and academia. The partners involved Airbus Defence and Space, Airbus Operations, Chalmers Tekniska Hoegskola, MTU Aero Engines, Politechnika Warszawska, Rolls Royce Deutschland Ltd., Siemens AG, Delft University of Technology, the University of Cambridge, and ARTTIC. The consortium was accompanied by a technical advisory board (TAB) of senior experts from industry and research including representatives from the German Aerospace Center DLR and ONERA, the French Research Lab.

The specific PFC configuration investigated in CENTRELINE (cf. Figure 1), features a twin-engine, turbo-electric PFC systems layout with the aft-fuselage BLI fan being powered through generator offtakes from advanced Geared TurboFan (GTF) power plants podded under the wing. When compared to a mechanical drive train concept such as focused on in the previous DisPURSAL project (cf. [37]), the FF electric drive approach facilitates the BLI propulsive device to be installed at the very aft-end of the fuselage body. Consequently, the aero-structural integration at the aft fuselage is simplified, while maximizing the wake-filling potential attainable from fuselage BLI. Aft-fuselage internal thermal shielding requirements are relieved, internal and external noise and vibration is reduced, and overall system maintenance costs decrease, as the third gas turbine engine used for the mechanical FF drive in DisPURSAL is omitted. At the same time, the design complexity due to relevant rotor burst scenarios is reduced.

Figure 1. Artist view of the CENTRELINE turbo-electric propulsive fuselage aircraft design.

Pursuing the conceptual proof, the main challenges associated with turbo-electric propulsive fuselage aircraft design (cf. [38]) were addressed in CENTRELINE. These included the obtainment of a thorough understanding of the aerodynamic effects of fuselage wake-filling propulsion integration, the development of suitable aero-structural design integration solutions for the BLI propulsor, the design elaboration of the FF turbo-electric drive train, as well as the multi-disciplinary systems design integration and optimization at aircraft level. As such, the CENTRELINE project aim was to maximize the benefits of fuselage wake-filling propulsion integration under realistic systems design and operating conditions. The high-level objectives at the beginning of the project stated a TRL goal of 3 to 4 at the end of the project, together with ambitions performance targets of 11% CO_2 and NO_x emission reductions against an advanced conventional reference aircraft equipped with aerodynamic, structural, power plant, and systems technologies suitable for a potential entry-into-service (EIS) year 2035.

1.3. Overall Methodological Approach

In order to address the identified key challenges for the PFC conceptual proof, in CENTRELINE, a set of problem-tailored analytical, numerical and experimental methods was employed. This included high-end and high-fidelity simulation techniques for the aerodynamics of the overall aircraft and the FF, for key structural elements as well as the components of the turbo electric drive train. For the purpose of initial experimental validation, low speed wind tunnel and BLI fan rig testing campaigns were performed. The work was organized in collaborative work packages handling the multi-disciplinary concept integration and design optimization, the detailed aerodynamic design simulation and testing, and the pre-design and integration of the FF turbo-electric power train. The TRL3 research activities were framed by a work package dedicated to deriving a realistic technology application scenario at the beginning of the project and the later critical system-level evaluation of the detailed research results. A visualization of the basic work logic followed is provided in Figure 2.

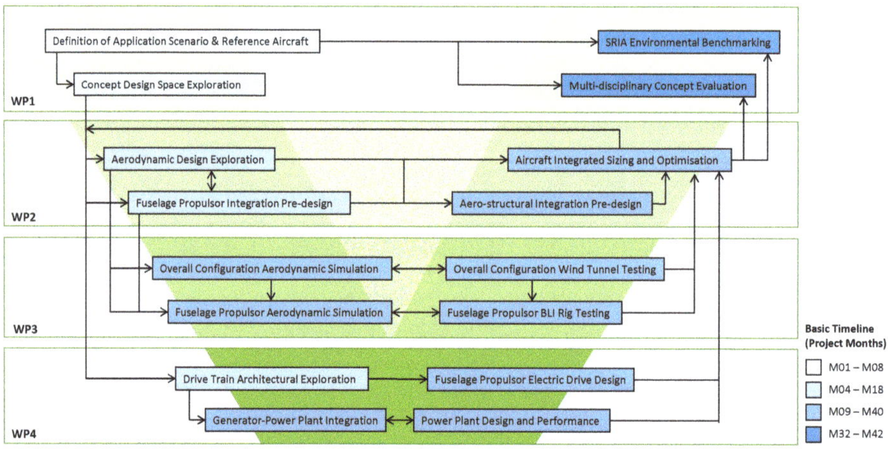

Figure 2. The basic CENTRELINE work logic.

Based on market outlook perspectives, anticipated socio-economic development trends and required transport capacities in the targeted EIS timeframe for a PFC aircraft, a most impactful market segment was identified and translated into Top Level Aircraft Requirements (TLARs). In order to allow for a rigorous evaluation of the PFC technology, two families of conventional reference aircraft were defined, reflecting year 2000 in-service aircraft, dubbed "R2000", and an advanced reference equipped with technologies suitable for an EIS year 2035, the "R2035". While the R2035 served as the immediate benchmark for the PFC technology, the R2000 represented the baseline for the PFC evaluation against the 2035 environmental targets set by the Strategic Research and Innovation Agenda (SRIA) [3] of the Advisory Council for Aeronautics Research in Europe (ACARE).

As part of an initial design space exploration, a most suitable aircraft layout for the PFC proof-of-concept was identified, and an initial aircraft target design was developed using simplified analytical and semi-empirical sizing methods. Completed within the first eight project months, the PFC aircraft target design served as a consistent starting point for the detailed multi-disciplinary design and analysis work, providing guidelines for key design parameters in both the numerical and experimental domains.

Starting from the derivation of unified performance bookkeeping standards for the BLI PFC and non-BLI reference aircraft, the aerodynamic design and analysis work included a comprehensive 2D numerical optimization of the axisymmetric bare PFC configuration, i.e., the isolated fuselage with installed aft-propulsive device, as well as a systematic 3D numerical analysis of the PFC aerodynamic performance properties and flow field characteristics. Based on the obtained PFC-specific FF inflow patterns, 3D numerical design for a distortion tolerant BLI fan was performed. All aero-numerical activities were closely accompanied by low-speed aerodynamic validation testing at the overall aircraft level as well as specifically for the BLI FF. The experimentally verified aerodynamic design and performance characteristics of the FF were integrated together with the 3D numerically refined aero-structural design for the key elements of the aft-fuselage propulsion installation. The FF turbo-electric power train design was elaborated under multi-disciplinary consideration, including the transmission system architectural definition, the electromagnetic and mechanical design of the involved electric machinery and the conceptual integration of the turbo-electric generators within the gas turbine environment. The impact of the significant generator power offtakes on the main engines' sizing and operational behavior was investigated through sophisticated design and performance modeling. Conceptual solutions for the thermal management of all power train components were developed and incorporated in the overall system design and performance simulation. The entireness of knowledge obtained from the detailed design and analysis activities were continuously incorporated for overall system sizing and optimization at the aircraft level. Finally, the optimized PFC aircraft family design was subjected to a comprehensive multi-disciplinary assessment against the R2035 and R2000 aircraft families.

1.4. Technology Application Case

The technology application case was tailored to maximize the leverage of the PFC efficiency potentials on the reduction of aviation's climate impact. To maximize the PFC technology impact at aircraft fleet level, existing forecasts of market and route development were analyzed in order to identify a most influential aircraft market segment with regard to reductions in fuel consumption. Similar to preceding analyses performed in the DisPURSAL project [25], the medium to long-range wide-body aircraft segment was determined to be particularly impactful. Judging the forecasted numbers of revenue passenger kilometers according to the specific regions, the highest demand for the aspired EIS in 2035 was identified for the Europe—Asia/Pacific inter regional connections, leading to a projected design range of 6500 nm. At the same time, a peak demand of aircraft installed seat for this mid-to-long range market segment was identified at 340 passengers [39]. A condensed list of the CENTRELINE TLARs is given in Table 1.

As a best suited in-service aircraft with reasonably similar design mission specifications the Airbus A330 was selected as a starting point for the creation of the CENTRELINE reference aircraft models. In order to obtain the year 2000 reference aircraft, R2000, the payload-range capacity of the A330 was slightly increased by a stretch of the fuselage length and a corresponding reinforcement of key structural components such as the landing gear. In order to maintain appropriate low-speed performance, the wing planform was geometrically scaled for the increased aircraft gross weight. The tail planforms were adapted to retain the aircraft stability and control characteristics. The propulsion system for the R2000 is based on the most common engine option for the A330, the Rolls Royce Trent 772B. The Trent 700 series is a three-spool turbofan featuring a long duct mixed-flow nacelle. The design and performance characteristics of the Trent 772B were reproduced and subsequently scaled for the R2000 thrust requirements using a design synthesis model created in Bauhaus Luftfahrt's in-house Aircraft Propulsion System Simulation (APSS) software [40–42].

Table 1. CENTRELINE top level aircraft requirements.

Parameter	Value
Range and PAX	6500 nmi, 340 * PAX in 2-class
TOFL (MTOW, SL, ISA)	≤2600 m
Second climb segment	340PAX, 100 kg per PAX, DEN, ISA + 20 K
Time-to-climb (1500 ft to ICA, ISA + 10 K)	≤25 min
Initial cruise altitude (ISA + 10 K)	≥FL 330
Design cruise Mach number	0.82
Maximum cruise altitude	FL410
Approach speed (MLW, SL, ISA)	140 KCAS

* Baseline family member; 296 PAX for shrink and 375 PAX for stretch version.

The R2035 advanced reference aircraft was directly derived from the R2000 aircraft. Therefore, a comprehensive technology scenario was devised, including advanced multi-disciplinary technological developments in the fields of aerodynamics, structures, systems and equipment considered realistic for aircraft product integration by 2035 [39,43]. Figure 3 shows a simplified three-view drawing of the R2035 with selected aircraft dimensions annotated. Compared to the R2000, the R2035 features a larger cabin cross section with a nine-abreast (two-five-two) economy seating arrangement. An obvious feature of the R2035 aircraft design is the slender wing featuring an aspect ratio of 12 enabled by an advanced composite design and improved aero-elastic tailoring capabilities.

In order to facilitate a realistic evaluation of the PFC technology, aircraft family design considerations were taken into account throughout the aircraft design and benchmarking process. The R2000 and R2035 aircraft were designed as families consisting of a baseline, shrink, and stretch version. The family design was conducted in accordance to common industry practice and the goal of sharing common components as empennage, landing gear and engines. The stretch and shrink versions of the baseline aircraft feature +10% and –15% payload capacity, respectively. The R2035 power plant systems are Ultra-High Bypass Ratio (UHBR) > 16 geared turbofan engines, sized to serve the entire aircraft family. Power plant thermodynamic cycle parameters and component design properties were selected appropriately to reflect advanced aerodynamics, materials, and manufacturing technologies for an EIS 2035 [42]. At typical cruise conditions, i.e., FL350, M0.82 and a lift coefficient $C_L = 0.5$, the baseline member of the R2035 family features a lift-to-drag ratio of 20.7, with an induced drag share of 28% and a wave drag share of 5%. The fuselage share of the total is approximately 26%. Further characteristics including aircraft weight component breakdowns and block fuel values for the R2000 and R2035 aircraft are provided in Section 7.3 in comparison to the corresponding PFC aircraft properties. More detailed descriptions of the CENTRELINE reference aircraft families were presented by [39].

In order to set a consistent basis for the more detailed design and analysis activities, in CENTRELINE, an initial PFC aircraft design was specified through qualitative configurational down-selection and a subsequent preliminary multidisciplinary design loop for the selected PFC aircraft layout. From an initial cloud of configurational candidates featuring alternative approaches to the aero-structural integration of the aft-fuselage BLI propulsive device as well as different empennage integration options, a PFC aircraft configuration comprising a T-tail arrangement with the ducted BLI fan integrated behind the vertical fin was selected as a most suitable basis for the further detailed studies in CENTRELINE (cf. [44]). With the basic configurational layout identified, an initial PFC aircraft target design was derived from a multi-disciplinary design study based on simplified analytical and semi-empirical methodology (cf. [43]). The design synthesis included estimated properties of the FF turbo-electric power train, PFC airframe structural weight and main power plant design implications, as well as performance targets for the PFC aero-shaping in order to meet the project's performance goal in terms CO_2 reduction. As a result, an operating empty weight (OEW) increase of 5.7% relative to the R2035 was predicted, which together with a target design block fuel improved of 11% yielded an almost identical maximum take-off weight (MTOW) as for the R2035 aircraft [43]. Key design features of the initial PFC aircraft design are illustrated in Figure 4.

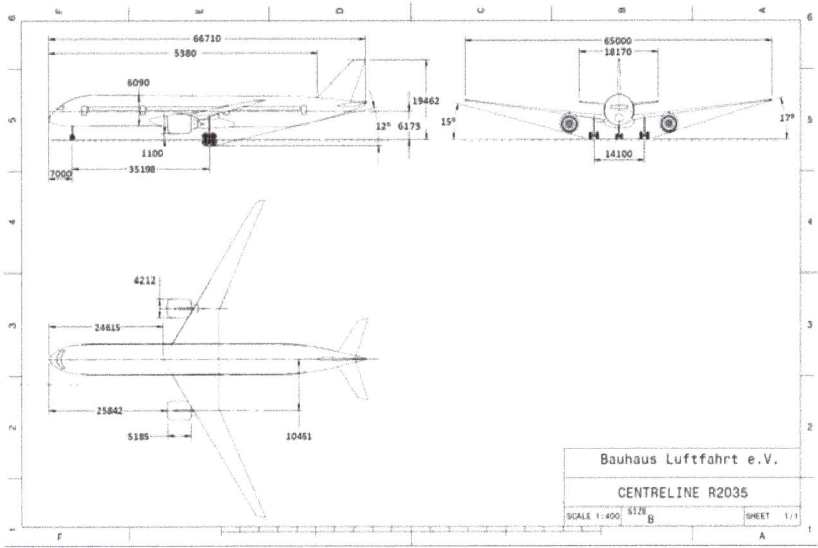

Figure 3. Schematic three-view of the CENTRELINE R2035 baseline aircraft.

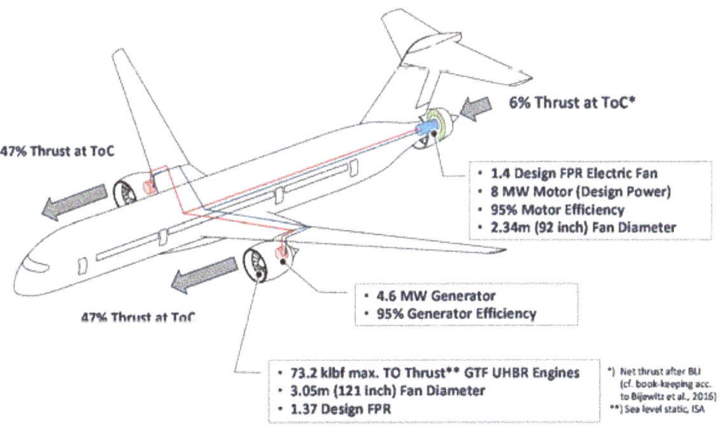

Figure 4. Main design features of the CENTRELINE initial propulsive fuselage target design (adapted from [43]).

2. Fuselage BLI Propulsion Aerodynamics

Key requisites for a meaningful evaluation of the fuselage BLI technology include a rigorous bookkeeping of the aerodynamic interaction between the BLI propulsion system and the airframe as well as a sufficiently refined aerodynamic shaping of the aircraft. This section discusses the performance bookkeeping standards followed in CENTRELINE and presents key aspects of the aerodynamic design and analysis activities at the overall configuration and the FF level.

2.1. Performance Bookkeeping and Efficiency Metrics

The assessment of aircraft concepts with highly integrated propulsion systems, such as the CENTRELINE configuration, requires adherence to rigorous bookkeeping standards and consistent performance assessment metrics. A comprehensive review of existing bookkeeping schemes and their applicability to aircraft concepts with a strong coupling of airframe aerodynamics and propulsion system can be found in [45]. In general, bookkeeping schemes can be classified by the quantity, which is conserved in a specified control volume. Approaches based on momentum conservation allow consistency with force bookkeeping, but are often neglecting the bi-directional effect of airframe aerodynamics and propulsion system performance. Integral energy methods use kinetic energy or exergy conservation. The "Power Balance" method introduced by Drela in 2009 [8], reflects the need for a holistic bookkeeping approach applicable to highly integrated propulsion system concepts. Similar to the "Exergy Balance" method, later developed by Arntz [46], the evaluation of BLI configurations requires the full resolution of the flow field, which is achieved through experimental test or CFD simulations.

Within the CENTRELINE project, two different bookkeeping approaches were employed: The power balance method and an integral momentum conservation approach. The application of the power balance method is very resource demanding, because it requires the full resolution of the flow field. Therefore, it was solely applied to analyze the wind tunnel particle image velocimetry (PIV) measurement results of a single PFC configuration at the end of the project in order to estimate the experimental BLI performance in detail [47,48].

For the aircraft-level sizing studies, an integral momentum conservation approach was deemed to be most practicable together with a distinction between the bare PFC configuration, i.e., the integrated assembly of the fuselage and the fuselage BLI propulsive device, and, all other adjacent aircraft components [45]. The control volume for the bare PFC configuration is pictured in Figure 5. It allows for the rapid evaluation of the bare PFC configuration aerodynamics and performance properties, e.g., based on a 2D axisymmetric shape definition (see Section 2.2).

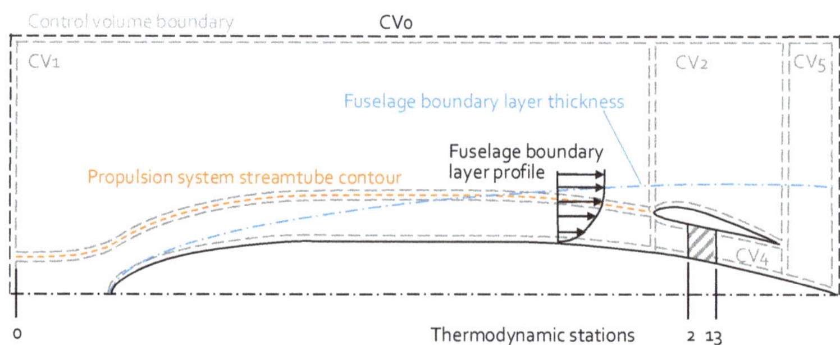

Figure 5. Control volume scheme for bare PFC bookkeeping (adapted from [45]).

As a key descriptor for the bare PFC configuration, the net propulsive force $NPF_{PFC,bare}$ is introduced, representing the net axial force acting on the bare PFC configuration. It is the result of an integral momentum conservation applied to the depicted control volume

$$NPF_{PFC,\ bare} = F_{FF,disc} + F_{Fus} + F_{Aftbody} + F_{FF,Nac} + F_{FF,Duct} = F_{FF,Disc} + F_{PFC,bare} \quad (1)$$

$F_{FF,Disc}$ represents the axial force produced by the FF (between the thermodynamic stations 2 and 13). F_{Fus} is the total aerodynamic force of the fuselage up to the BLI propulsion system air intake, $F_{Aftbody}$ is the total force on the (fuselage) aft-body behind the BLI propulsor nozzle exit, $F_{FF,Nac}$ is the total force of the FF nacelle external surface area, and, $F_{FF,Duct}$ represents the total force due to the FF duct internal flow. $F_{PFC,bare}$ is the sum of all surface forces acting on the bare fuselage-propulsor configuration. All total force components include both, viscous as well as surface pressure related axial forces [45].

The total aircraft drag force D_{tot}, i.e., the sum of all forces acting on the individual component surfaces, plus possible interference drag e.g., between the bare PFC and adjacent airframe components D_{int}, miscellaneous drag items such as due to protuberances and leakages D_{misc}. With potential flow buoyancy terms of the individual components assumed to be zero for the closed aircraft body, the D_{tot} for the PFC aircraft becomes

$$D_{PFC,tot} = F_{PFC,bare} + F_{PFC,res} + D_{PFC,int} + D_{PFC,misc} = F_{PFC,bare} + D_{PFC,res} \quad (2)$$

where $F_{PFC,res}$ represents the sum of aerodynamic forces acting on the surfaces of all aircraft component forces other than the bare PFC configuration, i.e., including the wing and empennage, any required fairings, as well as, the underwing podded nacelles and pylons. For convenience, the sum of $F_{PFC,res}$, $D_{PFC,int}$, and $D_{PFC,misc}$ may be written as $D_{PFC,res}$. It should be noted, that the total drag of a non-BLI reference aircraft $D_{Ref,tot}$ is directly obtained from Equation (4) by replacing $F_{PFC,bare}$ with the aerodynamic force acting on the reference aircraft fuselage $F_{Ref,fus}$, and, evaluating all other aerodynamic forces and drag numbers specifically for the reference aircraft.

Assuming steady level flight, the overall propulsion net thrust requirements F_N equal the aircraft total drag force. With the bare PFC net propulsive force defined in Equation (1), the overall net thrust requirement for the non-BLI main power plants of the PFC aircraft yields

$$F_{N,PFC,main} = D_{PFC,res} - NPF_{PFC,bare} \quad (3)$$

An overview of relevant figures of merit for the performance assessment of wake-filling propulsion system concepts—such as the CENTRELINE PFC configuration—is provided by Habermann et al. [45]. As a means of aerodynamic inter-comparison of alternative PFC designs, the bare PFC efficiency factor $f_{\eta,PFC,bare}$ is particularly relevant. It relates the net useful propulsive power of the bare PFC configuration, i.e., the product of $NPF_{PFC,bare}$ and the flight velocity V_0, to the isentropic power expended in FF disc, $P_{FF,\ Disc}$

$$f_{\eta,PFC,\ bare} = \frac{NPF_{PFC,\ bare} \cdot V_0}{P_{FF,disc}} \quad \forall P_{disc,FF} > 0 \quad (4)$$

For PFC aircraft design optimality considerations and performance assessment purposes against the conventional reference aircraft, integral mission figures of merits are used (see Section 7). This includes the design mission block fuel as well as the corresponding CO_2 reduction potential calculated based on jet fuel-specific emission factor of 3.150 kg CO_2/kg fuel postulated in [49]. For the assessment of PFC point performance against the non-wake-filling reference, the power saving coefficient, (*PSC*), originally introduced by [4], is applied

$$PSC = \frac{P_{Ref} - P_{PFC}}{P_{Ref}} \quad (5)$$

where P_{Ref} refers to the power required to operate the aircraft in the conventional, non-wake-filling case, and P_{PFC} represents the power requirement of the PFC configuration.

A detailed discussion of the *PSC* application to PFC aircraft performance evaluation is presented by [50]. An analytical formulation of the *PSC* coefficient for PFC aircraft in cruise is provided in Section 7.1.

2.2. Bare PFC 2D Aerodynamic Design Optimization

To find a feasible and well-refined aerodynamic design for the PFC, at first the aerodynamic design space for cruise conditions needs to be thoroughly explored. In order to accomplish this, a responsive and suitably accurate aerodynamic analysis method needs to be adopted. All key relevant parameters that influence the aerodynamic performance of the PFC need to be investigated. In order to maximize the gain of knowledge and to reduce the complexity at the early stage of the project, an optimization of the bare PFC configuration based on axisymmetric paradigms was focused on. The optimized axisymmetric aero-shaping formed the basis for the final three-dimensional design of the PFC.

2.2.1. Aerodynamic Design Space

The aerodynamic design space of the PFC is comprised of geometric and operational parameters. Examples of the former include the fuselage slenderness ratio, FF nacelle incidence angle and aft-body contraction ratio. Examples of operational parameters are the cruise flight level, flight Mach number and fan pressure ratio (FPR) of the FF. The PFC will be optimized for a given design cruise mission and benchmarked against a R2035 reference aircraft for the same mission. Nevertheless, it is also worthwhile to investigate the sensitivity of the PFC to the flight conditions (see Section 7.1).

The initial phase of the design phase consisted of design modifications of an initial aerodynamic model based on engineering judgement. Even though the initial design was improved significantly, the complex aerodynamic interactions halted further improvements. The last design revision based on iterative design is called "Rev05".

To explore the design space in a more systematic manner, a fully parametric model was developed capturing the axisymmetric shape of the bare PFC configuration. Using a combination of Non-Uniform Rational B-splines (NURBS) [51] and Bezier-Parsec [52], the parametric model describes the PFC geometry based on actual design parameters instead of free-floating control points: The aft fuselage geometry ahead of the aft propulsor was defined in terms of curvature and contraction ratio. The contour of the FF annular ducting was determined by cross-section area ratios as functions of the FF disc area. The nacelle contour was described through Bezier-Parsec parameterization. The boat tail shape follows from the overall fuselage length and the axial FF location. First order continuity between the various segments ensure smooth curvature transitions. Constraints were added, for example to limit the overall fuselage length and avoid very short or slender boat tail designs. Furthermore, it was ensured that the usable floor for each design was the same. In total 23 variables were used to describe the geometry, together with the following operating conditions: Mach number, altitude/flight level (FL), FPR [53].

2.2.2. Numerical Methods

To analyze the large variety of PFC designs, an aerodynamic solver is required which is able to capture even subtle differences in the design and at the same time have low or moderate computational demand to avoid bottlenecks in the assessment. In CENTRELINE, it was decided to use axisymmetric 2D Reynolds-averaged Navier–Stokes (RANS) CFD simulations, which are able to capture the mean flow characteristics and boundary layer development over the fuselage with appropriate accuracy. The axisymmetric 2D grids were sufficiently small to be able to run on a standard engineering workstation. In most cases, the generation of the grid consumed the majority of time required to obtain a CFD solution. In order to drastically reduce the person effort of geometry meshing, an automated mesh routine was developed. Using an open-source MATLAB-toolbox [54], a MATLAB-based framework was developed that executes all steps from the initial design vector to geometry-creation and meshing to post-processing of the results [53].

The structured hexahedral mesh was constructed using ICEM® by Ansys Inc. (Canonsburg, Pennsylvania, United States). In order to fully resolve the boundary layer up to the wall, it was ensured that the mesh satisfies the requirement $y^+ < 1$. To model turbulence, the k-ω shear stress transport (SST) model [55] was selected. The air was modeled as an ideal compressible gas with Sutherland's three-coefficient law. A pressure-coupled solver was employed to accelerate convergence of the simulation. The domain boundary was modeled using a pressure far field boundary condition. The simulations were performed in Fluent® 18.2 by Ansys Inc. (Canonsburg, Pennsylvania, United States).

To model the effect of the FF on the fluid, an actuator disc model was used based on source terms of momentum and energy. The source terms appear directly in the right-hand side of the momentum and energy equations. A dedicated volume in the mesh was constructed, representing the boxed-volume of the FF. Note that only axial momentum is added and no swirl is added to the flow. Since the fluid is modeled to be compressible, energy is added to the fluid based on the local work of the momentum source in axial direction. Verification and validation of the actuator disc model is discussed in brief in [50].

The volumetric integration of the momentum source term directly yields the force of the actuator disc provided to the fluid. As such, traditional drag numbers by integration of the wall shear force and pressure wall-normal component, can still be used to assess the drag force of the PFC. The latter is beneficial, as it is a very straightforward and unambiguous method for the bookkeeping of the aerodynamic forces.

2.2.3. Optimization Results

Finding the optimum bare fuselage design requires a systematic survey of the aerodynamic design space, sweeping as much of the available design space as possible. To do so, designed experiments based on a partially stratified sampling method [56] was selected to cover the aerodynamic design space. The average computation time for the simulation of a single sample, including pre- and post-processing, was approximately 10 min, on a state-of-the-art personal computer. In total, more than 9000 samples of the design space were evaluated. Of this initial number of samples, approximately one-third of the simulations converged and were used for further analysis. A 1D sensitivity study was carried out to evaluate sensitivities of each design parameter [53]. It was found that the following parameters, beside the operational parameters, drive the aerodynamic performance of the PFC the most:

- FF duct height (r_{tip}—r_{hub})
- FF nozzle exit to fan face area ratio (A_{18}/A_{12})
- FF hub-to-tip ratio (r_{hub}/r_{tip})
- FF relative axial position along the fuselage (x_{FF}/L_{fus})

Using the above principal design parameters, a surrogate model based on the technique presented by [57] was fitted to the results. From the pool of evaluated designs, the most promising design was selected for further optimization. Using a gradient-based solver from the MATLAB® Optimization Toolbox by The Mathworks Inc. (Natick, Massachusetts, United States), the design was optimized. As a second step, the design was further refined by including the knowledge of the sensitivity study for the parameters not included in the surrogate model. Verification in CFD confirmed that the new design "Rev06" was improved significantly over the previous design iterations. The evolution of the axisymmetric bare PFC performance is shown in Figure 6. Note that the aerodynamic performance is expressed as the ratio between the product of NPF with flight velocity and the ideal (isentropic) shaft power of the FF, $P_{FF,disc}$.

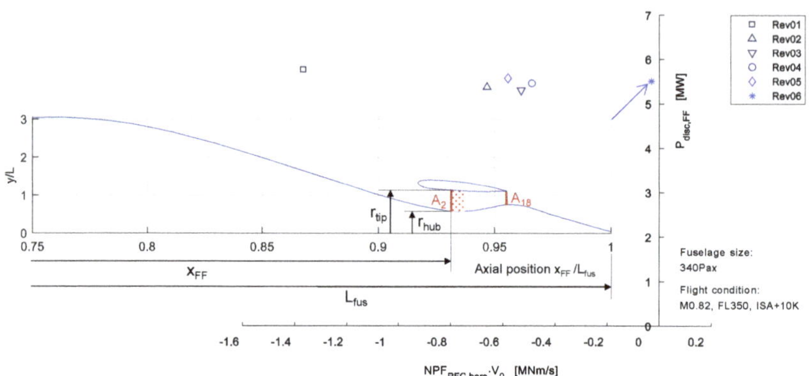

Figure 6. Evolution of the aerodynamic performance of the bare PFC fuselage (adapted from [50]).

To align the optimized design with the target values of the pre-design studies [43], the Rev06 optimization was modified to meet the following additional equality constraints:

- FPR equal to 1.40
- Fuselage diameter constrained to 6.09 m
- Hub-to-tip ratio equal to 0.51

Based on the obtained "Rev07" design, which represents the final axisymmetric bare fuselage design, three case studies were performed and simulated based on various targets of $P_{FF,disc}$. An overview of the variants of the Rev07 with different FF shaft power requirements is presented in Figure 7.

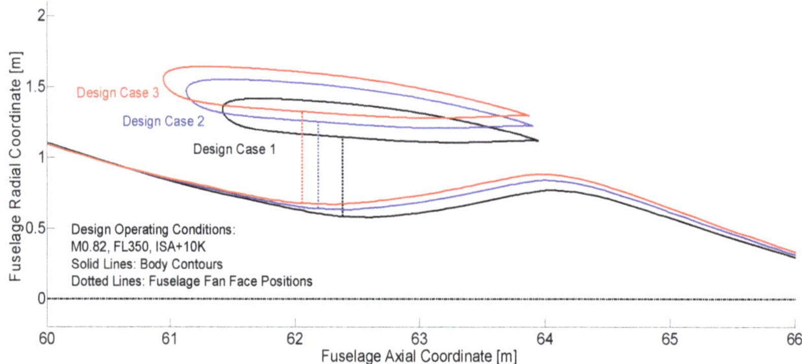

Figure 7. Changes in FF geometry of the Rev07 PFC design for different target FF shaft power.

An overview of the corresponding aerodynamic performance for each Rev07 design case study is presented in Table 2.

As can be observed from Table 2, the *NPF* force for the bare PFC becomes positive if the duct height of the FF is increased beyond a certain threshold. For a given FPR, the force exerted by the FF on the fluid is directly proportional to the area of the disc, which scales quadratically with the blade radius. Despite an increase of drag for the bare PFC for case 2 and 3, the BLI efficiency factor is increased for the design cases with a higher FF duct height. Note that the additional benefit for an increased FF duct height is decreasing, as the amount of additional momentum deficit that is ingested is diminishing towards higher duct heights of the FF (see also Section 7.1).

Table 2. Overview of the main aerodynamic performance parameters for three different case studies based on the Rev07 axisymmetric bare PFC design.

Parameter	Unit	Design Case 1	Design Case 2	Design Case 3
$F_{PFC,bare}$	kN	33.17	35.52	36.84
$F_{FF,disc}$	kN	32.44	39.32	44.02
$NPF_{PFC,bare}$	kN	−0.73	3.80	7.18
$P_{FF,disc}$	MW	5.53	6.80	7.76
$f_{\eta,PFC,bare}$	-	−0.033	0.139	0.229

2.3. *PFC Aircraft 3D Aero-Numerical Analysis*

Having obtained a feasible and much improved axisymmetric bare PFC design, a three-dimensional model of the PFC is constructed to analyze the aerodynamics of the PFC in more detail. The aim of the 3D CFD simulations is twofold, namely to obtain more detailed inflow conditions for the further development of the FF and to verify the main aerodynamic design of the PFC. In order to understand the effect of an increased level of model complexity on the aerodynamics of the PFC, a step-by-step approach was followed starting from the 3D simulation of the bare PFC configuration while successively adding the directly adjacent aircraft components, namely the wing including fuselage belly fairing and the vertical tail. The geometry of the full PFC aircraft model, which was aerodynamically studied is shown in Figure 8.

Figure 8. Side-view of the aerodynamic model of the PFC, including wing and vertical tail.

Note that the podded under-the-wing engines and horizontal tail have not been included in the aerodynamic model to reduce the complexity and size of the numerical grid. A symmetry boundary condition was applied to half the computational domain. The numerical setup has been kept as similar as possible to the bare axisymmetric CFD, as discussed in the previous section.

Since the increased model complexity is less suited for a structured hexahedral mesh, an unstructured tetrahedral mesh has been applied with mesh inflation layers to capture the boundary layer in an accurate manner. In total 30 inflation layers were applied, using an exponential growth law with $r = 1.15$. It was ensured that $y^+ < 1$. Two bodies of influence were added on the wing and the fuselage to refine the volume mesh along the bodies and near-wake. The grid size for the half-model of the PFC, as shown in Figure 8, exceeded 120 million elements.

Before proceeding with the 3D analysis of the PFC, the 2D axisymmetric results where compared with the 3D simulation of the bare axisymmetric PFC. It was found that the difference in drag was in the order of 3% due to numerical differences. Considering that a systematic mesh dependency study for the 3D mesh has not been part of this work, the difference is considered to be within the acceptable error margin.

2.3.1. Effect of Wing and Empennage

To investigate the effect of the wings and vertical tail on the inflow conditions to the FF, the wings and vertical tail are added step-by-step. To assess the inflow conditions, a total pressure coefficient is defined as

$$PC = \frac{p_{AIP} - \overline{p_{AIP}}}{\overline{p_{AIP}}} \tag{6}$$

where p_{AIP} is the local total pressure and $\overline{p_{AIP}}$ the mean total pressure across a specified area. The inflow conditions are measured at the aerodynamic interface plane (*AIP*), which is located at 60% of the inlet to the FF inside the FF duct. The results for the effect of wing and empennage are shown in Figure 9.

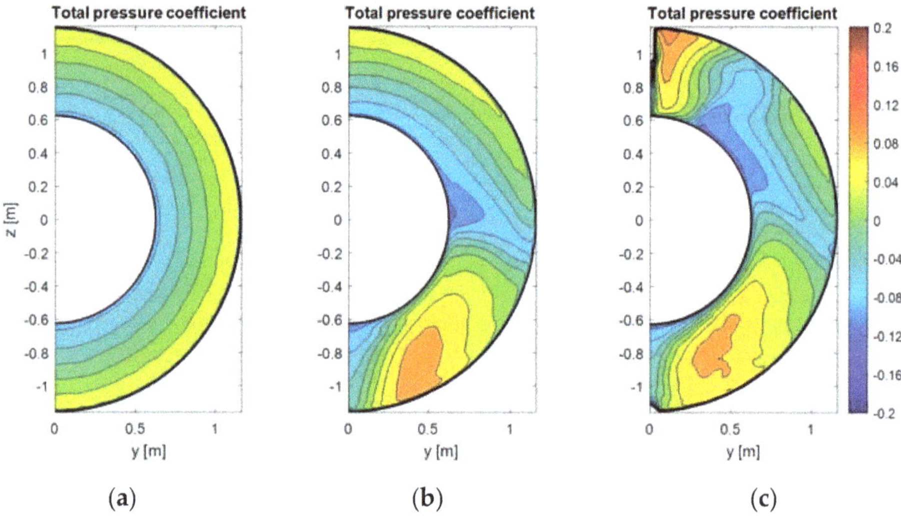

Figure 9. Contours of total pressure coefficient at the AIP. (**a**) bare PFC configuration, (**b**) bare PFC configuration + wing and (**c**) bare PFC configuration + wing + vertical tail (Source: [58]).

As can be observed, the inflow for the bare fuselage is relatively smooth, with the largest variation in total pressure coefficient near the hub. However, the addition of the wing appears to have a significant impact on the inflow conditions. In the lower half of the disc, there is a zone of significantly higher total pressure. The boundary layer along the fuselage hub, marked by lower levels of total pressure coefficient, is washed upwards. After adding the vertical tail, another zone of higher total pressure can be found on the upper side of the disc. Note that the AIP is intersected by the vertical tail.

Analysis of the external flow field revealed that there is flow separation on the lower side of the belly fairing. The separated flow forms a trailing vortex, which impinges on the nacelle of the FF. The core of the vortex is not ingested by the FF, however air is drawn in from the outer layers of the boundary layer and the freestream. As such, there is an inflow of air of higher total pressure to the duct of the FF. Other than causing a significant disturbance to the FF, the flow separation at the belly fairing is a source of additional drag. The belly fairing design has not been part of the aerodynamic shape refinement, hence it is expected that a redesigned belly fairing should reduce the disturbance to the FF considerably. Similarly, it was found that the zone of higher total pressure coefficient at the top of the AIP is caused by a vortex as well, stemming from the fuselage–tail junction. The latter is much more difficult to avoid; however, the horse-shoe vortex could maybe be reduced with additional fairing design [59].

2.3.2. Effect of Fuselage Upsweep

Although the axisymmetric aft fuselage is beneficial from an aerodynamic standpoint, it does limit the tail strike angle during take-off rotation of the airplane. To avoid excessive landing gear lengths, the introduction of an aft-fuselage upsweep was investigated. The upsweep was defined by a vertical offset of the FF hub by 600 mm. The fuselage contour was adapted to avoid severe flow separation, however, no iterative design changes have been made to fine-tune the aft-fuselage contour. As can be seen from Figure 10, the inflow pattern to the FF is similar, even though a more pronounced up-wash along the fuselage hub can be noted. Furthermore, it can be noted that the effect of the horse-shoe vortex is more pronounced as the contraction of the aft fuselage is less on the upper side for the PFC with aft-fuselage upsweep. Compared to the axisymmetric PFC aircraft, the fuselage upsweep increased the overall drag by 0.7%. Improvements to the design, such as a refined aft-fuselage shaping and adaption of the nacelle for non-axisymmetric flow could reduce the drag penalty. Detailed discussion of the performed 3D aero-numerical analyses is provided by van Sluis and Della Corte [58].

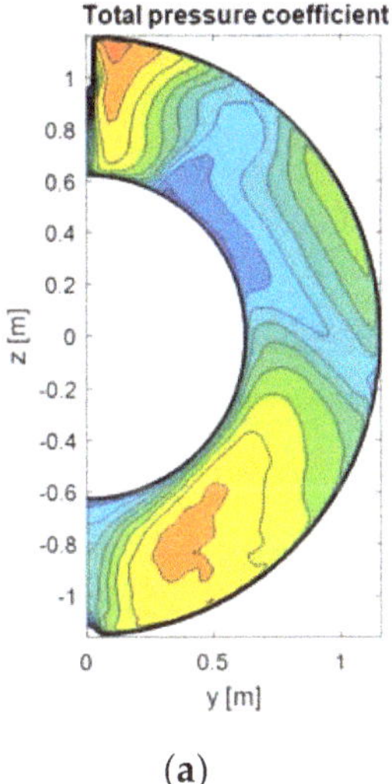

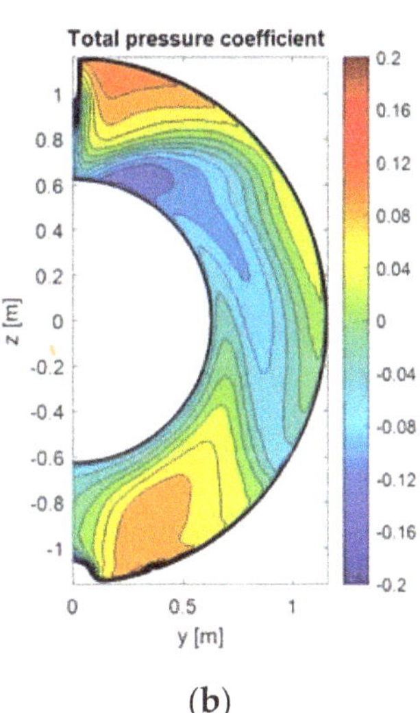

Figure 10. Contours of total pressure coefficient at the AIP. (**a**) PFC with axisymmetric aft fuselage, (**b**) PFC with aft-fuselage upsweep (Source: [58]).

2.4. Fuselage Fan Aerodynamic Design and Performance

The BLI fan at the fuselage aft-section has been designed and analyzed using 3D CFD. The solver used is the GPU-accelerated CFD code Turbostream [60]. It is a 3D, unsteady, RANS solver running on structured multi-block meshes. The one-equation Spalart–Allmaras turbulence model [61] was used for all simulations along with adaptive wall functions and y^+ of approximately 5 on all solid walls. The inlet and outlet boundaries were located one fan diameter away from the fan to allow the free interaction between the fan and the flow.

During the design phase, the flow at the inlet of the fan was assumed axisymmetric. Radially non-uniform profiles of stagnation pressure were extracted at the propulsor AIP for different operating points [58]. These correspond to continuous hub-low stagnation pressure distortion profiles. An example of the distortion profile found during cruise has been previously shown in Figure 9a. These profiles have been prescribed as inlet boundary conditions. An additional clean uniform inlet boundary condition was used to replicate the conditions at which a conventional podded fan would operate.

Two separate transonic fan stages have been designed and analyzed for the basic operating conditions listed in Table 3. The first fan stage, called Fan A, was designed for clean uniform flow at cruise aerodynamic design point (ADP) conditions. The second one, Fan B, was designed to match the severe continuous hub-low BLI inflow.

Table 3. Aerodynamic design point parameters of CENTRELINE FF.

Parameter	Unit	Value
Flight altitude	ft	35,000
Flight Mach number	-	0.82
Flow coefficient	-	0.69
Stage loading coefficient	-	0.45
Stage pressure ratio	-	1.4
Rotor inlet tip Mach number	-	1.24
Rotor inlet hub-to-tip-ratio	-	0.51
Running tip clearance (% span)	%	0.2
Number of rotor/stator blades	-	20/43

The main characteristics of Fan A are: midspan loaded work distribution, alignment of the leading edge with the flow for minimum pressure loss and controlled solidity distribution to minimize Lieblein's diffusion factor. More information on the design of Fan A can be found in [62]. Fan A has been tested under uniform and distorted inflow conditions to assess the effect of BLI and to derive the design changes required for an aft-section fan.

For a fixed mass flow, the deficit of axial velocity associated to hub-low distortion is balanced by an increase in axial velocity in the upper part of the span shown in Figure 11a. The decrease in velocity has a twofold effect: an increase of the local incidence into the rotor blade as visualized in Figure 11b and a rise of the local work load as indicated in Figure 11c. The opposite trends are found in the upper part of the span where there is an excess of mass flow. The overall effect of BLI distortion on the performance of the fan is driven by the upper part of the blade. Consequently, a consistent drop in work input is shown in Figure 12a for the 100% corrected speed line. The continuous operation of the blade at off-design results in an efficiency degradation displayed in Figure 12c. The combination of reduced work input and efficiency ultimately results in a reduction in stage pressure ratio shown in Figure 12b and thrust produced. Lastly, the flow capacity of the fan is reduced and the stability margin slightly increased.

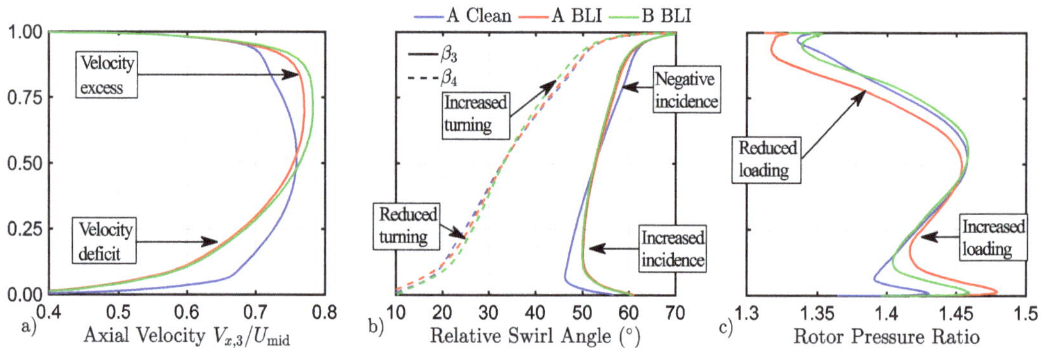

Figure 11. Spanwise distributions at the ADP of: (**a**) axial velocity, (**b**) relative swirl angle, (**c**) rotor pressure ratio.

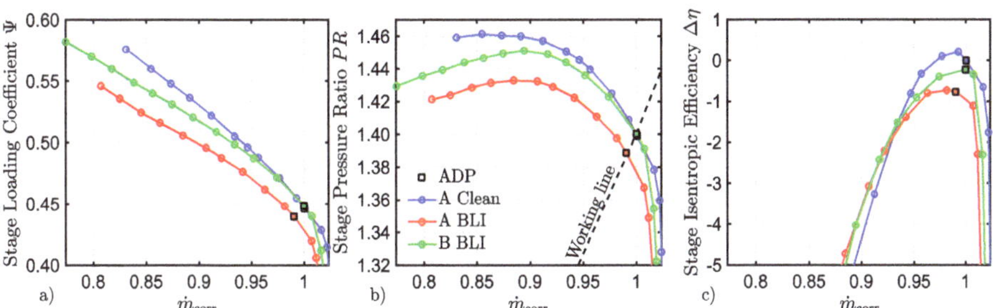

Figure 12. Effect of fuselage BLI on Fan A and Fan B at 100% corrected speed and cruise conditions: (**a**) stage loading coefficient, (**b**) stage pressure ratio, (**c**) stage isentropic efficiency (%).

To recover the performance lost due to BLI, a second fan stage, denominated Fan B, has been designed. The leading edge of the blade has been aligned with the flow for minimum pressure loss up to 75% span. Above 75% span the sections have been progressively restaggered to operate at negative incidence. This improves the operability of the blade in non-axisymmetric inflow. The magnitude of the negative restagger has been limited to minimize choking losses and maintain the flow capacity. The radial work distribution has been modified to restore the intended loading (cf. Figure 11c). Blade turning has been decreased near the hub and increased in the upper part of the blade as can be seen from Figure 11b. No changes to the solidity distribution were required to maintain the diffusion in optimum values. Note that low levels of rotor hub loading reduce the turning required by stator, maximizing the stage efficiency. The resulting design fulfils the performance requirements at the aerodynamic design point as shown in Figure 12 with a minimal reduction in efficiency and choking margin. Away from the design point, the work input drops. However, a significant rise in stable operating range can be observed.

Figure 13 presents the CFD-based fan map for cruise of Fan A and Fan B. Near ADP, Fan A with clean inflow and Fan B BLI inflow exhibit the same performance. However, away from ADP a drop in work input and efficiency (and consequently pressure ratio) is observed for Fan B. This is caused by the non-linear change in distortion profile with corrected mass flow. Despite the drop in pressure ratio generated, the stability margin of Fan B is greatly improved across the map. Potentially, improving its operability at off-design non-axisymmetric inflows.

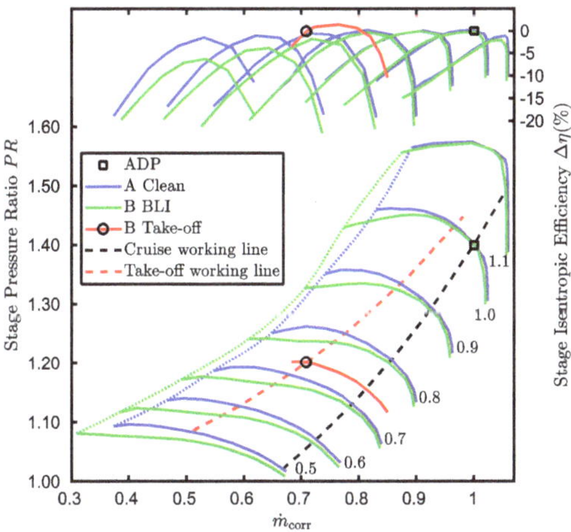

Figure 13. Comparison of cruise maps for Fan A and Fan B.

The performance of Fan B at take-off has been superimposed in Figure 13. This constitutes the most critical operating condition for the aft-section fan. As reported in [62] in detail, the level of distortion found during take-off is much lower than at cruise. Therefore, the distortion effectively becomes a pseudo tip-low distortion. The fan is driven by an electric motor; the maximum torque delivered by the motor limits (cf. also Section 7.3) the attainable pressure ratio of the fan. For this specific case, the fan is expected to operate only up to a corrected speed of approximately 72%, i.e., stage pressure ratio of 1.2 (cf. also Section 7.3). Additionally, the reduced ram-pressure during take-off moves the working line towards the stability margin, operating at reduced flow coefficient and increased incidence levels across the span. The design features of Fan B alleviate the high levels of incidence at the tip, ensuring stable operation [62].

3. Aerodynamic Validation Testing

The core of the conceptual proof for fuselage wake-filling propulsion integration in CENTRELINE is constituted by two experimental test campaigns aiming at obtaining a fundamental understanding of governing flow physics of both, the overall aircraft configuration and the fuselage BLI propulsor. For the overall configuration aero-validation testing, a wind tunnel model was developed to be tested in the low-speed test facilities at Delft University of Technology. The modular nature of the model allowed for an incremental analysis of the configurational effects on the PFC flow field, and more specifically, the FF inflow conditions. In order to verify the FF aerodynamics, the low-speed BLI fan rig facility in the Whittle Laboratory at the University of Cambridge was employed and modified in order to closely replicate the CENTRELINE PFC configuration.

3.1. Overall Configuration Wind Tunnel Testing

The experimental studies of the overall PFC aircraft were carried out at the Low Turbulence Tunnel facility of Delft University of Technology. This atmospheric closed-loop tunnel features a closed octagonal 1.8 m × 1.25 m test-section. The turbulence intensity is below 0.03% at a freestream velocity V_0 between 20 m/s and 40 m/s [63].

The goal of the wind-tunnel experiments was twofold: (1) investigating the effect of fuselage-BLI on the aircraft aerodynamic performance; and (2) understanding the governing aerodynamic interactions between the airframe and the BLI propulsor. The whole experimental campaign was designed with an incremental approach. Firstly, the 2D bare PFC shown in Figure 14 a was tested in simulated cruise conditions ($\alpha = \beta = 0°$) and without the influence of secondary elements, such as wing and tail surfaces. In this setup, the fuselage is directly connected to the external balance and the support structures covered with two symmetric fairings. Secondly, the 3D PFC aircraft including wing and vertical tail depicted in Figure 14 b was tested to study the PFC performance in representative conditions in on- and off-design phases.

Both setups featured an axisymmetric fuselage (fineness ratio $L_{fus}/2r_{fus}$ = 11.1) with a shape representative of the CENTRELINE PFC design Rev03 [50]. The fuselage aft-section was equipped with an integrated shrouded fan model. The 12-bladed fan stage was specifically designed to operate at the same conditions, defined by the flow and pressure rise coefficients, as the full-scale BLI fan [64]. The rotor was driven by an electric motor and the speed setting was directly controlled to match the required set point (±0.5 Hz). The shrouded fan could also be removed to obtain the performance of the unpowered PFC configuration.

A whole range of measurement techniques were employed to characterize the aerodynamic performance of the PFC configuration: (1) six-axis external balance measurements of the aerodynamic forces and moments; (2) total pressure measurements in the airframe viscous layers and far wake; (3) stereoscopic PIV measurements to quantify the 3D velocity field in various survey planes. During all experiments, turbulent transition was ensured on all the model surfaces with the use of turbolator strips. Key measurement results obtained from the performed test campaigns are characterised in the following.

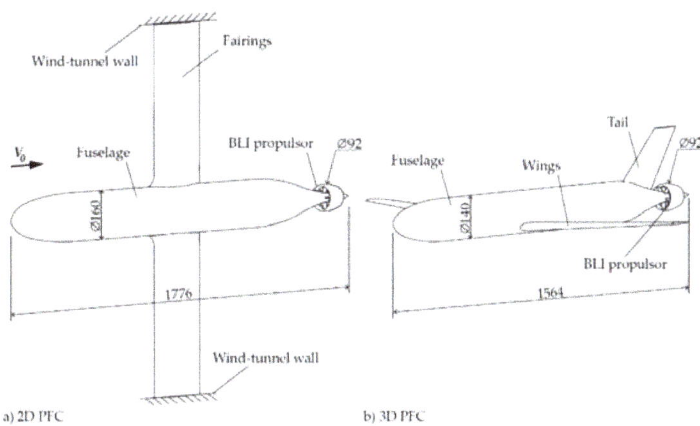

Figure 14. Schematics of the wind-tunnel models setup. Dimensions in mm.

3.1.1. Time-Averaged Flow Field

The velocity field was measured around the fuselage aft section for the 2D PFC setup according to Figure 14a through stereoscopic PIV. The time-averaged velocity magnitude field for the unpowered and powered PFC are reported in Figure 15a,b, respectively. The powered case is at axial equilibrium condition, $NPF = 0$, where NPF is the net axial force acting on the fuselage-fan assembly. Figure 15a shows the boundary layer development around the fuselage aft-cone. At the fan location ($x/L_{fus} = 0.94$) a boundary layer thickness $\delta = 0.8r_{fus}$ was measured, resulting in approximately 60% of the boundary layer flow ingested by the fan. In the powered PFC case of Figure 15b, the effect of the fan is twofold: first, it accelerates the boundary layer flow upstream of the inlet, especially in the lowest momentum region close to the fuselage wall; second, the fan slipstream develops and mixes with the outer part of the fuselage boundary layer creating a turbulent shear layer. Moreover, at the trailing edge of the fuselage, a region of turbulent separation is found.

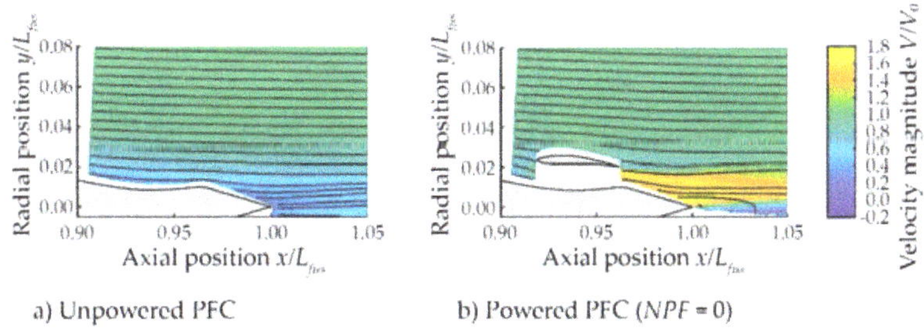

Figure 15. Time-averaged velocity field around the PFC aft-fuselage section of the 2D-PFC setup. Stereo-PIV measurements taken at $V_0 = 20$ m/s and $NPF = 0$.

3.1.2. Fuselage Fan Inflow Field

When integrated in the overall PFC aircraft configuration, the fuselage inlet flow field is not only dominated by the fuselage boundary layer, but it is also influenced by the flow around other aircraft elements, such as wing and tail surfaces. The resulting inlet distortion can potentially affect the fan aero-acoustic and aero-mechanical behavior and hence it needs to be known at all the relevant flight phases. The total pressure distribution at the fan inflow for cruise and off-design conditions was measured for the 3D-PFC setup. The results are reported in Figure 16.

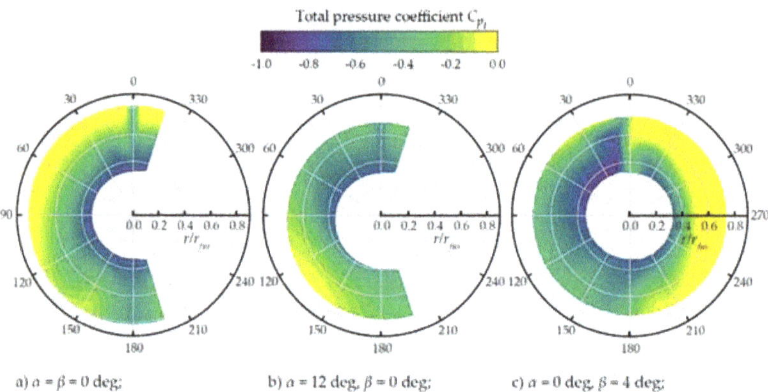

Figure 16. Total pressure field upstream of the FF air intake (x/L_{fus} = 0.89) in cruise and off-design conditions. Measurements taken on the 3D PFC setup at V_0 = 40 m/s.

In zero incidence conditions (at $\alpha = \beta = 0°$, Figure 16a), the main inlet distortion is due to the fuselage boundary layer which produces radial total pressure gradients. The vertical tail plane (centred at an azimuthal angle $\theta = 0°$) introduces two main distortions in the flow: first, the viscous wake of the tail, with a narrow extension ($\Delta\theta \approx \pm 5°$) and constant with r; second, the junction flow at the tail-fuselage intersection, with a larger distortion ($\Delta\theta \approx \pm 30°$). Despite a direct quantitative comparison cannot be drawn, the measured distortion patterns in Figure 16a reflect the full-scale high-speed predictions from the CFD simulations (cf. Figure 9c) in the good qualitative manner. This includes the radial distortion due to the fuselage boundary layer as well as relatively sharp vertical tail wake. Stronger differences between the experimental and numerical domains is found in the bottom section of the contours, due to the flow around the fuselage–wing junction fairing, as already discussed in Section 2.3.1.

At $\alpha = 12°$, representative of take-off rotation conditions, the increased incidence angle results in the onset of a crossflow around the fuselage section as can be observed from Figure 16b. The vertical tail introduces a low p_t disturbance for $-5° < \theta < 5°$ which is due to the viscous wake. Unlike the $\alpha = 0°$ case (cf. Figure 16a), the junction flow distortion is not clearly visible anymore close to the airframe.

Finally, at $\beta = 4°$ shown in Figure 16c, a strong distortion is introduced by the vertical tail in the sector $-30° < \theta < 30°$. A low p_t region is found on the leeward side (suction side of the vertical tail) presumably due to trailing edge separation or corner flow separation. On the windward side (pressure side of the vertical tail), the imprint of a horseshoe vortex can be identified, as a result of the junction flow developing at the tail-fuselage intersection. Furthermore, strong pressure gradients are found in the azimuthal direction around $\theta = 0°$.

All the measurements were taken at low speed, resulting in a Mach number in the range 0.06 to 0.12, and a Reynolds number of 0.5×10^6 based on the wing mean aerodynamic chord length. This clearly has an influence on the applicability of the presented results to the full-scale PFC. Even though the fan diameter was scaled to match the fuselage boundary layer at the model scale, other effects due to compressibility and different Reynolds number were not present in the measurements. For example, possible shock-waves at the shroud or tail surfaces could affect the full-scale PFC (see [58] for further analysis in high-speed conditions). More results of the presented wind-tunnel campaigns and detailed data analysis can be found in [48].

3.2. Fuselage Fan Aerodynamic Design and Performance

The aerodynamic behavior of the BLI fan has been tested in an experimental low-speed single-stage fan rig, known as the BLI rig. The facility, located in the Whittle Laboratory at the University of Cambridge, was purposely built for the aerodynamic analysis of inlet flow distortion fan interaction. Figure 17 presents the meridional view of the of the BLI fan rig. The annulus geometry replicates the geometry of CENTRELINE's aft-section FF.

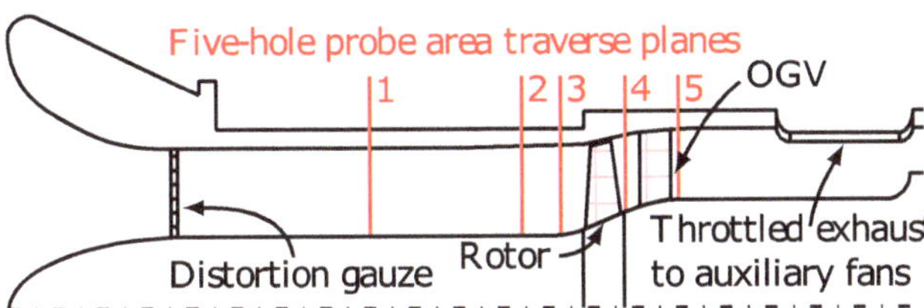

Figure 17. Meridional view of the BLI fan rig.

Two rotor blades designs have been manufactured and tested. The first blade, called Fan A, represents a conventional free-vortex fan designed for clean uniform inflow. The effect of a severe and continuous hub-low radial distortion has been investigated on this conventional blade. A series of design steps have been taken to match this blade to the BLI inflow. The resulting blade shares most of the design features of the transonic BLI optimized Fan B presented in Section 2.4. These are: leading edge aligned with the flow for minimum loss, mid-span loaded work distribution, controlled diffusion factor and increased operability of the tip section. The resulting blade design, called Fan B has been subsequently tested under axisymmetric and non-axisymmetric distorted inflows. The low-speed nature of the rig does not allow compressibility effects to be replicated. However, velocity triangles representative of the full-scale transonic FF have been obtained by matching the full scale flow coefficient and stage loading coefficient (cf. Table 3).

Detailed pressure measurements have been taken at five axial locations with a five-hole pneumatic probe area traverse system as indicated in Figure 17. The five-hole probe measures the time-average values of stagnation pressure, static pressure, swirl angle, and radial angle. Based on these flow properties and the incompressibility of the rig, the complete flowfield has been reconstructed. Further details of the experimental setup can be found in [64].

Experimental tests have been performed for clean and distorted inlet conditions. The baseline distortion chosen for the study is a scaled version of the axisymmetric but radially non-uniform hub-low stagnation pressure distortion calculated in Section 2.3 and used in Section 2.4 for the aerodynamic design point. In a subsequent study a full-annulus non-axisymmetric distortion representative of the CENTRELINE configuration comprising fuselage, wings, and vertical tail plane (cf. also Section 2.3) has been tested for Fan B.

To generate the target inlet velocity profiles, flow conditioning gauzes have been installed at the intake of the rig. The gauzes comprise thousands of small vanes with precisely controlled geometry and are additively manufactured. For the axisymmetric inflow (clean and radial stagnation pressure distortion) the gauzes were designed using the method proposed by Taylor [65]. The method makes use of CFD simulations in an iterative manner to obtain the shape of the vanes that compose the gauze. For the non-axisymmetric inflow, the method reported in [65] has been upgraded to replicate non-uniform distributions of stagnation pressure and swirl distortions. Figure 18 presents the flowfield measured downstream of three different gauzes. Axisymmetric clean and hub-low axial velocity measurements are shown in Figure 18a,b whilst a highly complex distortion pattern (stagnation pressure + swirl) associated to the operation of the aft-section fan with the full aircraft configuration is presented in Figure 18c.

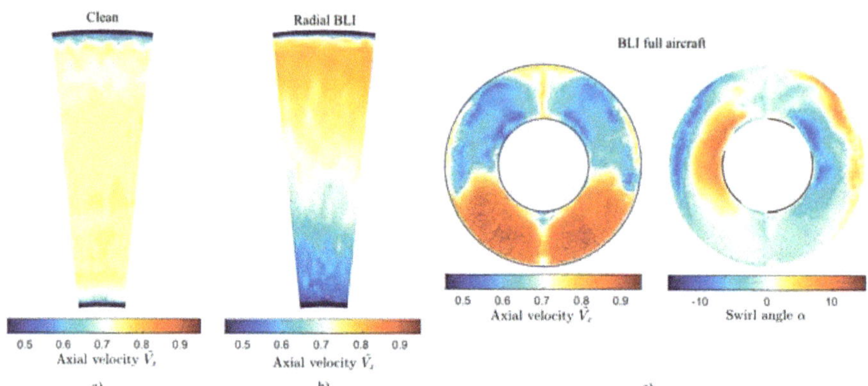

Figure 18. Measured inlet inflow behind the distortion gauzes: (**a**) Clean, (**b**) hub-low distortion, (**c**) full-aircraft configuration.

The low-speed nature of the rig does not allow to capture compressibility effect. However, most of the physics associated to the operation of the fan in a distorted inflow are still captured. To illustrate this, Figure 18 presented the measured spanwise distributions of axial velocity, relative swirl angle and angular momentum. Note that angular momentum is equivalent to work load for a purely axial inflow. The effect of the severe hub-low stagnation pressure distortion on Fan A is consistent to the one found for its transonic version in Figure 19. High incidence and increased work load are found near the hub whilst negative incidence and reduced loading are observed in the upper part of the span. The overall effect on both cases is a reduction in stage loading coefficient of the same magnitude. Additionally, the performance lost due to BLI is recovered applying the same design philosophy for the low- and high-speed versions of Fan B: tailored alignment of leading edge and controlled work and diffusion distributions. With exception of shock related losses, the low-speed rig is able to capture the physics of fan-distortion interaction. Making low-speed fan testing a quick and economical tool to validate the performance of numerically derived fan designs.

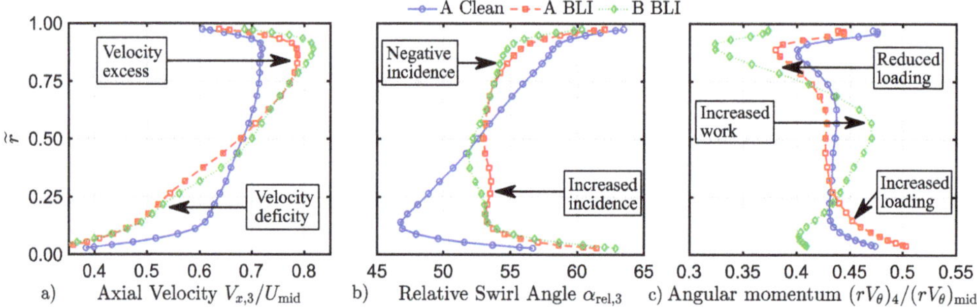

Figure 19. Measured spanwise distributions at the ADP of: (**a**) axial velocity, (**b**) relative swirl angle, (**c**) angular momentum.

4. Fuselage Propulsive Device Design and Structural Integration

This section is dedicated to the conceptual design integration of the fuselage BLI propulsive device. This covers parametric sizing and performance modeling of the FF module including key components of is mechanical design, as well as the aero-structural design integration at the aft-fuselage. Beside the fuselage nacelle integration, the structural design and analysis part also includes the key airframe components directly affected by the FF installation.

4.1. Fuselage Fan Conceptual Design Synthesis

The CENTRELINE FF propulsion system consists of a single-rotating ducted fan driven by an electric motor in a direct drive arrangement. The fan rotor design features fixed-pitch blades. The fan rotor blading and outlet guide vane geometry refers to the Fan B design described in Section 2.4. The fan ducting features a fixed nozzle. The developed design and performance models, however, feature immediate capability to appropriately emulate a nozzle variability device if required for operational flexibility during more detailed aircraft-integrated design optimization (cf. [42]). Nacelle mechanical complexity and structural mass is kept to a minimum as the fan cowling is not equipped with a dedicated thrust-reversing device. Instead, thrust reverse functionality—in case required from the overall aircraft design perspective—is intended to be realized through reverse operation of the electrically driven fan. The FF outer casing includes a rotor containment sized for the kinetic energy of a fan blade-off. The transfer of the aerodynamic, inertial and gyroscopic loads across the FF rotor plane is realized by the structural load path routed through the FF rotor hub. Supported by FEM (finite element method) structural analyses reported in [66,67], this design solution was down-selected from a number of options conceived and evaluated as part of the preliminary design work in the project (cf. [42,43]). This solution allows for the least number of structural items in the fuselage inflow streamtube and a complete structural decoupling from loads introduced by the empennage. For the FF bearing, electromagnetic bearing options such as proposed by Steiner et al. [5] were initially evaluated, but discarded for complexity reasons based on the trade-off between performance and overall system complexity. The design of the fan rotor bearing system is a rather conventional one, featuring a floating cylindrical roller bearing in the front. The rear bearing is a fixed ball bearing, designed to take the FF rotor thrust loads. The flanging of key component groups is tailored to enable orderly assembly and disassembly of the FF power plant system. Down-selected from a number of alternative concepts, a compactly folded forward mounting approach was identified as the best and balanced design solution for the electric drive motor attachment to the fan rotor. In Figure 20, a representation of the geometric arrangement of the FF rotor system including electric drive motor, bearings, and internal support structure is presented for the CENTRELINE Rev07 design case 1. The annotated figure also displays the integration of the auxiliary power unit (APU) in the fuselage aft-cone, as well as the forward and rearward attachment points for the structural load path across the FF rotor plane.

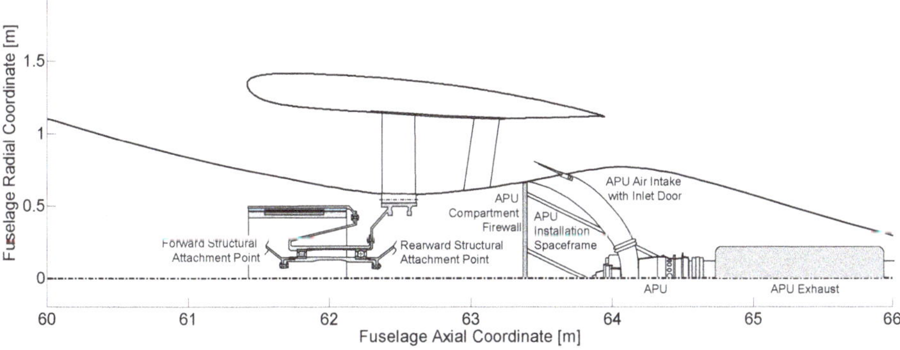

Figure 20. Concept sketch of the aft-fuselage power plant integration.

In order to ensure robustness against uncontained FF disk structural failure, the components of the FF power plant and all adjacent systems require positioning such that disk burst corridors do not interfere with critical functions of the vertical tail. A visualization of systems integration at the PFC aft-fuselage with annotation of spread angles as recommended in AMC 20.128A [68] for the shedding of fan blade fragments is provided in [69].

4.1.1. Flow Path Sizing and Performance Synthesis

The design and performance synthesis for the FF propulsive device was performed using the APSS framework (cf. also Section 1.4). During flow path sizing, the turbo component cross-sectional areas directly result from mass flow continuity, the local gas properties and prescribed axial design Mach numbers. Similarly, the nozzle exit area results from mass flow continuity and the nozzle exit velocity subject to the prevailing nozzle pressure ratio. A set of typical heuristics along with appropriate iteration strategies is included for cycle design and performance prediction according to [42,70,71]. For the FF propulsion system sizing, specific functional sensitivities were implemented as basic design laws including the mapping of fan tip speed as a function of design fan pressure ratio, nozzle thrust and discharge coefficients as functions of nozzle pressure ratio, as well as fan design efficiency as a function of corrected mass flow and mean stage loading. Flow path sizing was conducted for design cruise conditions with maximum climb power settings for the FF.

For FF operational performance simulation, the Fan B cruise map shown in Figure 13 was used. In order to be consistent with the map scaling standards in APSS, before adoption, the FF map was Reynolds-number scaled to ISA SL conditions. The mass flow and efficiency scaling was performed based on the Reynolds number index (cf. [72]). While a precise representation of FF performance behavior with full reflection of the varying fan inflow conditions—e.g., in cruise and take-off—would require an array of classic component maps (cf. [62]), even with the simplification of the single component map the APSS off-design performance results were found to be in good agreement with the values predicted by TU Delft's CFD simulations in cruise and take-off conditions. As expected, the operating line during take-off was situated at higher pressure levels and lower corrected mass flows, meaning reduced stability margins when compared to cruise conditions. With take-off stability margin still over-predicted due to the use of cruise component map, through a re-staggering of the fan blade tips acceptable stability margins in take-off were shown to be achievable for the CENTRELINE FF by Castillo Pardo and Hall [62]. Based on a preliminary assessment, sufficient stability margin in take-off was obtained at the loss of approximately 0.2% of isentropic efficiency at the aerodynamic design point [62].

As a pivotal point for the aircraft-level design studies, a FF polytropic design efficiency of 93.4% was chosen which corresponds to a 1% reduction in polytropic stage efficiency relative to the underwing podded reference engine fans. Taking into account the aerodynamic design efficiency penalties discussed in Section 2.4, this selected baseline design efficiency appears appropriately conservative. In order to account for 3D duct design and typical installations inside the fan cowling duct, the duct pressure losses predicted by the 2D CFD simulations were increased to yield pressure ratios similar to classic fan bypass ducts. A summary of key design parameters for the FF propulsive device is presented in Table 4.

For the actual fan sizing the reduced maximum flow capacity associated with the aerodynamic design of a BLI fan was taken in account. The design axial fan rotor inlet Mach number is 0.60, meaning a reduction of approximately 12% relative to the value assumed for the main engines fan operating without BLI effects. The reduced axial Mach number at the fan rotor inlet yields an increased fan face area, and thus, larger tip diameter in the order of 3% relative to CFD optimized Rev07 geometry cases, when a constant hub contour is retained. This modification was deemed to be achievable featuring a net neutral impact on the sum of surface pressure forces acting inside the fuselage nacelle and the fan disc [58]. The resultant reduced fan cowling profiles thickness of approximately 8% (cf. also Table 4) was assessed to remain acceptable for a proper nacelle aerodynamic and aero-structural design. The axial Mach numbers at the FF intake highlight and AIP were directly adopted from the CFD optimized Rev07 design cases 1 to 3. Assuming a limited upstream effect of the increased flow diffusion in front of the FF face at the given fan design pressure ratio of 1.40, the corresponding impact on the freestream total pressure recovery ratio ahead of the FF face was considered negligible. When operated at a FPR of 1.2 in take-off (cf. Section 2.4), the shaft power absorbed by the FF is increased by approximately 30% at a simultaneously reduced shaft rotational speed by 20%, relative to the flow path sizing conditions summarized in Table 4 above. In combination, this means an elevation of shaft torque by approximately 63%.

Table 4. Summary of results for actual FF flow path sizing (M0.82, FL350, ISA + 10K).

Parameter	Unit	Design Case 1	Design Case 2	Design Case 3
Fan isentropic shaft power ($P_{FF,Shaft}$)	MW	5.530	6.800	7.760
Fan pressure ratio (p_{13}/p_2)	-	1.40	1.40	1.40
Fan mass flow rate (W_2)	kg/s	212.6	261.4	298.3
Freestream total pressure recovery ratio (p_2/p_0)	-	0.831	0.841	0.848
Nozzle pressure ratio ($p_{18}/p_{s,amb}$)	-	1.795	1.817	1.832
Gross thrust	kN	62.59	77.70	89.23
Fan rotor inlet tip diameter (D_{FF})	m	2.349	2.585	2.745
Hub/tip ratio at rotor inlet	-	0.497	0.495	0.493
Rotational speed	1/min	2730	2481	2336
Bearing system inner diameter	[mm]	338	367	389
Cowling thickness at fan face	m	0.208	0.224	0.230
Cowling length	m	2.531	2.776	2.935

4.1.2. Mechanical Component Sizing and Mass Prediction

With the flow path hub and tip diameters determined through the flow path sizing, the more detailed parametric mapping of the FF flow path geometry was performed based on the methods described in [70]. For the prediction of the FF module weight, the masses of the fan key component group were mapped individually. These included the fan rotor blading, the outlet guide vanes, the rotor disk, the fan bearing, as well as the fan shaft and rotating bearing support structure. The rotor blading and outlet guide vane (OGV) masses are determined based on a simplified blade/vane geometry, i.e., cuboid bodies with prescribed aspect ratios and relative thickness distributions in the radial direction. The number of blades and vanes per row is controlled by a prescribed solidity value. The resultant total displacement volume per row was then scaled with a representative material density in order to obtain the total blading/vane mass. For the present study all input parameters for the blading and vane geometric properties were calibrated to the Fan B aero-shaping discussed in Section 2.4. For the rotor disk sizing, the disk design tool of GasTurb® Details 6 [73] was used. FF rotor bearing sizing was performed based on a method by [74] as discussed in [75]. In order to predict the required bearing size and corresponding mass, the method uses a correlation between the dynamic load rating and the bearing skin surface area, which was derived from existing SKF ball bearings [75]. The required bearing dynamic load ratings were calculated from the actual bearing load cases due to the fuselage rotating mass components under CS-25 regulation [76] and prescribed target service hours. The inner diameters of the bearing system obtained for a typical velocity index of 1.2×10^6 mm/min are listed in Table 4 above. For the fan rotor casing including blade containment, a simplified geometric description including axial length determination based on a constant spread was used. Containment thickness was derived from [77] as a function of blade kinetic energy. The individual components included in the mass budget of FF module are highlighted in Figure 21.

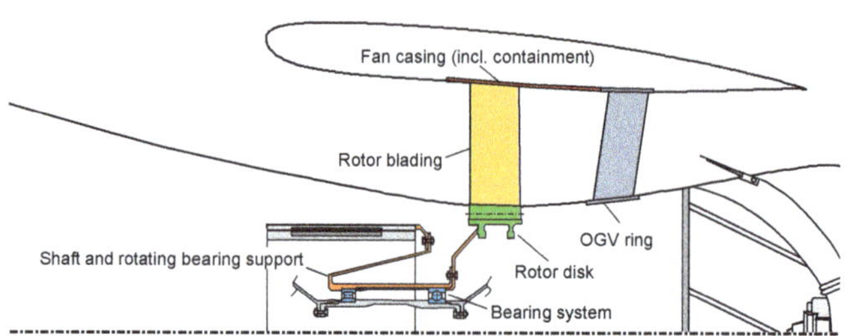

Figure 21. Visualization of control volume for FF module mass budget, individual component groups highlighted in color.

A synopsis of the individual component masses predicted for the FF modules for the Rev 07 design cases is provided in Table 5. The prediction of nacelle structural masses for the Rev 07 design cases involves the FF cowlings as well as all structural elements of the aft-fuselage section. Beside the outer shell structure this also includes the internal load beam routed through the inner diameter of the FF bearing system and the APU firewall and is discussed in the next section.

Table 5. FF component masses predicted for the CENTRELINE Rev 07 design cases.

Mass Terms (kg)	Design Case 1	Design Case 2	Design Case 3
Fan rotor blading	136	183	221
Rotating mass w/o blading	306	411	490
Stationary masses (OGV + fan casing)	164	206	235
Fan module total	606	800	946

As can be noted in Table 5, the FF module masses appear to be lower than expected for conventional fans featuring identical diameters. This is the consequent result of two important effects: Firstly, the hub-to-tip ratios of the CENTRELINE FF are considerably higher than for conventional fans (cf. Table 4). Secondly, in contrast to conventional fan face maximum mechanical speeds during maximum take-off conditions, the FFs maximum speed is at flow path sizing conditions, i.e., top of climb. As a result of these two effects, the AN^2 metric (cf. e.g., [70]) for the CENTRELINE FF design is approximately 35% reduced compared to typical conventional fans.

4.2. Fuselage Fan Aero-Structural Integration

The aero-structural design concept, analysis and sizing activities in CENTRELINE covered the most relevant airframe components for the investigated PFC aircraft layout. While a focus was placed on the structural integration of the fuselage aft-section including the FF propulsive device, also the entire fuselage structure was conceptually elaborated and sized as well as key components of the wing structure. In the following, important aspects of the developed structural design solutions will be introduced. The FEM simulation-based design and sizing strategy for relevant load cases according to CS-25 regulations [76] will be discussed and main results will be presented including structural mass prediction for the addressed airframe components.

4.2.1. Description of Structural Design and Integration Solutions

The aft-fuselage section including the FF nacelle and supporting structure was the central subject of the structural design work. Starting from a pool of candidate concepts considered as part of the pre-design phase in the first half of the project (cf. [66]), a suitable structural layout was elaborated for the principal conceptual integration of the FF propulsive device as presented in Figure 21. The overall structural concept is based on a ubiquitous use of carbon fibre reinforced polymer (CFRP) material.

The FF cowling is structurally decoupled from the vertical fin in order to avoid the complex aero-elastic interactions between the empennage, the nacelle and the FF rotor. As a result, all loads generated by the FF, as well as the aerodynamic, inertial and gyroscopic forces acting on and inside the rear of the fuselage and FF nacelle are transferred in a single structural path through the FF rotor hub (cf. Figure 21). Important boundary conditions for the design of the load carrying structure were imposed by the Rev07 optimized aerodynamic shaping, as well as by a defined arrangement and positioning of the main internal components, including the FF electric drive motor, the rotor and stator and the aircraft APU. The geometric space available for the internal load-carrying structure was particularly limited by the bore diameters of the FF bearing system (cf. Table 4). A basic visualization of the final aft-fuselage aero-structural integration is provided in Figure 22.

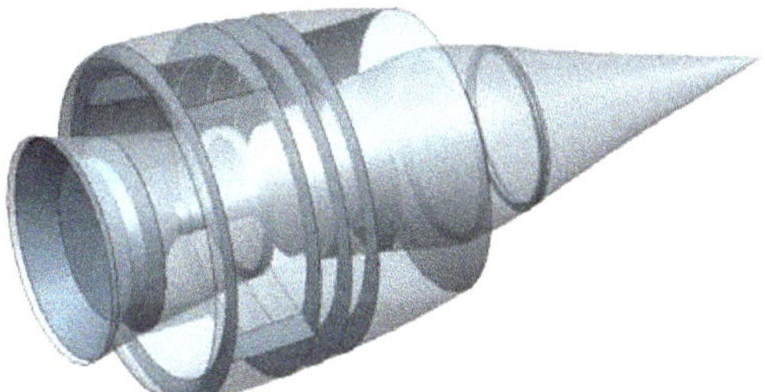

Figure 22. Basic visualization of CENTRELINE aft-fuselage aero-structural integration concept.

The most loaded structural element is the front fixing frame, which transfers the loads to the main structure of the aircraft. This main frame is conically shaped and reinforced to provide the required strength and rigidity. Both, on the rear part of the fuselage and the nacelle, a monocoque structure is used for the transfer of loads through the outer shell. In addition to its load carrying outer shell, the fuselage aft-section is equipped with smaller forming frames, two main fixing frames and a fire wall providing the structural attachments for the space frame holding the APU. The FF OGVs are used as a multifunctional element. Beside the aerodynamic purpose, the OGVs serve as the sole structural connecting element between the FF cowling and rear-section of the fuselage. The load transfer across the FF rotor is implemented through the hollow cylindrical beam connecting the aft-fuselage installation to the front fixing frame. Important aspects for the aero-structual integration of the PFC aft-fuselage are described in [67].

Beyond the FF installation structure, a concept for the entire fuselage main structure was developed. For both aircraft, the PFC and the R2035, the fuselage is made of a hybrid structure utilizing sections with a geodetic layout combined with conventional structure sections. In the cylindrical parts—except for the fuselage-wing junction section—a geodetic structure is adopted, while in the remaining parts a conventional structure is considered more well-suited for manufacturing purposes. As in the case of the PFC aircraft a T-tail empennage arrangement was selected, additional frames are added to its local fuselage structure in order to strengthen the vertical tail attachment area. A PFC aircraft side view featuring basic visualization of the main elements of the fuselage structural layout is provided in Figure 23. Not shown in the figure, are the window and door cut-outs, which however have been taken into account during the structural analysis.

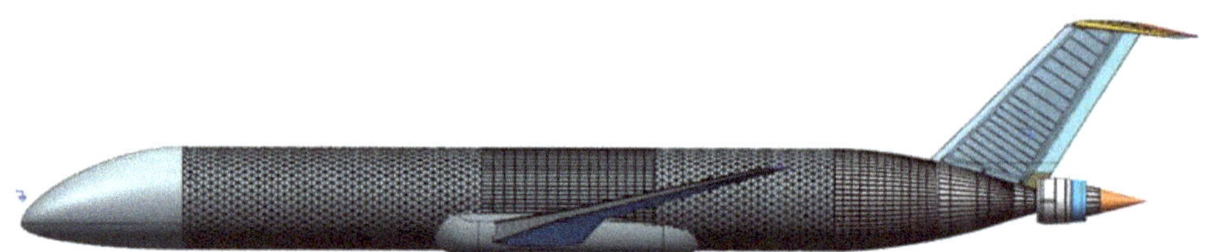

Figure 23. PFC aircraft side view including basic visualization of the main elements of the fuselage structural design concept.

As part of the structural design activities, a concept of the primary load-carrying structure of the PFC aircraft wing was prepared. Accordingly, the wing consists of the skin, two spars (main and auxiliary), ribs, and stringers. The wing skin is divided into zones with different fabric layouts adapted to the loads. The centre wing box features innovative design approach based topologically optimized zones featuring laminates with different layouts, individually adjusted to the local load distribution, allowing for a significant mass reduction while maintaining appropriate mechanical properties of the individual parts [78]. The fuselage and wing structural concepts and their analyses are described in more detail in [79,80].

4.2.2. FEM-Based Structural Sizing Method

For the effective sizing of the structural concepts described above, 3D Computer Aided Design (CAD) models of the CENTRELINE aircraft geometries were developed and analyzed through FEM using the Siemens NX® software by Siemens PLM Software Inc. (Plano, Texas, United States). Based on the defined external geometry, the internal load-carrying structure of the aircraft was built, which was later converted to a 2D surface model as shown in Figure 24 and subjected to FEM analysis afterwards. The strength analysis was repeated iteratively until the specified requirements were met in terms of strength, stiffness, and weight.

Figure 24. Finite element mesh for (**a**) internal structure and (**b**) skin of the aft-fuselage section.

For the PFC nacelle and aft-fuselage integration, the main sources of loads affecting the load-carrying structure include inertial loads coming from the structural weights, torque due to the FF rotor drive, aerodynamics loads on the nacelle surface and gyroscopic loads due to the FF and APU rotating parts. Considering the unconventional placement of the BLI propulsor as well as the nature of the operation of the entire assembly, the CS-25 regulations [76] were analyzed. Table 6 summarizes the load cases identified and used during the FEM structural analysis.

Table 6. Fuselage fan integration structural load cases according to CS-25 (cf. [76]).

Case Number	Case Description	Regulation Chapter
1	Emergency landing	CS 25.561
2	Side load	CS 25.363
3	Engine and APU loads	CS 25.361
4	Gyroscopic loads	CS 25.331
5	APU acceleration	CS 25.361
6	APU gyroscopic loads	CS 25.371

The overall fuselage structure must be able to withstand a variety of load limitations acting individually and in combination. Rationally analysing the regulations, nine critical load cases were selected and presented in Table 7. Cases of hard landing were not explicitly analyzed, as this would require a detailed design of the undercarriage attachment structure. However, the obtained results can be considered valid since the undercarriage attachment structure was included in another component of the used mass estimation method.

Table 7. Fuselage fan integration structural load cases based on CS-25 (cf. [76]).

Case Number	Description of Loads
1	1 g + cabin pressurization
2	−1 g manoeuvre + cabin pressurization
3	2.5 g manoeuvre + cabin pressurization
4	2.5 g manoeuvre
5	1.33 times cabin pressurization (over pressurization)
6	1 g + elevator deflection downward
7	1 g + elevator deflection upward
8	Lateral gust + cabin pressurization
9	Lateral gust + cabin pressurization

The PFC structural analysis and sizing was performed for all Rev07 design cases as introduced in Section 2.2. In each of the three cases, the load-carrying structure was adapted to the different external geometries (see Figure 7) and internal component masses (e.g., Table 5). The iterative structural refinement process was applied in every case until all structural requirements—such as predefined minimum strength, stiffness, and buckling resistance—were reached.

4.2.3. Key Structural Design Results

A structural displacement map for the aft-fuselage installation according to Rev07 design case 2 under CS 25.561 loads is presented in Figure 25. Assuming a fixated front frame, the maximum vertical displacement at the fuselage aft-tip remains within approximately 57 mm.

The buckling analysis performed for the final sizing of the Rev07 design case 2 structural design showed the onset of local wrinkling at loads more than 60% above the ultimate loads, indicating sufficient structural safety and buckling resistance. The FEM structural analyses, furthermore, confirmed the load carrying capability of the FF OGVs as intended by the structural design concept.

Beside the principal proof of the PFC aero-structural feasibility, the purpose of the structural design activities in the project was to deliver weight predictions for key aircraft components suitable for the integrated aircraft design sizing and performance evaluation. With the primary load carrying structure of the fuselage designed and sized based on the previously described FEM-analysis based approach, other structural mass components including the fuselage pressure bulkheads, cabin floor supports, doors, windows cargo hold structures, wheel bays, wing carry-through structural elements, as well as the attachment structures for the wing, tail, and landing gear were estimated using semi-empirical methods as described in [79,80]. Table 8 provides a synopsis of the fuselage integration and overall fuselage structural weights determined for the R2035 aircraft as well as the three Rev07 design cases for the PFC aircraft. Following a family design strategy, all structural components have been sized for the largest member of each aircraft family.

Figure 25. Displacement map for aft-fuselage nacelle and integration structure of PFC Rev07 design case 2 under CS 25.561 loads. Color legend shows vertical displacement in mm.

Table 8. Overview of fuselage and FF integration structural weight results.

Component	R2035	PFC (Rev07 Case 1)	PFC (Rev07 Case 2)	PFC (Rev07 Case 3)
Aft-fuselage integration (kg)	n/a	490	646	775
Main fuselage (kg)	22,286	22,742	22,889	23,027

5. Turbo-Electric Fuselage Fan Power Supply

In this section, the pre-design of the turbo-electric drive train system is introduced and explained. The turbo-electric drive train is the system transmitting the mechanical power from the podded engines to the FF in the aft fuselage section. The discussion includes the overall system architectural layout, the design integration of the main power plants, the involved electric machinery and system thermal management aspects.

5.1. Transmission System Architectural Definition

The CENTRELINE PFC aircraft features a partial turbo-electric propulsion system, in which the majority of the core engine excess power used to drive the under-wing main fans. The fraction of power required to drive the BLI FF is transmitted electrically, through power conversion by electric generators installed on the main engines and the FF electric drive motor. Due to the low energy density, batteries as used in partial serial-electric power train arrangements have not been considered for the CENTRELINE systems layout.

The electric components in the system must be sized for the maximum mechanical power, which occurs on ground at high power levels. The fuselage power scheduling for the various phase of the mission is determined by the overall aircraft sizing and performance synthesis (cf. also Section 6.2). As a result, the actual power profile differs from conventional fans throughout the mission. This translates into a variable power split between the podded engines and the FF. Compared to a fixed power split the adjustment to the FF power to the current mission segment allows to avoid excessively high sizing power requirements for the electric components to cover the take-off performance.

The turbo-electric drive train can be realized by different electric architectures. The most relevant configurations are the direct current (DC) architecture, where the mechanical power extracted from the engines is converted by a generator and subsequently rectified to DC electric power. The rectification allows to decouple the rotational speeds of the engine generators and the FF. On the other hand, power electronic devices are required to rectify and invert the electric power, which results in additional losses and component masses. Moreover, the system must be controlled actively, which increases the development effort and adds failure cases to the propulsion system.

An alternative solution would be an alternating current (AC) architecture, where the extracted mechanical power is converted to electric power, but it is not rectified afterwards. The AC electric power is supplied to the FF drive motor, where it is directly converted back to mechanical power. This architecture would not decouple the rotational speeds of the engine generators and the FF. From a functional point of view, this transmission can be compared to a gearbox with a fixed ratio. AC transmission systems are used in hybrid-electric ships to maximize efficiency as the power electronic devices are disconnected from the system and the diesel engine drives the propellers "directly". With an AC transmission, a variable power split between the engines and the FF can only be realized with a variable pitch mechanism in the FF. The advantages are the higher efficiency and no need for active control, which increases the system efficiency and does not require an active input. The downsides of this system are the higher cable masses and the danger of desynchronization of the electric machines. The cables of the AC transmission must be sized to transport the reactive power, which results in higher cable masses and higher losses. As there is no active control of the system, the electric machines can transmit a limited torque, which must no be exceeded to prevent desynchronization. The desynchronization of such a propulsion system is considered catastrophic for the aircraft as the mechanical forces will destroy the electric machines and lead to significant structural damage.

A DC transmission has been selected for the PFC aircraft to enable the variable power split without using a variable pitch fan and to avoid any desynchronization problems typically related to AC transmission systems. The rated DC link voltage is derived from commercially available semiconductor switches, possible inverter topologies and a voltage derating due to increased cosmic radiation at cruise altitude. Since the cosmic radiation intensity at a flight altitude of approximately 12 km, or FL390, increases by a factor of 300 compared to sea level, the voltage rating of the semiconductor switches is reduced to 55%. A positive effect of reducing the voltage rating is the significantly reduced failure rate of the semiconductor switches. The considered voltage ratings of commercially available switches are 900 V, 1200 V, 1700 V, 3300 V, and 6500 V. Switches with lower voltages are available, but the derating and the power level do not favor switches with lower voltage ratings. Finally, two-level, three-level, and five-level converter topologies are considered for this application. The combination of the three factors results in the possible DC link voltages listed in Table 9 below.

Table 9. Suitable DC link voltages including de-rating (Source: [81]).

Voltage Rating (V)	Two-Level Converter	Three-Level Converter	Five-Level Converter
6500	3575	7150	14,300
3300	1815	3630	7260
1700	935	1870	3740
1200	660	1320	2640
900	495	990	1980

Suitable values for the DC link voltage are highlighted in green. Lower values lead to high currents and high cable masses. Higher voltages will increase the insulation effort and affect the cooling performance of the motor and the power electronics. As a reference, a five-level inverter with 1200 V switches is considered, which results in a DC link voltage of 2640 V has been selected based on a system-level sensitivity analysis including the electric component weights and efficiency effects.

To complete the propulsion system architecture pre-design, a basic safety assessment is conducted. The results show that the expected failure rates for the electric component do not match the level, which is expected to be required by a long-range aircraft with ETOPS (Extended-range Twin-engine Operation(al) Performance Standards) rating. To reduce the failure rate for the loss of the turbo-electric drive train, the DC transmission lines from the engine generator to the FF drive motor are split into multiple parallel transmission lines, which are considered to operate independently. Like this, the loss of the entire transmission system is very unlikely, however, increasing the number of components reduces the system availability as the chance of a component failure increases. In a trade-off, the number of parallel transmission lanes is selected to four. This design allows to reduce the failure rate of a partial loss of power of less than 75% to less than 1×10^{-9} 1/h. Figure 26 shows the resulting propulsion system architecture. The two podded engines on the left and the turbo-electric drive train with four transmission lanes for each generator are depicted on the right.

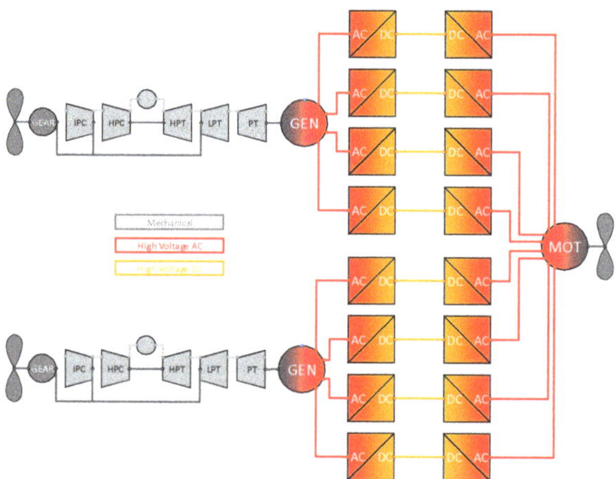

Figure 26. Overall propulsion system architecture (Source [81]).

5.2. *Electric Machinery Pre-Design*

The design of electric machines is driven by the maximum rotational speed and the level and duration of the required maximum torque. The maximum rotational speed and the maximum torque levels do not necessarily appear at the same time. Hence, all major mission points must be considered to size the electric machines and their power electronic devices correctly. The FF performance requirements stipulated by the CENTRELINE initial PFC aircraft target design (cf. Section 1.4) include maximum rotational speed to appear at flow path sizing conditions at Top of Climb (ToC), while maximum torque is occurring during take-off as outlined in Section 4.1.1. As a result, the motor needs to be designed for a mechanical power level going clearly beyond the maximum rated power during operation. Consequently, the fuselage drive motor resulting from the requirements according to initial PFC aircraft target design (Figure 4) would be able to deliver 11.5 MW but is operated at a maximum power of 8 MW during take-off. In constrast, the relevant rotational speeds for the TEPT generators allow for a more straight-forward sizing in take-off.

The basic machinery type selected for the TEPT electric motor and generator applications in CENTRELINE is a permanent magnet radial flux electric machine. While the FF drive motor is an out-runner machine to maximize power density, the gas turbine integrated generators are in-runner machines in order to better to shield the magnets on the rotor from the high ambient temperatures.

For the electric machine computation, the commercial software package SPEED®, PC-BCD 12.04 version 12.04.010 by CD-adapco® (Melville, New York, United States) was used. The basic scheme followed for the electric machine design and sizing is visualised in Figure 27. As can be seen in the figure, the system sizing and performance requirements, including the required torque and rotational speeds during key mission phases, are used to scale an ab-initio machine with approximate radius, length, current, slot and yoke width, etc. under the constrains of physical limits, e.g., the maximum current density in copper under the certain cooling conditions. The obtained geometric and electromagnetic characteristics of the machine are translated to SPEED®-specific input settings in order to compute the machine performance properties. If the prescribed machine design is identified as valid based on the performance calculated by SPEED®, the underlying machine model is used as an initial point to optimize the exact geometry of the slots, teeth, magnet volume in an iterative way. Machine mass is calculated from the 2D geometry with the 3D components such as the winding heads included based on the winding scheme. The best machine configuration in terms of power to weight with a minimum required efficiency is saved as a SPEED® design.

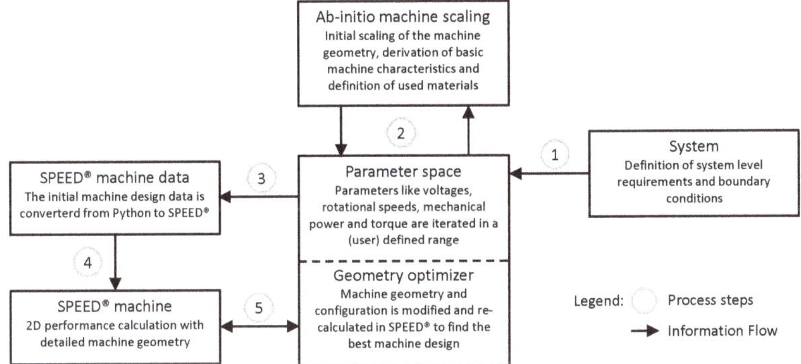

Figure 27. Basic electric machine design scheme.

In CENTRELINE, this machine scaling and optimization procedure was implemented to run automatically for every unique set of input requirements [81]. In order to support the aircraft-level systems design activities in a parametric manner, the procedure was employed for the FF electric drive motor and the main power plant integrated generators using systematic variations of maximum rotational speed, torque, DC voltage level and electric frequency. Beside the obtained mass, geometric and efficiency properties, also the resulting external driving and cooling parameters required to operate the electric machinery, such as the current and voltage per phase provided by the inverter or the cooling power by the heat exchanger, were provided back to the system [81].

The design calculation for the inverters was done for two-level, three-level, and five-level topologies and individually for each machine design. The result of these design calculations were the masses of all involved parts. Connection and mounting devices were not considered in this pre-design phase since they depend strongly on the detailed mechanical design. It is highlighted that the mass obtained from this calculation reflects only a net mass of the 'active' parts of the inverter. By incorporating all other parts in order to obtain the inverter gross mass, the net mass needs to by multiplied by a factor between 1.5 to 2. The mass values for power switches, their drivers and the DC-link capacitors were calculated by the help of an in-house database with collected state of the art devices. If a certain device was not explicitly found in the database, required key properties were derived by interpolation and curve fitting approached from similar devices of the same voltage class.

The cooling device scales with the power electronic base plate area. The housing was scaled with the volume of all above mentioned parts. After the parts were designed, the conducting and the switching losses and hence the efficiency was evaluated. This was realized by a numeric electrical simulation with a PLECS® (PLECS® is a registered trademark of Plexim GmbH, Zurich, Switzerland) model [81]. Table 10 summarizes the sizing results for the detailed settings of the initial PFC aircraft design provided by the initial PFC target design in the project (cf. [43]). For the aircraft-level investigations, the evaluation of electric component sizing and performance was incorporated using response surface models based on the produced parametric data for each component (see also Section 6.1).

Table 10. Component mass breakdown for the turbo-electric power train (Sources: [82,83]).

Component	Unit	Engine Generator	Fuselage Fan Motor
Converter	kg	152	182
Converter cooling system	kg	71	80
AC cables (5 m)	kg	30	100
Electric machine	kg	434	905
Electric machine cooling system	kg	165	280
Total mass	kg	852	1475

Not included in Table 10 is the mass of the DC cabling required for the TEPT, which was assessed to be 690 kg. A parametric analysis of the DC cable sizing is provided in [81]. The TEPT component efficiencies for the performance requirements at the beginning of cruise are listed in Table 11.

Table 11. Component efficiencies of the turbo-electric power train in cruise (Source: [84]).

Component	Unit	Value
Generator	%	96.5
Rectifier	%	98.5
Inverter	%	99.0
Motor	%	96.5
Total system efficiency	%	90.8

5.3. *Component Cooling and System Waste Heat Management*

One key issue of electric propulsion systems is the thermal management. In aircraft, the only way to reject waste heat is to transfer the heat to the ambient air. The advantage of classic gas turbine engines is that most of the waste heat is generated at a high temperature level and ejected via the hot exhaust gas flow. Electric machines and power electronics operate at much lower operating temperatures, which increases the effort to cool the components, as the available temperature difference to the ambient air is much smaller. Moreover, the losses are generated in solid material and not in a gas flow that is intrinsically released to ambient air. This implies that the heat must be transferred to the ambient air via multiple transfer mechanisms, which include thermal resistances and reduce the cooling performance compared to gas turbine engines. Moreover, designing an electric machine for maximum power density means that the amount of material is minimized. This also minimises the volume losses are generated within, and, the areas available for the external heat rejection. Hence, an electric machine with high power density requires a high cooling effort and smart cooling methods.

The electric machines designed for this project operate at high current density so that direct air cooling is not possible. Hence, a fluid cooling system is required to extract the waste heat from the electric machines. The generator, which is integrated downstream of the engine LPT requires a complex cooling system to sustain the high ambient temperatures. The casing is cooled with an internal jacket cooling and the permanent magnets on the rotor are cooled with oil running on the inner shaft wall. The fuselage drive motor requires a direct cooling of the stator windings.

The significant amount of electric machinery waste heat contained in the cooling fluid needs to be rejected to the ambient air at a minimised drag penalty. Beyond the classic approach of heat rejection via ducted radiators, the utilization of existing aircraft surfaces as an interface to the external heat sink was explored in CENTRELINE. Therefore, the on-board fuel was used to transfer the heat from the fluid cooling circuits of the electric machines to the lower wing surface, in order to be cooled down by the forced convection on the outer wing surface during as the aircraft is in motion. A preliminary estimation of fuel temperatures at the end of diversion, i.e., the most critical conditions during the design mission, showed that for the present case the fuel temperature can be kept well below the ignition temperature limit even with low fuel quantities remaining in the tank [84]. This result is in good agreement with independent studies of fuel sink systems and electric waste heat rejection through existing aircraft surfaces published in the very recent past [85,86].

The power electronic devices are designed for a low current density on the switches, which enables air cooling and increases efficiency. This comes at the price of many semiconductor switches and a high component price. Moreover, the additional drag from cooling fins must be considered and traded carefully against oil cooling of the power electronic devices. The converters of the engine generator are located at the inner wall of the engine nacelle to enable air cooling, like the cooling of the gearbox oil. The advantage of this location is that an air flow is available as soon as the engine (and the generator) are rotating. The drawback is the high sensitivity of the engine specific fuel consumption on pressure losses in the bypass section. The power electronics for the FF motor may be oil cooled and the heat may be transferred to fuel circulation system built-in for the aircraft's trim tank capability. For anti-icing purposes, incurring electric waste heat may be directly rejected though the aircraft external surface areas near the leading edge of the FF cowling and the fuselage section in front of the BLI propulsive device, whenever required.

5.4. Main Power Plant System Design Integration and Performance

Similar to the R2035 power plants, the basic architecture of the PFC main engines is based on a two-spool, boosted, GTF layout featuring a short duct separate flow nacelle. However, aside from the residual net thrust requirements (cf. Section 2.1), the main power plant systems of the turbo-electric PFC aircraft also need to deliver the power required for the FF electric drive. These power offtakes exceed the level of typically considered customer offtakes, even if all electric subsystems are considered. Together with the overall design power offtake level comes the operational requirement of relative independence between the thrust produced by the main propulsion system nozzles and the shaft power extracted by the generator. More specifically, an optimal operation of the PFC aircraft (see also Section 7.3) means the ratio of generator power offtakes to main engine thrust to be relatively low during take-off, however particularly high in cruise.

5.4.1. Cycle Design and Performance Modeling

Propulsion system design and performance synthesis in the project was conducted using the APSS framework (cf. Sections 1.4 and 4.1.1). Flow path sizing was performed for ToC conditions at maximum climb (MCL) rating. For the cycle design definition, a set of typical heuristics according to [71] and [70] along with appropriate iteration strategies was adopted along with appropriate iteration strategies. Specifically, fan tip speed was mapped as a function of the design outer FPR based on data given in [71]. Turbo component design efficiencies were modeled with functional sensitivity to component size and Reynolds number effects, the aerodynamic loading conditions and, if applicable, cooling air insertion [42]. For off-design performance, component maps from the GasTurb™ map collection [87] were employed. Maximum rating settings including maximum allowable temperature levels and component mechanical speeds for take-off and climb were correlated to the turbine cooling air demand based on the multi-point sizing strategy as described by Bijewitz et al. [42]. Lower part power operation was facilitated by handling bleed laws as proposed by Seitz [70]. For the mapping of the PFC-specific power offtakes in design and off-design, a set of relative descriptors was used [42] which—together with the overall model parameterization—allow for broad ranges of power offtake in a Cartesian input space. Specific details for the R2035 and PFC main engine weight estimations used for the aircraft-level design and performance evaluation presented in Section 7 are provided in [69].

5.4.2. Engine Architectural Integration Options

The TEPT impact on the PFC main power plants under nominal and abnormal operating conditions needs to be limited by design as much as possible. Therefore, two main propulsion system design options have been investigated within the project combining the considerations of cycle performance, mechanical design and integration aspects, as well as operability in all flight modes.

Figure 28 shows the flow path layout of the two main propulsion system design options, which can be distinguished by the source of the power extraction from a free power turbine (FPT) shaft or from the LPT shaft [88]. Both options have the generator integrated behind the final turbine stage in the hub section. This position, firstly, could satisfy sufficient circumferential speeds and space for the generator design with direct drive; secondly, it is the best position with respect to the weight balance of the propulsion system; thirdly, it enables a disassembly on wing [89].

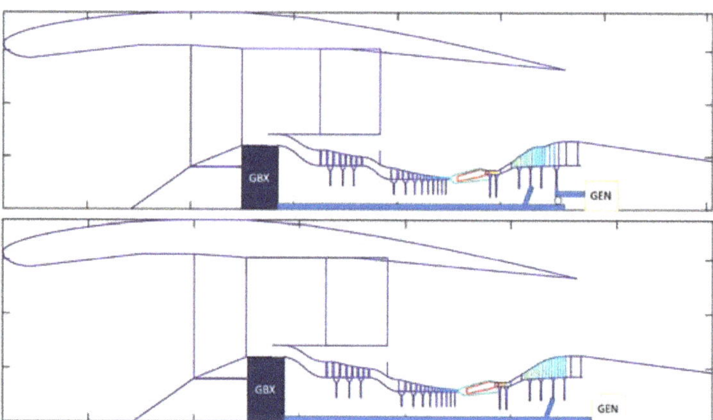

Figure 28. Flow path layout of the main power plant design options, free power turbine power offtake (**top**) and low-pressure turbine power offtake (**bottom**).

While the LPT shaft power extraction was considered as a default case since the intitial PFC aircraft target design, the FPT-based architecture was investigated as an alternative in order to explore its potential benefits from the mechanical decoupling of the generator from the main engine core and propulsive device and the intrinsic possibility to select a suitable design rotational speed for the FPT and generator.

The design point performances of both design options are nearly the same: A slightly better TSFC for the configuration with direct LPT power extraction at normal operating points can be attributed to a better performance of the LPT compared to the variable geometry FPT. An efficiency penalty of 0.5% at design point and a efficiency correlation derived from publicly available resources [90,91] were included for the effects incurring from the FPT variability. The power output of the FPT at a given flight condition is basically controlled by two speeds, the fan speed and its own speed. The former dominates the power output potential at a certain operating condition whilst the latter has a marginal influence mainly given by its effect in the FPT efficiency variation. In order to increase the flexibility of FPT power extraction, a variable turbine nozzle has been considered for the special thrust/power extraction requirements as mentioned above. At a constant main engine thrust setting, a variation of the FPT nozzle area allows to shift powers between the FPT and the LPT without the necessity of the varying the FPT rotational speed. For normal operations, such as mid cruise and take-off, an FPT nozzle variation of ±30% could satisfy the need for flexibility. However, operating points, featuring low main engine thrust settings at simultaneously high generator power offtake requirement such as end of cruise pose signification challenges to the FPT design: Firstly, the FPT efficiency penalty increases considerably when the turbine nozzle area is reduced below 70% of the design area [88]. Second, with the electric power demand remained constant, the mechanical loading of the FPT disc increases despite the de-throttling of the core engine.

Beside the performance implications, the FPT design would add about 1.5% more weight to the main propulsion system compared to the LPT design. The additional weight is mainly caused by the variable geometry of the power turbine. The weight estimation is based on the layout given in Figure 28, which has no casing between the LPT and FPT. If a casing must be included, a 3.5% additional weight is expected. A detailed discussion of the design and operational implications including impacts on turbo component stability margin at relevant operating conditions for both main engine design options is provided by [88].

5.4.3. Architectural Selection Based on Abnormal Modes

The final selection of the PFC main propulsion system architecture is dominated by the considerations of abnormal flight conditions. Therefore, cases without nominal TEPT power offtake from the PFC main engine need to be considered. Key abnormal conditions include the shortfall of FF power absorption, as well as a case in which the generator has to be stopped from rotation due to an internal short circuit [89].

For the FPT design variant, both problems are critical, because the FPT would accelerate and exceed safe speed limits as it is coupled aerodynamically to the rest of the core engine. To prevent the FPT from overspeeding, in both cases a number of counter measures are conceivable, including using the TEPT generators for the subsystems power supply, rotating the turbine inlet vanes to a position in order the power delivered by the free power turbine [92], and, bypassing a fraction of the core mass flow around the FPT through bleed valves located between the LPT and the FPT [89].

For the LPT direct drive design, the former case is not considered to be a major problem as the main fan still consumes the majority of the mechanical power generated from the LPT. The latter condition, however, would require a mechanism to disconnect the generator from the LP spool, such as predetermined breaking point in the shaft connecting the LPT and the generator rotor. In case the generator could continue rotating despite a short circuit in one of its winding systems, such as disconnection mechanism would not be required. Corresponding technologies that are currently being investigated are considered to be available for an EIS in 2035 [89].

In conclusion, both architectural options are conceivable; however, under the occurrence of TEPT failure cases, the LPT direct drive design is considered significantly less complex than the FPT variant. Together with its slight weight and performance advantages discussed in the previous section, the LPT direct drive design was selected as the final architecture for the CENTRELINE PFC aircraft.

6. Multi-Disciplinary Aircraft-Level Design Synthesis

This section describes the overall aircraft-level design and multi-disciplinary knowledge integration performed in the project. Therefore, the developed collaborative framework and data integration solution are presented, before key aspects of the PFC and R2035 aircraft design and performance synthesis methodology is introduced. To round off, the partial system safety assessment conducted as part of the aircraft design integration activities is discussed.

6.1. Collaborative Framework and System-Level Knowledge Integration Approach

The foundation of the aircraft design and performance studies presented in this paper was formed by the multi-disciplinary, multi-partner collaborative research conducted within CENTRELINE. This section provides a brief overview of the interfacing and data handling processes established in order to facilitate efficient collaboration between the project partners as well as the consistent propagation of the multi-level, multi-disciplinary design and performance information to the level aircraft sizing and operational assessment.

The knowledge on PFC aircraft design was incrementally refined during the CENTRELINE project involving multiple levels of modeling fidelity. Incremental results from the various detailed design and analysis tasks need to be evaluated at aircraft-level in a continuous manner in order to provide adequate design guidance from an overall system optimality perspective. Therefore, a rapid-responding aircraft-level sizing and optimization setup was required, featuring robust parametric sensitivity for key design parameters from all relevant system components and disciplines; flexibility and extensibility in the parametric interfacing between the various design aspects; and the capability of zooming by a quick inter-changeability of disciplinary models.

Given the nature of this design problem, a fully-coupled multi-disciplinary, multi-partner design optimization process featuring the direct execution of the specialized disciplinary models seems widely impractical. Instead, decomposition of the overall system design and optimization problem was performed. The individual disciplinary and component design optimization subtasks were decoupled from the aircraft level, by imposing local objectives and constraints directly derived from the aircraft level design and optimization task. Such an approach, commonly known as the "Bi-Level Integrated System Synthesis (BLISS) technique [93,94], allows the handling of complex design optimization problems based on a relatively small number of top-level variables effectively shared by the various subtasks. For the interfacing between the system-level design optimization and the local design optimization activities, a problem-oriented set of direct exchange of parametric data and tailored surrogate modeling techniques was adopted. Due to their extremely fast response times, surrogate models enable rapid design space exploration and a quick gain of system behavioral knowledge. Surrogate model application intrinsically enforces quality assurance measures such as expert checks prior to the system level integration of subsystem analysis result data (cf. e.g., [70]). Accordingly, the design and performance characteristics of the main power plants systems as well as the BLI FF power plant are integrated using feedforward neural networks (FNN), trained and validated by Latin hypercube sampled (LHS) [95] data as described in [70]. PFC aerodynamic performance properties, structural design characteristics as well as the design and efficiency properties of the turbo-electric power train components were integrated based on custom-developed non-linear regression models. The detailed structural design results in terms of component weights and geometric properties were transferred through customized parametric data fittings. The turbo-electric powertrain design and performance information was incorporated through multi-dimensional data interpolation.

In order to facilitate an efficient, collaborative, distributed, multidisciplinary analysis and design process, a highly effective set of tools, infrastructure and processes is necessary. Therefore, basic requirements, common conventions, processes and infrastructure had been derived already at the beginning of the project, based on the organizational structure and the type of planned activities. The areas of data security, semantically correct data integration, consistency and traceability of results had been identified as of central importance. In result, a secure git server was set up for data exchange between the project partners. Versioning and branching guidelines were developed at the beginning of the project and refined throughout the project in order to facilitate convenient data traceability and consistency. A consistent set of suitable formats for data exchange was defined, in order to support an efficient exchange of data. The developed infrastructure, policies and procedures for the efficient and secure data handling in CENTRELINE is reported by Shamiyeh [96]. The multi-partner, multi-disciplinary workflow is discussed in detail by Troeltsch at al. [97]. A detailed overview and discussion of the system-level knowledge integration for the overall aircraft design including specifications of the individual data and surrogate model parametric characteristics is provided by Habermann et al. [69].

6.2. Aircraft Sizing and Performance Modeling

The overall sizing process for the R2035 and PFC aircraft families in CENTRELINE was implemented using a customized version of the commercial aircraft preliminary design (APD) software within the modeling environment Pacelab Suite [98]. The APD software offers a set of handbook methods for aircraft conceptual design mostly based on Torenbeek [99]. As a starting point for the present activities, a customized version of the framework featuring comprehensively supplemented methodology based on Bauhaus Luftfahrt (BHL) in-house developed semi-empirical and analytical methods was employed (cf. [70,100–102]). During the CENTRELINE project, the baseline methods were systematically replaced by the surrogate models and datasets produced from the in-depth analyses of the PFC-specific design and performance aspects as described in the previous sections.

The R2035 and PFC aircraft sizing was performed using typical family sizing conditions, i.e., with the three members of the family (shrink, baseline, and stretch) sharing a majority of components. Each of the shared components is sized for the most critical requirement with in the aircraft family. The stretch and shrink members of both families of the aircraft are derived from the baseline variant by adding or removing common barrel sections from the baseline fuselage, respectively. The commonly used empennage accords to the shortest family member (shrink version). The wing, landing gear, main propulsion system, pylons, and other aircraft subsystems are sized for the largest family member (stretch). The geometry of the FF propulsion system is aerodynamically optimized for the baseline aircraft. The sizing of the FF PT is driven by the take-off power requirements of the stretch version. A compact overview of the specific component sizing laws followed for the R2035 and PFC aircraft families is provided in Table 12.

The core of the integrated PFC aircraft performance modeling in CENTRELINE is the point performance evaluation scheme depicted in Figure 29. The total aircraft thrust demand FN_{req} is calculated from the aerodynamic performance of the aircraft at a specific point of the flight envelope, for which the operational parameters altitude, ISA temperature deviation and flight Mach number are prescribed. The FPR schedule for the FF (see below) determines the instantaneous FPR for the given operating condition. All parameters serve as an input to the bare PFC aerodynamic model presented in [69], which predicts the NPF and the required ideal FF shaft power as introduced in Section 2.1. The NPF is subtracted from the total net thrust FN_{req} in order to calculate the thrust requirement for the main engines, $FN_{MainPPS}$ discussed in Section 5.4. The FF performance model (cf. Section 4.1) delivers the actual shaft power required by the FF, P_{Shaft} and the corresponding shaft speed N_{Shaft} for the operating point. Subsequently, the power requested from the main engine generators is calculated in the FF power train model. With the generator power offtake taken into account, the main power plant model returns the fuel flow W_{fuel} for the operating point.

The FF is operated at its design FPR of 1.4 during cruise. In take-off, FPR is limited to 1.2 as a best and balanced trade-off between overall propulsion system performance and the component weights of the turbo-electric FF power train. During climb, FF performance is controlled by a linearized schedule of FPR versus flight altitude. While flow incidence angles relative to the fuselage longitudinal axis are assumed to be small during high-speed operation in clean aerodynamic configuration. Bare PFC performance effects during take-off rotation and initial climb with high angles of attack are taken into account based on a dedicated 3D CFD simulation at low-altitude and low-speed featuring an incidence angle 10°.

The aircraft design mission simulation includes a step cruise profile with three steps targeting a maximum specific air range for each cruise point. Standard climb is conducted using a 250/300 KCAS schedule. Design ToC is at M0.82, FL350, ISA +10K. For the determination of design loaded fuel, international reserves with 200 nmi diversion, 30 min hold, 5% final reserves are considered.

Table 12. Overview of specific component sizing laws for the R2035 and PFC aircraft families.

Component	Characterization of Key Sizing Conditions
Wing	• Wing span: Variable, with fixed aspect ratio AR = 12.0. • Wing loading: Scaled to meet required landing field length. • High-lift: $C_{L,max}$ scaled to meet approach speed requirements.
Empennage	• Sizing based on tail volume coefficients. • PFC T-tail: 5% reduced relative to R2035 with conventional tail. • Potential stability and control enhancements due the FF cowling neglected, in the first instance (cf. [69]).
Fuselage	• Sizing: Design payload in 2-class, nine-abreast seating arrangement in economy class (6.09 m diameter). • Family design via common barrel sections. Family members share fuselage nose and tail (incl. FF in PFC case).
Landing gear	• Main gear extended length determination for required tail strike angle. • Constant load distribution between nose and main gear.
Main power plants	• Flow path sizing in order of meeting take-off field length and time-to-climb to initial cruise altitude requirements. • Identical cycle design settings at flow path sizing point (MCL@ToC).
Fuselage fan *	• Flow path sizing for design cruise conditions. • FPR during take-off limited by trade-off between FF performance and TEPT weight.
Turbo-electric power train *	• FF electric drive motor sizing for max. required torque (take-off) and speed (top-of-climb). • Power electronics and electric generator sizing for max. power demand (take-off).

* Applicable to PFC aircraft family only.

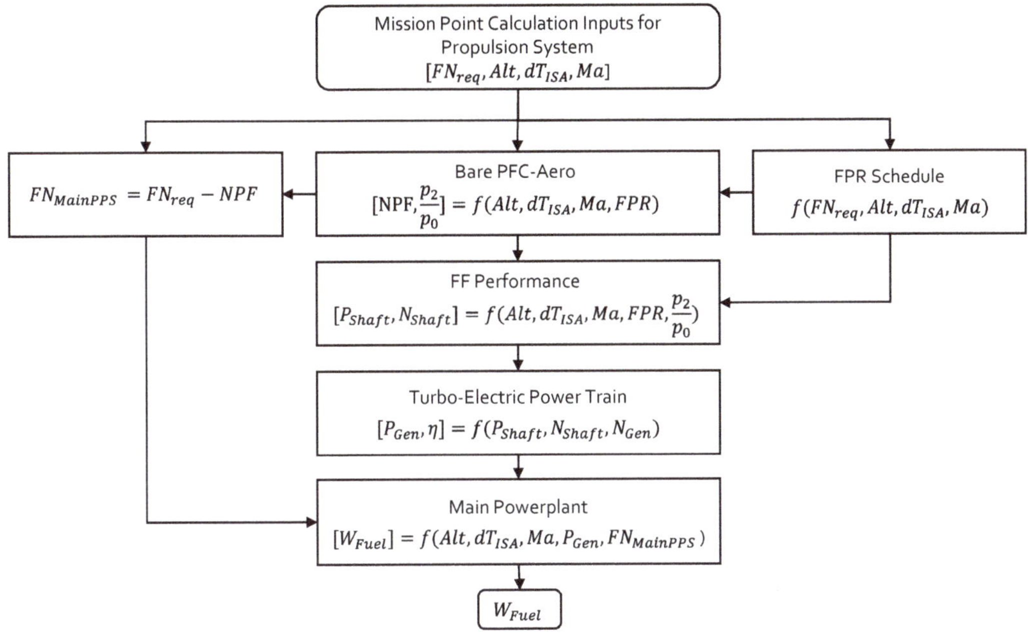

Figure 29. Logic for PFC point performance evaluation within aircraft sizing framework (adapted from [97]).

6.3. Preliminary System Safety Analysis

As part of the aircraft-level design integration effort, a partial system safety assessment was performed. The basis for the safety assessment process (SAP) activities was formed by the CS-25 [76], SAE ARP4761 [103], and SAE ARP4754 [104] standards. The execution of the complete SAP as defined in the ARP4761 was out of scope for the CENTRELINE project. However, the large number of conventional components in the overall technology set, allowed the focus of the SAP activities to be placed specifically on the parts directly connected to the technological innovations of the project. In specific, the actions carried out were a partial aircraft and system level function hazard assessment (FHA), a partial aircraft level failure tree analysis (FTA), and partial a preliminary system safety assessment (PSSA) together with its system level FTA.

The first step of the aircraft level FHA was the selection and definition of the relevant aircraft top level functions. From the examples given by SAE ARP4754, only the "Engine Control" was selected to be analyzed in the SAP activities. Subsequently, the aircraft level FHA process was conducted with the steps defined by SAE ARP4754:

1. Identification of related failure condition(s).
2. Identification of the effects of the failure condition(s).
3. Classification of each failure condition based on the identified effects (i.e., catastrophic, hazardous/severe–major, major, minor, or no safety effect) and assignment of the necessary safety objectives, as defined in AC 25.1309-1A and AMJ 25.1309 extended to include the no safety effect classification.
4. Identification of the required system development assurance level.
5. A statement outlining what was considered and which assumptions were made when evaluating each failure condition (e.g., adverse operational or environmental conditions and phase of flight).

The steps in this process were executed and key information connected to the identified failure conditions was collected. Therefore, the template provided by Kritzinger [105] was used. From the resulting failure conditions, the hazardous and catastrophic rated conditions as listed in Table 13 were analyzed further for their most critical flight phase.

Table 13. Identified hazardous and catastrophic aircraft level failure conditions specific to the CENTRELINE PFC aircraft.

ID	Failure Condition	Phase	Severity
2.1.2.a	Unannounced partial loss of ability to provide thrust	Take-off	Hazardous
2.1.3.a	Complete loss of ability to provide thrust	Take-off	Catastrophic
2.1.4.a	Unforeseen vibration frequencies and/or amplitudes	Take-off	Hazardous
2.1.5.a	Ignition of fire in the electric components of the FF	Take-off	Hazardous
2.1.6.a	Ingestion of foreign object or ice in the FF	Take-off	Hazardous
2.2.1.a	Announced loss of command authority for thrust control	Take-off	Hazardous
2.2.2.a	Unannounced loss of command authority for thrust control	Take-off	Catastrophic
2.2.3.a	Announced erroneous thrust control	Take-off	Hazardous
2.2.4.a	Unannounced erroneous thrust control	Take-off	Catastrophic

Each of these failure conditions was used as a top event in a specific FTA diagram. The top event probability was defined by the allowable quantitative probability in average probability per flight hour of failure occurrence, i.e., 10^{-9} for the as catastrophic and 10^{-7} for the as hazardous classified failure conditions (cf. CS-25 [76]). The accumulated probability of the events leading to the top event must not exceed the probability of the top event. The probabilities of an AND gate, meaning that both events have to occur for the higher event to occur, are multiplied. The probabilities of an OR gate, meaning that both events have to occur for the higher event to occur, are added. The probabilities of the events leading to the top event were assigned by either known data or experience. As an example, the FTA for the failure condition "Complete loss of ability to provide thrust" is shown in Figure 30.

The bottom events of these aircraft level FTAs were then distributed amongst the partners responsible for the respective systems and the process was repeated at system level, thereby executing the PSSA as a part of a partial system level safety assessment. The system level PSSA was successfully executed for all PFC specific component failure modes and target levels for maximum allowable failure probabilities were determined along the associated functional chains. In no occasion, concerns were raised by the expert partners regarding the criticality of achieving the required levels of failure probabilities for the involved basic events. The component specific safety goal obtained from the SAP provided valuable guidelines in the design decision-making processes in CENTRELINE.

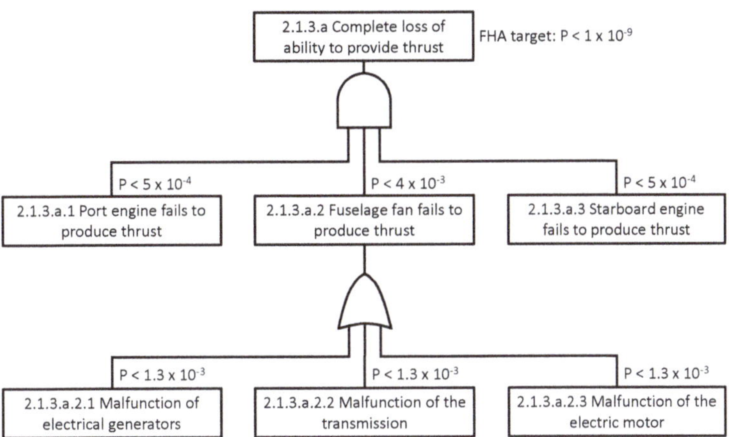

Figure 30. Example FTA for complete loss of ability to provide thrust for the CENTRELINE PFC aircraft.

7. Aircraft-Level Evaluation Results

An overview of key results and insights gained from the PFC aircraft design and technology evaluation is provided in the following. This include a brief discussion of fundamental aspects for optimum fuselage BLI aircraft design. A summary of the current maturity of the PFC technology together with an outlook towards further technology development steps is presented. Eventually, the final design and performance benchmarking for the CENTRELINE PFC aircraft is presented and discussed together with the results from the multi-disciplinary technology evaluation including noise, emissions and cost assessments.

7.1. Fundamentals of Optimum Fuselage BLI Aircraft Design

Before the integrated CENTRELINE PFC aircraft design results will be presented and benchmarked against the R2035 reference aircraft family, the discussion of a few fundamental aspects on fuselage BLI aircraft design optimality for maximum power savings is warranted. Therefore, a multitude of refined and optimized 2D aero-shapings of the bare PFC configuration as obtained from the EU-funded DisPURSAL and CENTRELINE projects have been analyzed with regard to the $f_{\eta,PFC,bare}$ metric by Seitz et al. [50]. The aerodynamic data basis of the analysis was formed by 2D-axisymmetric RANS CFD simulations conducted for typical cruise conditions at zero angle of attack. A common heuristic for $f_{\eta,PFC,bare}$ as a function of $P_{disc,FF}$ was derived from the best aero-shaping cases of both projects.

A more generalised version of the heuristic is presented in the following. Therefore, the FF isentropic power values for all given bare PFC configuration cases were non-dimensionalized by reference fuselage drag powers, i.e., the products of reference fuselage drag values $D_{Ref,fus}$ and given flight velocities V_0, yielding the non-dimensional FF isentropic powers $P_{disc,FF,non\text{-}dim}$

$$P_{disc,FF,non-dim} = \frac{P_{disc,FF}}{D_{Ref,fus} \cdot V_0} \tag{7}$$

The correspondingly upgraded BLI efficiency factor analysis chart featuring a non-dimensionalized abscissa is presented in Figure 31.

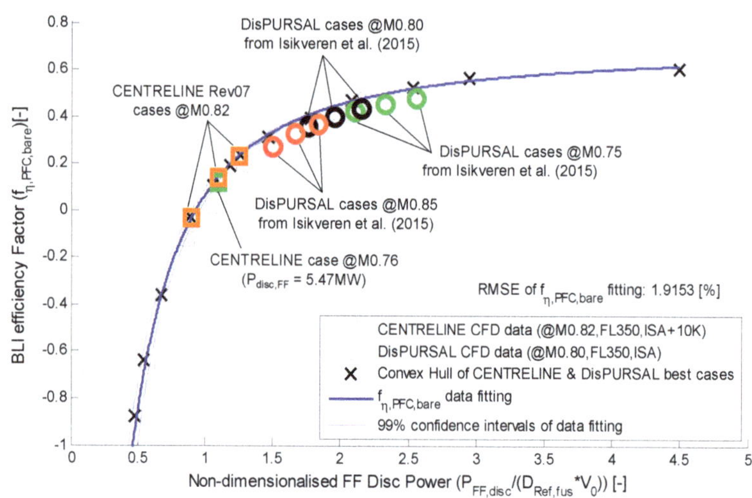

Figure 31. Non-dimensional analysis of BLI efficiency factor for CENTRELINE and DisPURSAL aero-shapings (adapted from [50]).

With the original $f_{\eta,PFC,bare}$ versus $P_{disc,FF}$ chart and its validation discussed in detail by Seitz et al. [50], inspection of Figure 31 reveals that the basic shape of the Pareto front formed by the best bare PFC 2D aero-shaping cases is retained. The generalised form of the data-fitting curve for the best designs originally introduced by Seitz et al. [50] yields

$$f_{\eta,PFC,bare} = 0.6863 - 0.7321 \cdot (P_{disc,FF,non-dim}[MW] + 0.1177)^{-1.5334} \tag{8}$$

Beyond the bare PFC designs contained in the original study, in Figure 31 above, a number of additional cases is added. These involve the final PFC 2D aero-shapings of CENTRELINE (Rev07, cf. Table 2) as well as cases for a range of different cruise Mach numbers. Beside a CENTRELINE design featuring a $P_{disc,FF}$ of 5.47 MW that was optimized for a reduced cruise Mach number of 0.76, an array of cases from a speed sensitivity performed as part of the DisPURSAL project has been incorporated from [25]. Despite the slight sub-optimality of the addition DisPURSAL cases when compared to the Pareto front of best designs, it is apparent that even for varying cruise speeds between approximately M0.75 and M0.85 the non-dimensional $f_{\eta,PFC,bare}$ heuristic is a good indication for 2D bare PFC performance. Accordingly, lower cruise Mach numbers lead to higher $P_{disc,FF,non\text{-}dim}$ and correspondingly increased $f_{\eta,PFC,bare}$ values. Higher cruise Mach number yield opposite trends along the heuristic curve. This observation made for the DisPURSAL speed sensitivities is basically confirmed by the M0.76 design case featuring almost identical $P_{FF,disc}$ as the lowest power case of the Rev07 designs for M0.82. The displacement to higher $f_{\eta,PFC,bare}$ and $P_{FF,disc,non\text{-}dim}$ directly along the heuristic curve is visible as a first order effect in Figure 31.

For steady level flight, the heuristic provided in Equation (8) can be directly used in order to calculate optimum power saving potentials for PFC aircraft based on the analytical formulation derived by Seitz et al. [50]

$$PSC = 1 - \left(\frac{D_{PFC,res}}{D_{Ref,tot}} + \frac{P_{disc,\ FF}}{V_0 \cdot D_{Ref,tot}} \cdot \left(\frac{\eta_{pd,eff}}{\eta_{PT,FF} \cdot \eta_{pol,FF}} - f_{\eta,PFC,bare} \right) \right) \tag{9}$$

where $D_{PFC,res}$ refers to the sum of drag components of the PFC aircraft apart from the bare PFC configuration and $D_{Ref,tot}$ represents the total drag of the non-BLI reference aircraft at the given operating conditions. The efficiency figure $\eta_{pd,eff}$ refers to the effective propulsive device efficiency for the conventional non-BLI power plants of the PFC and the reference aircraft, describing the ratio of net propulsive power ($FN \cdot V_0$) to the effective core engine excess power $P_{co,eff}$ as defined in Seitz et al. [50]. The aerodynamic efficiency of the fuselage and the transmission efficiency of the FF power train—i.e., the actual FF shaft power related to the power off-take from the main engine turbine—are reflected by $\eta_{pol,FF}$ and $\eta_{PT,FF}$, respectively. The evaluation of Equation (9) for the CENTRELINE aircraft application scenario based on the bare PFC efficiency heuristic from Equation (8) immediately allows for the identification of maximum power saving potentials and correspondingly optimum FF isentropic powers $P_{FF,disc}$ as shown in Figure 32.

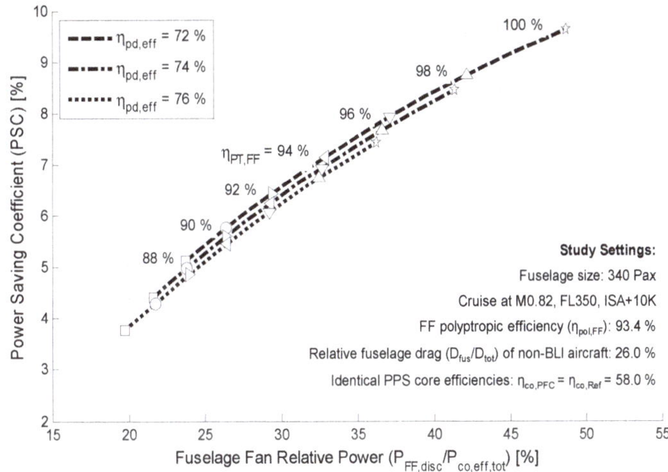

Figure 32. Identification of optimum power savings and corresponding FF cruise power settings for combinations of power train and non-BLI propulsive device efficiencies.

For the CENTRELINE specific settings and $\eta_{pd,eff}$ = 74% for the main engines, under ideal power transmission conditions, i.e., $\eta_{PT,FF}$ = 100%, a PSC of approximately 8.5% might be achieved with a corresponding optimum FF power share $P_{disc,FF}/P_{co,eff}$ of 41%. For a turbo-electric power transmission as given in Table 11, the obtainable PSC value is 5.3% as optimum $P_{disc,FF}/P_{co,eff}$ = 25%. A potential mechanical transmission scenario featuring $\eta_{PT,FF}$ = 98%, would yield PSC = 7.7% for an optimum $P_{disc,FF}/P_{co,eff}$ of 37%. It is obvious that larger transmission losses make a shift of power to the aft-fuselage BLI fan less attractive for maximum vehicular efficiency. The reduction of transmission losses is key to enabling high power saving potentials.

7.2. Technology Readiness Level Assessment

The maturation of the PFC technology to TRL 3—i.e., to perform the proof-of-concept for the promising technical approach to fuselage BLI propulsion integration—was one of the prime objectives for the research presented in this paper. In order to evaluate the maturity of the technical results developments, a TRL assessment was conducted for all critical components.

The first part of this section provides a compact overview of key technology component of the CENTRELINE PFC aircraft design and their TRLs at the end of the project. The overview is based on dedicated TRL assessments performed as part of the final deliverables of the individual tasks in CENTRELINE. Table 14 below lists the condensed assessment of the current TRL including component-specific justifications of the claimed TRLs.

Table 14. Assessment of the current TRL of the PFC critical technologies.

Aspect	TRL	Justification
Overall vehicle aerodynamic design for fuselage BLI & wake-filling	3–4	The CENTRELINE PFC aircraft configuration was investigated in a laboratory environment (low-speed wind tunnel) at relevant flow incidence angles [48]. The low-speed scale-model experimental results were extrapolated to full speed and scale based on extensive CFD numerical analyses [58].
FF aerodynamic design and performance	3–4	The CENTELINE FF was 3D numerically designed and tested in a laboratory environment (low-speed fan rig) at relevant operating conditions [106].
Aero-structural integration of FF propulsive device	3	Numerical simulation of CS-25 load cases were executed, The primary structures were sufficiently conservative designed and analyzed with FEM using the certification relevant load cases [80]. A partial SAP was conducted and a sufficiently conservative design was derived.
Turbo-electric power transmission system	3	The architectural definition, component design the thermal management specifications for the overall transmission were performed system under consideration of realistic system redundancy requirements and failure modes [81].
FF electric drive	3	Advanced numerical methods have been used for the pre-design of the electric machines in the drive train. The electro-magnetic and structural design of the electric machines was conducted in compliance with the geometric, structural and thermal boundary conditions in the aft-fuselage environment. The applied technology has been validated by experiments outside of the CENTRELINE project [83,84].
Electric generators	3	The electro-magnetic and structural design of the generator was conducted in compliance with the geometric, structural and thermal boundary conditions in the podded engine environment [82,89].
Underwing-podded power plants	3	A sufficiently conservative design based on a comprehensive investigation of the impact of significant power offtakes on cycle and engine operating behavior was conducted [88].
Overall aircraft design	3	A sufficiently conservative aircraft sizing with incorporation of detailed numerical aerodynamic and key electrical component data was executed. The baseline aircraft was integrated in an aircraft family designed by common industrial practices [69].

At the heart of the PFC conceptual proof, the aerodynamic experiments on the overall configuration and the BLI propulsor validated the core technology of the PFC. Therefore, TRL 3 is claimed. However, even though the current research is considered at TRL 3, some aspects of the wind-tunnel and fan rig experiments are approaching TRL 4. The numerical and analytical methods used to predict the PFC aerodynamic performance were validated against the experimental data. The design of the aero-structural components, all key elements of the propulsion system and the overall aircraft design synthesis was performed under realistic operating conditions, compliant with common industrial practice and by the adoption of sufficiently conservative design assumptions.

Further actions for increasing the TRL go from more elaborate numerical analysis and design optimization taking into account more detailed boundary conditions and physical effects, to experiments with increasing approximation of the relevant environmental conditions, leading up to sub and full-scale flight testing. An intermediate step in between experiments in commonly available wind tunnels and a scaled flight demonstrator would be a test campaign in a representative wind tunnel, able to recreate relevant flow conditions with sufficient accuracy. Tests in such facilities (e.g., the European Transonic Windtunnel) are very expensive. An alternative option, either in addition or as a substitute, could be the application of a 'flying wind tunnel'. The European funded demonstrator "Breakthrough Laminar Aircraft Demonstrator in Europe" (BLADE) testbed is an example of this concept [107]. BLADE employs an Airbus A340 [108] refitted with wing tips aiming at the assessment of natural laminar flow on a representative scale. Several configurational options are possible and are shown in Figure 33. The testbed could be fitted with wing-tip devices, representing a fuselage with and without a FF in different integration levels. A measurement of the occurring drag by the necessary compensation of yaw moment, or by a force scale at the junction of the test section, can provide information of the effectiveness of the FF under very representative environmental conditions. In addition, flow conditions could be monitored from the junction fairings, as is done for the BLADE tests. Alternative configurations are shown on the right side of Figure 33, featuring either an attachment on top of the fuselage or the replacement of an engine as is ofter done for engine program tests.

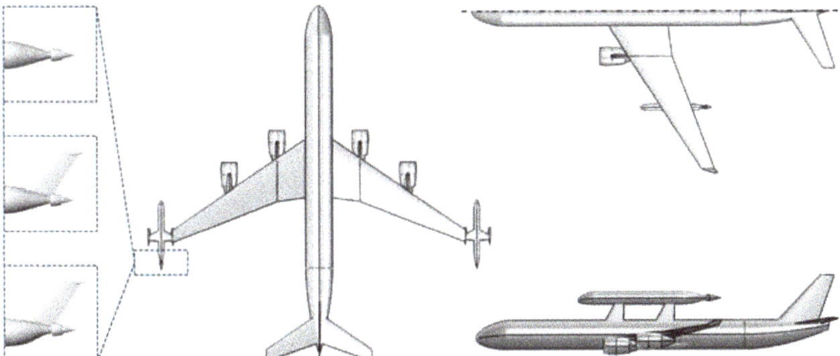

Figure 33. Basic schematic of PFC flying lab similar to the BLADE demonstrator aircraft.

Data acquired with such a testbed would also be applicable to different propulsion train architectures as, e.g., a turbo-mechanical drive train. Naturally, the last stage of the roadmap foresees the integration on a full-scale PFC configuration and consequently the progression of all technology maturities, sufficient to be included in the program of a production aircraft. More detailed discussion documentation of the technology maturation results within the CENTRELINE project as well as a detailed list of the necessary actions to advance the critical PFC technology aspects to target TRL6 is provided [109].

7.3. Design and Performance Benchmarking

The aircraft integrated sizing and performance benchmarking of the PFC technology versus the R2035 was performed as a family design exercise as described in Section 6.2. In the following, key characteristics of the final PFC baseline aircraft are presented and compared to the R2035 reference aircraft. Table 15 shows a comparison of important design properties for both aircraft.

Table 15. Key design characteristics of the PFC and R2035 baseline aircraft.

Category	Parameter	Unit	PFC	R2035	Delta (%)
General	PAX (2-class)	-	340	340	0.00
	Design range [a]	nmi	6500	6500	0.00
	MTOW/S_{ref}	kg/m^2	666	679	−1.91
	SLST/MTOW	-	0.32	0.33	−3.03
	MLW/MTOW	-	0.77	0.75	2.67
	OEW/MTOW	-	0.55	0.53	3.77
	Design block CO_2	t	196.6	206.2	−4.66
Wing	Reference area	m^2	355.3	346.4	2.57
	Aspect ratio	-	12.0	12.0	0.00
	Quarter chord sweep	°	29.7	29.7	0.00
	Span	m	65.8	65.0	1.23
Fuselage	Total length	m	67.0	66.7	0.43
	Diameter (centre section)	m	6.09	6.09	0.00
Main power plants	Design net thrust [b]	kN	54.6	59.8	−8.70
	Design specific thrust [b]	m/s	86.0	86.0	0.00
	OPR [b]	-	61.0	61.0	0.00
	T_4 [b]	K	1780	1780	0.00
	Relative HPT cooling air	%	20.0	20.0	0.00
	Fan tip diameter	m	3.21	3.36	−4.46
	Design bypass ratio [b]	-	14.5	16.4	−11.6
	SLS net thrust	kN	376.6	343.6	−8.76
	Mid-cruise net thrust [c]	kN	35.5	46.7	−24.0
	Mid-cruise TSFC [c]	g/kN/s	17.4	14.0	+24.3
	Mid-cruise total fuel flow [c]	kg/s	1.24	1.31	−5.52
Main landing gear	Extended length	m	5.70	4.70	21.3
Turbo electric power train	Total efficiency (take-off)	-	0.914	n/a	n/a
	Total efficiency (cruise)	-	0.919	n/a	n/a
	Total specific power [d]	kW/kg	2.10	n/a	n/a
	Electric motor sizing power (output)	MW	10.0	n/a	n/a
	Generator sizing power (output per unit)	MW	3.67	n/a	n/a
Fuselage fan	Tip diameter	m	2.26	n/a	n/a
	Rotor blade height	m	0.57	n/a	n/a
	Design isentropic disc power [b]	MW	5.00	n/a	n/a
	Design pressure ratio [b]	-	1.40	n/a	n/a
	Max. take-off shaft power [e]	MW	7.00	n/a	n/a
	Max. take-off pressure ratio [e]	-	1.20	n/a	n/a

[a] LRC (long range cruise), ISA, International allowances, 200 nmi diversion. [b] MCL@ToC: FL 350, Ma 0.82, ISA + 10 K. [c] FL370, M0.82, ISA + 10 K at 50% design loaded fuel. [d] based on the FF design isentropic shaft power. [e] SL, M0.23, ISA + 10 K.

As can be seen from Table 15, key aircraft properties of the sized turbo-electric PFC aircraft differ from the R2035: The additionally installed propulsion system components and the associated structural design cascade effects yield an increased operating empty weight (OEW) by 3.8%. Maximum wing loading is reduced by 1.9% relative to the R2035, as a result of its higher maximum landing weight (MLW) fraction of MTOW. The maximum sea level static thrust (SLST) loading is reduced by 3.0% due to the PFC-specific thrust lapse characteristics during take-off. The PFC design block CO_2 emissions are reduced by 4.7% relative to the R2035.

For a fair evaluation of the PFC technology, the wing aspect ratio was kept identical at 12.0 for both aircraft, the PFC and the R2035. It should however be noted, that with the given R2035 wing span of 65.0 m, the PFC wing span exceeds the limit posed by ICAO

Annex 14 Code E (52 m < b < 65 m) [110] by 0.8 m. Enforcing compliance with the Code E span limit would yield an increase of PFC design block CO_2 emissions by 0.6%.

The total transmission efficiency of the turbo-electric FF power train is 91.9% during cruise. The overall turbo-electric power train specific power is 2.1 kW/kg based on the power extraction from the main engines for the FF design isentropic shaft power of 5.0 MW which means a FF rotor tip diameter of 2.26 m. The FF shaft power is limited to 7.0 MW, allowing for stable operation of the FF at an FPR of 1.2 in take off but minimizing electric system oversizing, and thus excess weight, with regard to the transmission power needed during cruise. The shape-optimized PFC overall fuselage length exceeds the R2035 fuselage by only 0.3 m. The extended landing gear length of the PFC aircraft is increased by 1.0 m in order yield the same tail strike angle as the R2035. This additional landing gear length is smaller than half of the FF diameter due to the backwards shift of the aircraft center of gravity for the PFC aircraft and the associated longitudinal positions of the wing and the main landing gear.

The thermodynamic cycle settings for the R2035 power plants at flow path sizing point, namely design specific thrust, OPR, T_4 and relative HPT cooling are selected for minimum R2035 design block fuel (cf. [69]). To allow for a most straight-forward comparison, identical cycle design settings are adopted for the PFC main engines. As a result, both engine types feature the same maximum rated temperature levels for take-off and climb. While the R2035 power plants are sized for the time-to-climb requirement, which is typical for ultra-high BPR engines, the PFC main engine sizing is driven by the take-off field length (TOFL) requirement. This is a direct result of the FF maximum shaft power being limited to approximately 1.3 times the shaft power in cruise or top-of-climb conditions, which reduces the FF share of the total fan power in take-off considerably when compared to typical cruise. As a result, the ratio between take-off and cruise thrust is increased relative to the R2035 aircraft (cf. Table 15). With maximum rating settings identical to the R2035, the allowable TOFL becomes the active sizing constraint for the PFC main power. As a result, the PFC aircraft features a slightly short time-to-climb than the R2035, and, the PFC main engines operate at slightly lower power settings in cruise when compared to the R2035 engines.

Figure 34 presents the three view of the resultant PFC baseline aircraft. Dimensions are given in mm and rounded to the nearest 5. The airfoils used for the lifting surfaces are the same as the airfoils used for the R2035 reference aircraft.

Table 16 presents the mass breakdown of the baseline family members of the R2000, R2035, and PFC aircraft family. Additionally, changes of the PFC aircraft compared to the R2035 aircraft are shown. A synopsis of key weight terms for all nine members of the CENTRELINE R2000, R2035, and PFC aircraft families is provided by Habermann et al. [69].

Compared against the R2000, all structural component weights are considerably reduced for the PFC aircraft. Most of the weight reduction is due to the technology advancement assumptions projected for the year 2035. In a direct comparison with the R2035, the PFC fuselage structural weight is increased by 1.8%, due to the additional bending moment introduced by the BLI propulsion system. Endplate and lever arm effects inherent to the T-tail arrangement, offer sizing benefits for the PFC empennage relative to the conventional tail arrangement of the R2035. However, the weight of the VTP is increased over the R2035 due to the necessity to transmit the loads of the HTP through the fin. The HTP area is reduced, resulting in a 4.3% weight benefit. The increased MTOW and wing size yield a 1.1% rise in wing structural weight for the PFC. The increased landing gear length required for the PFC aircraft translates to an 11% increased weight of the undercarriage. The added weight of FF and TEPT constitute 17.62% of the total aircraft propulsion system weight. Payload, operational items, furnishing, and aircraft systems weights are kept unchanged between the R2035 and the PFC.

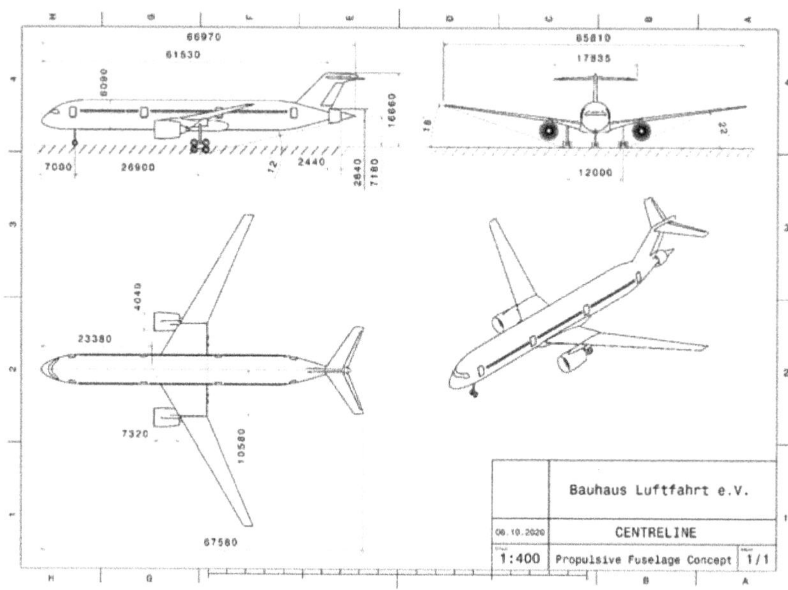

Figure 34. Three-view of the baseline member of the CENTRELINE PFC aircraft family, dimensions in mm.

Table 16. Mass breakdowns of the CENTRELINE R2000, R2035 and PFC baseline aircraft.

Mass Term (kg)	R2000	R2035	PFC	PFC vs. R2035 (%)	PFC vs. R2000 (%)
Structure Total	97,388	76,280	78,364	2.73	−19.5
Wing (incl. surface controls)	46,770	37,075	37,484	1.10	−19.9
Fuselage	27,854	22,286	22,683	1.78	−18.6
Horizontal stabilizer	3144	1835	1757	−4.25	−44.1
Vertical fin	2540	1936	2402	24.1	−5.43
Undercarriage	12,482	10,304	11,450	11.1	−8.27
Pylons	4598	2844	2588	−9.00	−43.7
Propulsion system	23,167	18,637	21,253	14.0	−8.26
Main power plants	23,167	18,637	17,509	−6.05	−24.4
Bare engines	16,217	12,232	11,742	−4.01	−27.6
Underwing nacelles	6950	6405	5767	−9.96	−17.0
Fuselage fan propulsive device	n/a	n/a	958	n/a	n/a
Fan module	n/a	n/a	427	n/a	n/a
Nacelle and integration structure	n/a	n/a	531	n/a	n/a
Turbo-electric power train	n/a	n/a	2787	n/a	n/a
Main generators	n/a	n/a	618	n/a	n/a
Fuselage fan drive motor	n/a	n/a	700	n/a	n/a
Cabling	n/a	n/a	586	n/a	n/a
Power electronics	n/a	n/a	434	n/a	n/a
Cooling system	n/a	n/a	449	n/a	n/a
Aircraft systems	10,151	11,803	11,802	−0.01	16.3
Operational items	13,213	10,248	8354	−18.5	−36.8
Furnishing	8154	8354	10,248	22.7	25.7
Operating empty weight	152,072	125,322	130,021	3.75	−14.0
Design payload	32,300	34,000	34,000	0.00	5.26
Design reserves fuel	10,986	10,570	10,416	−1.46	−5.19
Design landing weight	195,358	169,892	174,438	2.68	−10.7
Design trip fuel	97,829	65,192	62,158	−4.65	−36.5
Maximum take-off weight	293,187	235,084	236,596	0.64	−19.3
Design block fuel	97,951	65,453	62,403	−4.66	−36.3

Key characteristics of the PFC baseline aircraft during the cruise phase of the design mission are presented in Figure 35. With a constant flight Mach number and FF FPR in cruise, the three flight level changes are directly visible in the trends of the flight altitude, ideal fuselage shaft power and the main engine fuel flow. The reduction of the shaft power absorbed by the FF is an immediate reflection of the reduction in air density as flight altitude is increased. The main engine fuel flow behavior results from the superposition of continuous loss in aircraft gross weight along the mission the power peaks associated with the en-route step climb maneuvers. The NPF of the bare PFC configuration stays fairly invariant when fuselage FPR is retained constant.

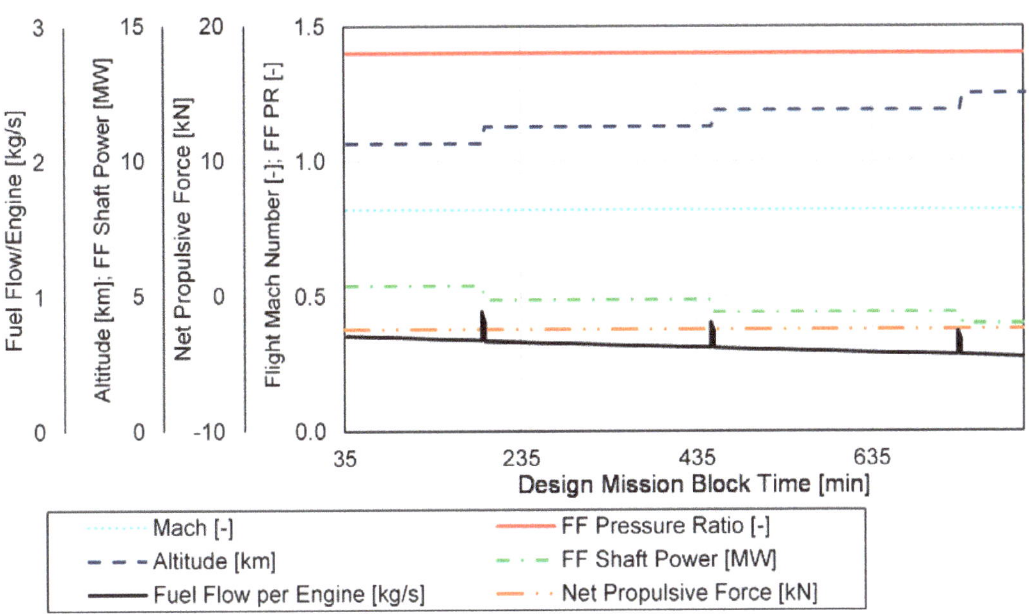

Figure 35. Simulated PFC aircraft specific performance characteristics recorded for the design mission cruise phase (adapted from [69]).

Due to the inherently different wing loadings, the family members feature different optimum initial cruise altitudes as well as individual optimum schedules for flight level changes during cruise. Figure 36 shows the step cruise mission profile of all PFC family members, targeting maximum specific air range for each cruise point. In specific, flight altitude and FF shaft power are plotted, again, with a constant FF PR of 1.40. As expected from the previous figure, the cruise altitude has a direct influence on the required ideal FF design shaft power. Despite the different stepping schedules, the FF power absorption at a given flight level; however, does only differ marginally by the order of 1% between the different members of the PFC aircraft family. This small difference at identical FPR is rooted in the change in actual fan mass flow due to the slightly different boundary layer profiles depending on the wetted fuselage area ahead of the FF for the individual PFC aircraft family members. In fact, the fuselage mass flow rate, and thus the fan power absorption, is the highest for the shrink family member. More detailed analyses of PFC mission performance characteristics including transversal flight phases are presented in [69].

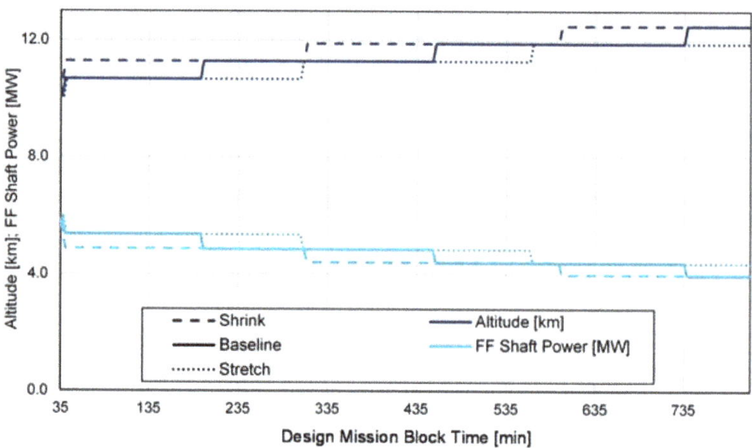

Figure 36. Cruise altitude and FF ideal shaft power profile for the PFC family members at Ma 0.82, ISA+10 K (adapted from [69]).

7.4. *Propulsive Fuselage Multi-Disciplinary Evaluation*

To complete the PFC technology assessment, in this section a more detailed analysis of the main design and performance aspects that influence PFC aircraft fuel efficiency is provided together with assessments regarding NO_x emissions, certification noise, and operating economics.

7.4.1. Stepwise Analysis of Effects on PFC Fuel Efficiency

In order to better resolve the key design and performance effects driving the fuel efficiency of the CENTRELINE PFC aircraft relative to the R2035 aircraft, an incremental analysis approach was followed. Therefore, a series of successive modifications to the aircraft platform were defined and evaluated in a stepwise manner using a simplified aircraft sizing and performance evaluation methodology as described in [111]. The individual steps defined for the incremental analysis approach are characterized in Table 17.

Table 17. Individual steps for incremental analysis of effects on PFC fuel efficiency.

Step	Characterization of Key Sizing Conditions
1	Baseline—Idealized PFC design without cascade effects: • FF cruise polytropic efficiency identical to underwing podded main fans • Idealised efficiency of FF power train (losses identical to LP spool of main engines) • Tail configuration identical to reference (conventional tail) • FF power train specific power identical to main engines • Adoption of pure optimum 2D-Aero heuristic for bare PFC aerodynamics • No aircraft design scaling considered
2	Baseline + Aircraft scaling switched on (sizing cascade effects enabled)
3	Step 2 + Turbo-electric power train losses adopted from CENTRELINE turbo-electric PFC aircraft design (cf. Table 14)
4	Step 3 + Turbo-electric power train specific weight adopted from CENTRELINE turbo-electric PFC aircraft design (cf. Table 14)
5	Step 4 + PFC-specific component weight implications—inclusion of fuselage structural mass penalty
6	Step 5 + PFC-specific component weight implications—inclusion of landing gear mass penalty
7	Step 6 + Update of empennage configuration from conventional to T-tail arrangement
8	Step 7 + Inclusion of FF efficiency penalty due to PFC-specific 3D inflow distortion (cf. Table 14).
9	Step 8 + Preliminary estimate of PFC 3D aerodynamic design implications

Steps 1 through 8 in the table above represent the scope of design integration results discussed in Section 7.3. In Step 9, the result of a preliminary assessment of the PFC aircraft performance impact due to 3D aerodynamic integration effects in cruise is included. The assessment was performed based the 3D RANS simulation results for an integrated PFC configuration with axisymmetric PFC aft-fuselage section, featuring attached wings, belly fairing, and vertical tail plane as presented by van Sluis and Della Corte [58]. A detailed discussion of the 3D performance assessment is provided by Seitz and Engelmann [111]. Here, the results of nine conceivable scenarios for the PFC 3D aerodynamic performance benchmarking were compared against the corresponding PFC aircraft performance based on the 2D aerodynamic design heuristic (cf. Section 7.1). The identified 3D aerodynamic discounts in terms of *PSC* ranged between 1.0 and 1.8%, with a nominal scenario considered to the most comprehensive and realistic one out the nine scenarios yielding a *PSC* penalty of 1.5%. This means an actual all integrated *PSC* value of 3.6% [111]. The 3D aerodynamic delta effects were incorporated within PFC aircraft sizing and performance evaluation through a calibration of the BLI efficiency factor heuristic (cf. Equation (8)) via constant offset values for the individual 3D aerodynamic benchmarking scenarios.

Important results gained from the incremental analysis of effects on PFC fuel efficiency are presented in Figure 37. Here, key figures of merit for the PFC resulting aircraft including the *PSC* metric, OEW, MTOW, and design mission fuel burn relative to the R2035. Also shown in the figure are the FF design isentropic shaft powers, which have been optimized for minimum PFC design mission fuel burn at each of step of the incremental analysis.

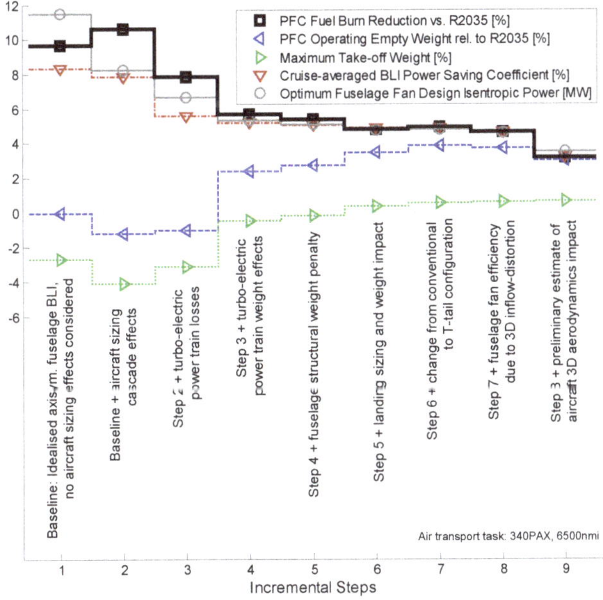

Figure 37. Stack-up of effects analysis on fuel burn reduction of CENTRELINE turbo-electric PFC against R2035 aircraft.

In the idealized analysis case without aircraft weight and sizing effects considered, namely Step 1, the highest power savings are obtained together with the highest FF design power, yielding a cruise-averaged PSC value of 8.6% and corresponding PFC fuel benefit of 9.9% for the given design range of 6500nmi. With the unaffected OEW in this case, the PFC MTOW is reduced by 2.7% relative to the R2035. The $P_{disc,FF,des}$ value of 11.9 MW corresponds to a 43% share of $P_{co,eff}$ for the PFC aircraft. With aircraft scaling effects and propulsion system component weight evaluation included in Step 2, the overall aircraft sizing cascade starting from MTOW benefit given by Step 1, yields mass reductions for key airframe structural components, in particular the wing. The additive mass of the fuselage propulsive device is partially set off against a decrease in fuel optimum $P_{disc,FF,des}$ and a correspondingly reduced power saving potential. The resultant 1.1% net reduction in OEW and the associated lower aircraft gross weights during cruise is compounded by the fuselage drag share being boosted due to the reduced wing size and induced drag. On top of the pure PSC value, this yields an idealized fuel burn benefit for the CENTRELINE PFC aircraft of 10.7% relative to the similarly advanced R2035 aircraft without fuselage BLI technology.

The introduction of the transmission losses and the component masses associated with the FF turbo-electric power train in Steps 3 and 4, both, the power saving potentials and PFC OEW are penalized noticeably. In effect, leading to a reduction in PFC fuel benefit of approximately 5% relative to the idealized case in Step 2. The particularly strong discount in fuel efficiency due to the transmission losses directly emanates from the reduced PSC. While in this case, aircraft gross weight effects remain mostly compensated by a decreased fuel optimum $P_{disc,FF,des}$, the turbo-electric power train weight effect has a major impact with regard to an increased OEW. The airframe structural component mass penalties specific to the PFC, i.e., the reinforced fuselage (Step 5) and the resized landing gear (Step 6), have a combined effect of less than 1% on the PFC fuel benefit. Under consideration of the empennage sizing implications, the introduction of the T-tail is beneficial for the PFC aircraft (Step 7). This is the case even without a full accounting of the aerodynamic and aero-structural interaction a conventional tail would have with the aft-fuselage BLI propulsive device. The penalty in PFC fuel benefit due to the BLI impact on FF efficiency as introduced in Step 8 is approximately 0.3%, yielding a net PFC fuel burn benefit against the R2035 of 4.7%, which directly corresponds to PFC aircraft design result presented in the previous section of the paper. The PSC discount according the to nominal PFC 3D aerodynamic design and performance scenario translates into a 1.5% reduction of the PFC fuel benefit, leaving an all-integrated design mission fuel burn improvements of 3.2% for the CENTRELINE turbo-electric PFC aircraft, relative to the R2035. When assuming the most pessimistic from the conceived scenarios [111], the PFC fuel benefit yields 2.8%, while the most optimistic from the conceived 3D aerodynamic scenarios would show a 3.7% PFC fuel benefit.

7.4.2. NO_x and Noise Assessments

The environmental impact of the CENTRELINE PFC, as well as that of the two reference aircraft R2000 and R2035, has been evaluated at system level based on the ICAO certification regulations in terms of NO_x and noise emissions. For the NO_x emissions assessment, a standard ICAO landing and take-off (LTO) cycle is used with a suitable combustor technology assumed for the corresponding EIS year. For R2000, a correlation developed by AECMA (European Association of Aerospace Industries) which provides a good estimation of the NO_x emissions of large engines around the year 2000 is used [112]. For the advanced R2035 and PFC applications targeting an EIS in 2035, a Lean Direct Injection (LDI) concept is assumed. A correlation found in [113] is adopted. The correlation is based on experimental data with a similar total pressure (p_3) level and marginally lower total temperature (T_3) level as occurring at the HPC exit of R2035 and PFC main engine sea level static take-off operation. For the noise assessment, noise emissions from aircraft and engine components at the ICAO certification points (sideline, cutback, and approach) are calculated from component-based noise source modeling [114]. The total effective perceived noise level (EPNL) is then computed from the perceived noise level accounting for spectral irregularities and duration, as presented in the ICAO Noise Certification Workshop [115].

Results of the NO_x and noise assessments are given in the charts below based on [116]: The LTO NO_x emissions data presented in Figure 38 are normalized by Trent 772B NO_x

emissions given in [117]. As can be seen, a 40% reduction in LTO cycle total NO_x emissions is predicted for the R2035, and a reduction of 41% could be expected for the PFC, compared to the R2000 aircraft. This is mainly contributed by the radical improvement of the combustor technology and fuel burn reduction. On the other hand, due to the much higher OPR and turbine entry temperature of the main engine, R2035 and PFC actually have a much higher $EINO_x$ at high ratings than R2000.

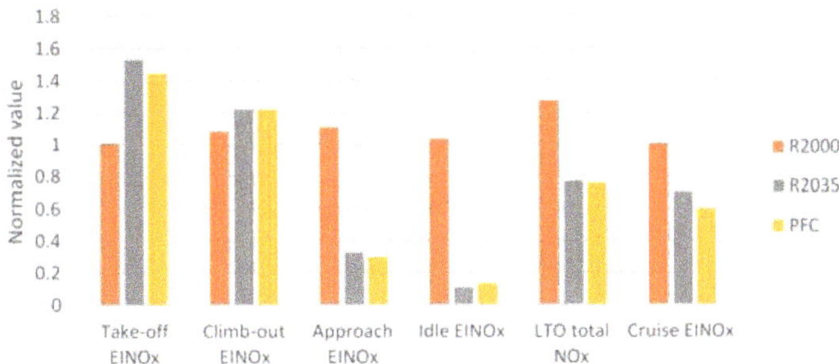

Figure 38. NO_x emissions assessments results for the CENTRELINE R2000, R2035 & PFC aircraft.

When comparing the PFC main engines against the R2035 power plants in cruise, the slightly lower part power settings of the PFC main engines (cf. Section 7.3) lead to a reduction in cruise NO_x emissions of 20% relative the R2035. Combining the low fuel consumption and low $EINO_x$ at cruise, which is the mass of NO_x emitted per unit of mass of fuel burned, the cruise NO_x emissions of the PFC would be as low as 36% of R2000.

Looking at the noise footprint shown in Figure 39, a considerably lower EPNL may be expected from the R2035 and PFC aircraft when compared to the R2000. A cumulative reduction of about 12 dB is observed. The key drivers for the lower noise level are the smaller size of the wing, the low FPR and low specific thrust design main engine fan. Between the R2035 and the PFC, the difference is, however, small. Uncertainty exists due to the FF inlet distortion. Several research activities focusing on BLI effects on noise generation, such as ONERA NOVA concept [34] and NASA D8 [118], have indicated a possible cumulative noise penalty up to 18 dB. Nevertheless, these study objects are different from the CENTRELINE 360° BLI concept for which the FF is not the dominating source of noise. Assuming an increase of 10 dB in sound pressure level at the predicted directivity angle and predicted frequency for the entire reception time, the increase of FF EPNL is 6.98, 6.15, and 1.52 for sideline, cutback, and approach, respectively. Though the estimation is similar to the NASA D8 BLI penalty estimation [118], it increases the total cumulative EPNL by only 0.83.

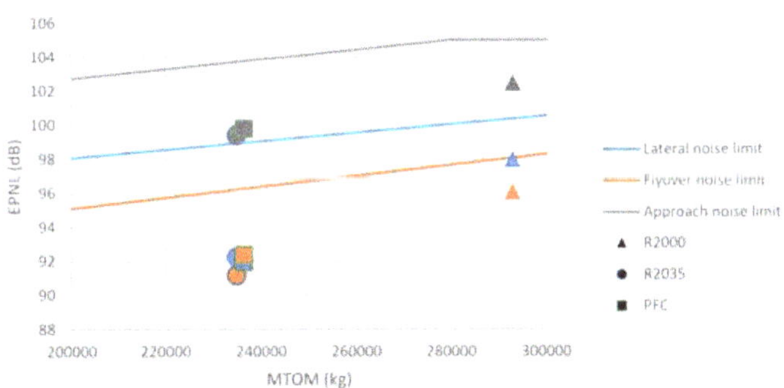

Figure 39. Noise emissions assessments results for the CENTRELINE R2000, R2035 & PFC aircraft.

7.4.3. Cash Operating Costs Estimation

For the purpose of an initial assessment of the operating economics of the CENTRELINE turbo-electric PFC technology, aircraft Cash Operating Costs (COC) were evaluated using ALiCyA (Aircraft Life Cycle Assessment), Bauhaus Luftfahrt's in-house cost assessment model. Initially based on methods introduced by [119], the model has been methodologically extended and customized (cf. [120,121]) and applied to PFC technology already in previous assessments [25].

In order of facilitate a fair comparison between the turbo-electric PFC and R2035 aircraft in CENTRELINE, specific model enhancements were developed to adequately resolve differences between the two configurations. A key aspect in this was the estimation of the direct maintenance costs (DMC) for both aircraft. Based on a reference DMC breakdown according to chapters defined by the Air Transport Association (ATA), individual scaling approaches for the cost of key chapters affected by the CENTRELINE technology were introduced. Moreover, emission charges suitable for a EIS 2035 scenario was adopted. D detailed discussion of the CENTRELINE COC model is provided in [111].

The actual COC assessment was performed for the PFC and R2035 aircraft as presented in Section 7.3 and reported in detail in [111]. Accordingly, the PFC aircraft is predicted exhibit by 9.5% increased DMC relative to the R2035, due to the additional FF and the turbo-electric power train and the increased weights of key airframe components (cf. Table 15). Airport charges are increased due to the noise characteristics according to Section 7.4.2. The reduced fuel consumption and emissions of the PFC aircraft have positive effect of COC which is amplified when aspects such as possible fuel taxation or carbon pricing and offsetting cost for and EIS 2035 are taken into account. The result of a parametric study of design mission COC against a range of fuel prices is presented in Figure 40.

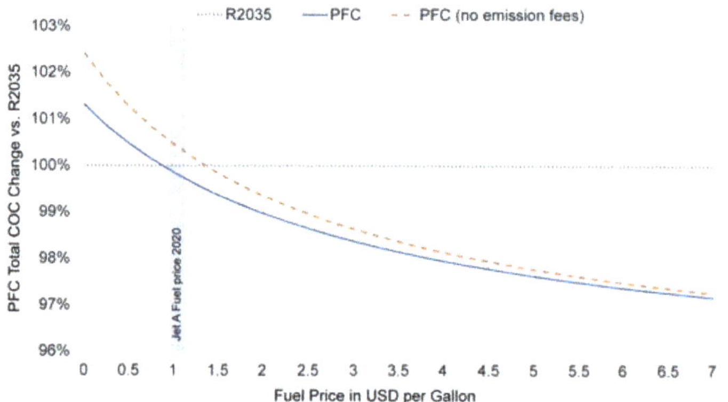

Figure 40. Fuel price study of CENTRELINE PFC versus R2035 design mission COC (adapted from [111]).

It can be observed that the PFC features reduced COC relative to the R2035, in case the fuel price of increases beyond US$ 1.5 per gallon. If a scenario of possible fuel taxation or carbon pricing and offsetting cost for and EIS 2035 are taken into account, this threshold drops to approximately 1.0 US$ per gallon, which corresponds an average Jet-A1 price in the year 2020. With future fuel prices, especially for sustainable aviation fuels, expected to rise considerably above current values, the PFC aircraft may be expected to feature lower COC than the R2035.

7.4.4. Preliminary Assessment of More Advanced PFC Power Train Technology

To round off, a preliminary study of alternative FF power train conceptual paradigms was conducted at a lower TRL basis. Two basic technology scenarios were considered: A turbo-electric power transmission using high temperature superconducting (HTS) technology and FF mechanical drive train. For the HTS technology case, a scenario with a cryogenic heat sink was considered to be available on-board the aircraft. This could be in the form of liquid hydrogen (LH_2) fuel, used as an energy source for the aircraft subsystems (cf. [25]) or as a primary fuel for propulsion purposes. Given the superior specific weights and dramatically reduced losses enabled by the HTS-specific current and flux densities in electrical machinery [122,123], target settings for an HTS turbo-electric transmission system were stipulated including a total power train efficiency of $\eta_{PT,FF,HTS} = 96\%$ and a total specific power of $SP_{PT,FF,HTS} = 5$ kW/kg. The high design complexity and the system behavioral patterns specially in abnormal modes of operation involved with the application of HTS technology have not been considered. At the same time, further potential synergy effects between the PFC and hydrogen fuel technology, such as emanating from an increased fuselage wetted area when LH_2 fuel is stored inside the fuselage, have not been accounted for, in the first instance. For the case of a mechanical FF drive train, efficiency, and specific power targets were set to $\eta_{PT,FF,mech} = 98\%$ and $SP_{PT,FF,mech} = 10$ kW/kg. Specific architectural implications such as a detailed assessment of the aero-structural integration of the aft-fuselage BLI propulsive device have not been undertaken as part of this preliminary study. A specifically elaborated conceptual solution was presented to TRL 2 as part of the DisPURSAL project (cf. [25]). In Table 18, the aircraft-integrated results obtained for the HTS and mechanical FF power train scenarios are compared to the baseline turbo-electric transmission developed to TRL 3 in CENTRELINE. The aircraft-level evaluation for the results in the table refers to the approach introduced in Section 7.4.1. Specifically, Steps 8 and 9 of the incremental analysis are presented.

Table 18. PFC assessment results for more advanced power train technology.

Parameter	Unit	Baseline Turbo-Electric	HTS Turbo-Electric	Mechanical
$\eta_{PT,FF}$	-	0.919 *	0.96	0.98
$SP_{PT,FF}$	kW/kg	2.1 *	5.0	10
Fuel benefit (Step 8)	%	4.7	7.2	8.4
Fuel benefit (Step 9 **)	%	3.2	5.2	6.2

* Cf. Table 15. ** Nominal scenario acc. to Section 7.4.1.

The results of the preliminary evaluation of the considered more radical technology concepts for the FF power supply indicate improvements in the all integrated PFC fuel benefit (Step 9) over the R2035 aircraft of additional 2.0% for the HTS and 3.0% for the mechanical scenarios relative to the baseline turbo-electric design. For the fuel assessment at the level of 2D PFC aerodynamic design (Step 8), the evaluated reduction potentials against the R2035 yield 7.2% and 8.4% for the HTS and mechanical cases, respectively. These results are in basic agreement with results obtained during the DisPURSAL project [124].

8. Conclusions and Further Work

Novel propulsion technology and propulsion-airframe integration play a key role in enabling aviation's long-term sustainability. Strong improvements of vehicular propulsive efficiency can be achieved by fuselage wake-filling propulsion integration. The presented research aimed at maximizing these benefits under realistic systems design and operating conditions. A key objective of the design, analysis and experimental testing activities discussed in this paper was to perform the proof of concept for a most straightforward approach to fuselage BLI—the so-called PFC. The investigated PFC configuration specifically featured a turbo-electric transmission in order to drive the aft-fuselage BLI fan through power offtakes from the underwing podded main engines. The interdisciplinary team of researchers addressed all main challenges associated with a turbo-electric PFC aircraft. A thorough understanding of the aerodynamic effects of 360° fuselage BLI has been developed through extensive aero-numerical simulations and low-speed experimental testing. Optimized aerodynamic pre-designs have been produced for the FF as well as for the fuselage and FF nacelle bodies. Conceptual solutions for the aero-structural integration of the BLI propulsive device and the TEPT have been elaborated. Important heuristics for PFC aircraft design have been deduced from multi-disciplinary design optimization. The developed design solutions have been either analytically or experimentally verified to demonstrate TRL 3 for the PFC technology. Initial steps towards experimental validation in a laboratory environment have been taken.

Assessments of fuel burn, NO_x emissions and noise were presented for the PFC aircraft and benchmarked against advanced conventional technology for an EIS in 2035. The PFC design fuel burn benefit for a fully integrated multi-disciplinary aircraft design and performance synthesis based on the 2D optimized PFC aero-shaping is 4.7% against a similarly advanced aircraft without fuselage BLI propulsion. The all-integrated PFC fuel benefit including a preliminary assessment of the PFC 3D aerodynamic design impact yields 3.2%. The NO_x emission assessment performed for LDI combustor technology shows emission benefits for the PFC aircraft relative to the R2035 at all high power conditions, including take-off, climb, and cruise. The high-level PFC NO_x reduction potentials were predicted as 64% relative to the year 2000 reference. PFC certification noise were found to be similar to the R2035, meaning a cumulative noise reduction of 12EPNdB compared to the R2000. The design mission COC for the PFC aircraft was assessed to be lower than for the R2035 as soon as the fuel price increases beyond US$ 1.5 per gallon, even if fuel taxation or carbon pricing and offsetting cost for an EIS 2035 are neglected. Important steps to be taken by follow-on research and innovation activities have been outlined that would facilitate a speedy maturation of the PFC technology towards a target TRL 6.

Author Contributions: Conceptualization: A.S., F.P., F.T., A.C.P., B.D.C., M.K., X.Z., J.B. and G.W.; Methodology, A.S., A.L.H., F.P., F.T., A.C.P., B.D.C., M.v.S., M.K., X.Z., T.G., J.B. and G.W.; Validation: A.S., A.L.H., F.P., A.C.P., B.D.C., M.v.S., M.K., X.Z., J.B. and G.W.; Formal analysis: A.S., A.L.H., F.P., M.K., X.Z., J.B. and G.W.; Investigation: A.S., F.T., A.C.P., B.D.C., M.v.S., M.K., X.Z., J.B. and G.W.; Data curation: A.S., A.L.H., F.P. and F.T.; Writing—original draft preparation: A.S., A.L.H., F.P., A.C.P., B.D.C., M.v.S., M.K., X.Z. and G.W.; Writing—review and editing: A.S., A.L.H., A.C.P. and J.B.; Supervision: A.S., Z.G. and T.G.; Visualization: A.S., A.L.H., F.P., A.C.P., B.D.C., M.v.S., M.K., X.Z. and G.W.; Project administration, A.S.; Funding acquisition: A.S., Z.G., T.G., J.B., and G.W. All authors have read and agreed to the published version of the manuscript.

Funding: This research was funded by the European Union's Horizon 2020 research and innovation programme under grant agreement no. 723242.

Acknowledgments: This paper is based on the work performed by the CENTRELINE project consortium comprising Airbus, Bauhaus Luftfahrt, Chalmers Tekniska Hoegskola, MTU Aero Engines, Politechnika Warszawska, Rolls Royce, Delft University of Technology, University of Cambridge, and ARTTIC. The authors would like express their gratitude to the following individuals for their support during the CENTRELINE project (in alphabetical order): Anders Lundbladh, André Denis Bord, Arvind Gangoli Rao, Bartlomiej Goliszek, Cesare Hall, Daniel Reckzeh, Frank Meller, Jasper van Wensween, Lars Jørgensen, Marc Engelmann, Mark Voskuijl, Markus Nickl, Martin Dietz, Maxime Flandrin, Michael Lüdemann, Michaël Meheut, Michael Shamiyeh, Olaf Brodersen, Olivier Petit, Philipp Maas, Rasmus Merkler, Sebastian Samuelsson, Sophie Rau, and Stefan Biser.

Conflicts of Interest: The authors declare no conflict of interest. The funders had no role in the design of the study; in the collection, analyses, or interpretation of data; in the writing of the manuscript, or in the decision to publish the results.

Nomenclature

Abbreviation/Acronym	**Description**
ACARE	Advisory Council for Aviation Research and Innovation in Europe
AC	Alternating Current
ADP	Aerodynamic Design Point
AECMA	European Association of Aerospace Industries
AIP	Aerodynamic Interface Plane
ALiCyA	Aircraft Life Cycle Assessment
APD	Aircraft Preliminary Design
APSS	Aircraft Propulsion System Simulation
APU	Auxiliary Power Unit
AR	Aspect Ratio
ATA	Air Transport Association
BLADE	Breakthrough Laminar Aircraft Demonstrator in Europe
BLI	Boundary Layer Ingestion
BLISS	Bi-Level Integrated System Synthesis
CAD	Computer-Aided Design
CENTRELINE	ConcEpt validatioN sTudy foR fusElage wake-filLIng propulsioN integration
CFD	Computational Fluid Dynamics
CFRP	Carbon Fiber Reinforced Polymer
COC	Cash Operating Costs
CS	Certification Specification
CV	Control Volume
DC	Direct Current
DEN	Denver International Airport
DLR	Deutsches Zentrum für Luft- und Raumfahrt
DMC	Direct Maintenance Costs
EC	European Commission
$EINO_x$	Nitrogen Oxide Emission Index
EIS	Entry Into Service
EPNL	Effective Perceived Noise Level
ETOPS	Extended-range Twin-engine Operation(al) Performance Standards
FEM	Finite Element Method
FF	Fuselage Fan
FHA	Function Hazard Assessment
FL	Flight Level
FNN	Feedforward Neural Network
FPR	Fan Pressure Ratio
FPT	Free Power Turbine
FTA	Failure Tree Analysis
GTF	Geared TurboFan
HTS	High Temperature Superconducting
ICAO	International Civil Aviation Organization
ISA	International Standard Atmosphere
HPC	High Pressure Compressor
HTS	High Temperature Superconducting
IPC	Intermediate Pressure Compressor
LDI	Lean Direct Injection
LHS	Latin Hypercube Sampling
LP	Low Pressure
LPT	Low Pressure Turbine
LRC	Long-Range Cruise
MIT	Massachusetts Institute of Technology
MLW	Maximum Landing Weight

MLW		Maximum Landing Weight
MTOW		Maximum Take-Off Weight
NPF		Net Propulsive Force
NURBS		Non-Uniform Rational B-splines
OEW		Operating Empty Weight
OGV		Outlet Guide Vanes
ONERA		Office National d'Etudes et de Recherches Aérospatiales
OPR		Overall Pressure Ratio
PAX		Passengers
PFC		Propulsive Fuselage Concept
PIV		Particle Image Velocimetry
PSC		Power Saving Coefficient
PSSA		Preliminary System Safety Assessment
RANS		Reynolds-Averaged Navier-Stokes
SAP		Safety Assessment Process
SFC		Specific Fuel Consumption
SL		Sea Level
SLST		Sea-Level Static Thrust
SKF		Svenska Kullagerfabriken
SRIA		Strategic Research and Innovation Agenda
SST		Shear Stress Transport
TAB		Technical Advisory Board
TLAR		Top Level Aircraft Requirements
ToC		Top-of-Climb
TOFL		Take-Off Field Length
TRL		Technology Readiness Level
TU Delft		Delft University of Technology
UHBR		Ultra-High Bypass Ratio
WUT		Warsaw University of Technology
Symbol	**Unit**	**Description**
α	°	Angle of attack
β	°	Sideslip angle
δ	m	Local boundary layer thickness
$\Delta\eta$	-	Change in stage isentropic efficiency
$\eta_{pd,eff}$	-	Effective propulsive device efficiency
$\eta_{pol,FF}$	-	FF polytropic efficiency
$\eta_{PT,FF}$	-	FF power train efficiency
Θ	°	Azimuthal angle
A_{12}	m	FF fan face area
A_{18}	m	FF nozzle exit area
AR	-	Aspect Ratio
$C_{L,max}$	-	Maximum lift coefficient
D	kN	Drag
D_{FF}	m	FF rotor inlet tip diameter
$D_{Ref,fus}$	kN	Reference fuselage drag
F	kN	Force
$F_{PFC,bare}$	kN	Bare PFC surface forces
$F_{FF,disc}$	kN	FF disc force
F_N	kN	Net thrust
$f_{\eta,PFC,bare}$	-	Bare PFC efficiency factor
L_{fus}	m	Fuselage length
$\dot{m}_{corr}$	kg/s	Corrected mass flow
$NPF_{PFC,bare}$	kN	Bare PFC Net Propulsive Force

p_0	Pa	Freestream total pressure
p_{AIP}	Pa	Local total pressure at FF AIP
$\overline{p_{AIP}}$	Pa	Mean total pressure at FF AIP
p_{18}	Pa	Mass flow averaged total pressure at FF nozzle throat
p_2	Pa	Mass flow averaged total pressure at FF rotor inlet
p_3	Pa	Total pressure at HPC exit
$p_{s,amb}$	Pa	Ambient static pressure
$P_{co,eff}$	MW	Effectvie core engine excess power
$P_{disc,FF}$	MW	FF disc power
$P_{FF\ Shaft}$	MW	FF shaft power
PC	-	Total pressure coefficient
PSC	-	Power saving coefficient
r_{fus}	m	Radius of fuselage centre section
r_{hub}	m	FF hub radius
r_{tip}	m	FF tip radius
$SP_{PT,FF}$	kW/kg	FF power train specific weight
T_3	K	Total temperature at HPC exit
V_0	m/s	Freestream velocity
W_2	kg/s	FF mass flow rate
x_{FF}	m	FF axial position
y	m	Radial coordinate from fuselage centreline

References

1. European Commission. *Flightpath 2050 Europe's Vision for Aviation–Report of the High Level Group on Aviation Research*; European Commission: Luxembourg, 2011.
2. Air Transport Action Group. *Recuding Emissions from Aviation through Carbonneutral Growth from 2020*; Working Paper Developed for the 38th ICAO Assembly September/October; Air Transport Action Group: Geneva, Switzerland, 2013.
3. Advisory Council for Aviation Research and Innovation in Europe (ACARE). *Strategic Research and Innovation Agenda (SRIA)–Volume 1*; ACARE: Brussels, Belgium, 2012.
4. Smith, L.H. Wake ingestion propulsion benefit. *J. Propuls. Power* **1993**, *9*, 74–82. [CrossRef]
5. Steiner, H.-J.; Seitz, A.; Wieczorek, K.; Plötner, K.; Isikveren, A.T.; Hornung, M. Multi-disciplinary Design and Feasibility Study of Distributed Propulsion Systems. In Proceedings of the 28th International Congress of the Aeronautical Sciences, Brisbane, Australia, 23–28 September 2012.
6. Smith, A.M.O.; Roberts, H.E. The jet airplane utilizing boundary layer air for propulsion. *J. Aerosol Sci.* **1947**, *14*, 97–109. [CrossRef]
7. Goldschmied, F.R. A Theoretical Aerodynamic Analysis of a Boundary Layer Controlled Airship Hull. *Goodyear Aircr. Rep. GER-6251* **1954**.
8. Drela, M. Power Balance in Aerodynamic Flows. *AIAA J.* **2009**, *47*, 1761–1771. [CrossRef]
9. Betz, A.; Ackeret, J. Verfahren zur Verminderung des Widerstandes eines Körpers in Flüssigkeiten oder Gasen. German Patent No 513116, 5 September 1923.
10. Heppner, F.A.M. Improvements relating to Jet-Propelled Aircraft. GB Patent GB577950, 31 December 1941.
11. Cerreta, P.A. *Wind-Tunnel Investigation of the Drag of a Proposed Boundary-Layer Controlled Airship*; David Taylor Model Basin Aero Report 914; Cornell University: Ithaca, NY, USA, 1957.
12. Atinault, O.; Carrier, G.; Grenon, R.; Verbecke, C. Numerical and Experimental Aerodynamic Investigations of BLI for Improving Propulsion Efficiency of Future Air Transport. In Proceedings of the 31st AIAA Applied Aerodynamics Conference, San Diego, CA, USA, 24–27 June 2013.
13. Lv, P.; Ragni, D.; Hartuc, T.; Veldhuis, L.; Rao, A.G. Experimental Investigation of the Flow Mechanisms Associated with a Wake-Ingesting Propulsor. *AIAA J.* **2017**, *55*, 1–11. [CrossRef]
14. Uranga, A.; Drela, M.; Greitzer, E.M.; Titchener, N.; Lieu, M.; Siu, N.; Huang, A.; Gatlin, G.M.; Hannon, J. Preliminary Experimental Assessment of the Boundary Layer Ingestion Benefit for the D8 Aircraft, AIAA 2014-0906, AIAA SciTech 52nd. In Proceedings of the Aerospace Sciences Meeting, National Harbor, MA, USA, 13–17 January 2014.
15. Gunn, E.J.; Hall, C.A. Aerodynamics of Boundary Layer Ingesting Fans. In Proceedings of the ASME Turbo Expo 2014, Dusseldorf, Germany, 16–20 June 2014.
16. De LaRosa, E.; Hall, C.; Crichto, D. Special session–Towards a silent aircraft challenges in the silent aircraft engine design. In Proceedings of the 45th AIAA Aerospace Sciences Meeting and Exhibit, Reno, NE, USA, 8–11 January 2007.

17. Felder, J.L.; Kim, H.D.; Brown, G.V. Turboelectric Distributed Propulsion Engine Cycle Analysis for Hybrid-Wing-Body Aircraft. In Proceedings of the AIAA 2009-1132, 47th AIAA Aerospace Sciences Meeting Including The New Horizons Forum and Aerospace Exposition, Orlando, FL, USA, 5–8 January 2009.
18. Bolonkin, A. A high efficiency fuselage propeller ('Fusefan') for subsonic aircraft. In Proceedings of the World Aviation Conference, San Francisco, CA, USA, 19–21 October 1999.
19. Drela, M. Development of the D8 Transport Configuration. In Proceedings of the AIAA 2011-3970, 29th AIAA Applied Aerodynamics Conference, Honolulu, HI, USA, 27–30 June 2011.
20. Seitz, A. Flugzeugantriebssystem. German Patent DE200810024463, 21 May 2008.
21. Stückl, S.; van Toor, J.; Lobentanzer, H. VoltAir—The All Electric Propulsion Concept Platform—A Vision for Atmospheric Friendly Flight. In Proceedings of the 28th International Congress of the Aeronautical Sciences (ICAS), Brisbane, Australia, 23–28 September 2012.
22. Bradley, M.K.; Droney, C.K. Subsonic Ultra Green Aircraft Research Phase II: N+4 Advanced Concept Development. NASA/CR-2012-217556, 1 May 2012.
23. Welstead, L.; Felder, J. Conceptual Design of a Single-Aisle Turboelectric Commercial Transport with Fuselage Boundary Layer Ingestion. In Proceedings of the AIAA-2016-1027, 54th AIAA Aerospace Sciences Meeting (SciTech), San Diego, CA, USA, 4–8 January 2016.
24. Welstead, J.; Felder, J.; Guynn, M.; Haller, B.; Tong, M.; Jones, S.; Gray, J.S.; Ordaz, I.; Quinlan, J.; Mason, B.; et al. Overview of the NASA STARC-ABL (Rev. B) Advanced Concept. NF1676L-26767, 22 March 2017.
25. Isikveren, A.T.; Seitz, A.; Bijewitz, J.; Mirzoyan, A.; Isyanov, A.; Grenon, R.; Atinault, O.; Godard, J.-L.; Stückl, S. Distributed propulsion and ultra-high by-pass rotor study at aircraft level. *Aeronaut. J.* **2015**, *119*, 1327–1376. [CrossRef]
26. Gray, J.S.; Mader, C.A.; Kenway, G.K.; Martins, J.R.R.A. Modeling Boundary Layer Ingestion Using a Coupled Aeropropulsive Analysis. *J. Aircr.* **2018**, *55*, 3. [CrossRef]
27. Kenway, G.K.; Kiris, C.C. Aerodynamic Shape Optimization of the STARC-ABL Concept for Minimal Inlet Distortion. In Proceedings of the AIAA 2018-1912, AIAA/ASCE/AHS/ASC Structures, Structural Dynamics, and Materials Conference, Kissimmee, FL, USA, 8–12 January 2018.
28. Yildirim, A.; Gray, J.S.; Mader, C.A.; Martins, R.R.A.J. Aeropropulsive Design Optimization of a Boundary Layer Ingestion System. In Proceedings of the AIAA 2019-3455, AIAA Aviation Forum, Dallas, TX, USA, 17–21 June 2019.
29. Martínez Fernández, A.; Smith, H. Effect of a fuselage boundary layer ingesting propulsor on airframe forces and moments. *Aerosp. Sci. Technol.* **2020**, *100*, 105808. [CrossRef]
30. Budziszewski, N.; Friedrichs, J. Modelling of a Boundary Layer Ingesting Propulsor. *Energies* **2018**, *11*, 708. [CrossRef]
31. Provenza, A.; Duffy, K.P.; Bakhle, M.A. Aeromechanical Response of a Distortion-Tolerant Boundary Layer Ingesting Fan. *J. Eng. Gas Turbines Power* **2018**, *141*, 011011. [CrossRef]
32. Kratz, J.L.; Thomas, G.L. Dynamic Analysis of the STARC-ABL Propulsion System. In Proceedings of the AIAA 2019-4182, AIAA Propulsion and Energy Forum, Indianapolis, IN, USA, 19–22 August 2019.
33. Heinlein, G.; Chen, J.; Bakhle, M. Aerodynamic Behavior of a Coupled Boundary Layer Ingesting Inlet–Distortion Tolerant Fan. In Proceedings of the AIAA 2020-3780, AIAA Propulsion and Energy Forum, Virtual Event, 24–28 August 2020.
34. Romani, G.; Ye, Q.; Avallone, F.; Ragni, D.; Casalino, D. Numerical analysis of fan noise for the NOVA boundary-layer ingestion configuration. *Aerosp. Sci. Tech.* **2020**, *96*, 105532. [CrossRef]
35. Wernick, A.R.; Chen, J.-P. Rotor Blade Design Optimization for Boundary Layer Ingesting Inlet Fan. In Proceedings of the AIAA 2020-0131, AIAA Scitech Forum, Orlando, FL, USA, 6–10 January 2020.
36. Celestina, M.L.; Long-Davis, M.J. Large-scale Boundary Layer Ingesting Propulsor Research. In Proceedings of the ISABE Conference 2019, Canberra, Australia, 24–27 September 2019.
37. Seitz, A.; Bijewitz, J.; Kaiser, S.; Wortmann, G. Conceptual investigation of a propulsive fuselage aircraft layout. *Aircr. Eng. Aerosp. Technol.* **2014**, *86*, 464–472. [CrossRef]
38. DisPURSAL Consortium. *Report on the Technology Roadmap for 2035*; Public Project Deliverable D1.2, Grant Agreement No. 323013; DisPURSAL Consortium; European Commission Directorate General for Research and Innovation: Brussels, Belgium, 2015.
39. Peter, F.; Bijewitz, J.; Habermann, A.; Lüdemann, M.; Plötner, K.; Troeltsch, F.; Seitz, A. *Specification of the CENTRELINE Reference Aircraft and Power Plant Systems*; Virtual Event; Deutscher Luft- und Raumfahrtkongress: Garching, Germany, 2020.
40. Kaiser, S.; Seitz, A.; Donnerhack, S.; Lundbladh, A. Composite Cycle Engine Concept with Hectopressure Ratio. *J. Propuls. Power* **2016**, *32*, 1413–1421. [CrossRef]
41. Seitz, A.; Nickl, M.; Stroh, A.; Vratny, P.C. Conceptual study of a mechanically integrated parallel hybrid electric turbofan. *Proc. Inst. Mech. Eng. Part G J. Aerosp. Eng.* **2018**, *232*, 2688–2712. [CrossRef]
42. Bijewitz, J.; Seitz, A.; Hornung, M. Power Plant Pre-Design Exploration for a Turbo-Electric Propulsive Fuselage Concept. In Proceedings of the AIAA 2018-4402, AIAA Propulsion and Power Forum 2018, Cincinnati, OH, USA, 9–11 July 2018.
43. Seitz, A.; Peter, F.; Bijewitz, J.; Habermann, A.; Goraj, Z.; Kowalski, M.; Castillo Pardo, A.; Hall, C.; Meller, F.; Merkler, R.; et al. Concept validation study for fuselage wake-filling propulsion integration. In Proceedings of the 31st Congress of the International Council of the Aeronautical Sciences (ICAS), Belo Horizonte, Brazil, 9–14 September 2018.

44. Meller, F.; Kocvara, F. *Specification of Propulsive Fuselage Aircraft Layout and Design Features*; Public Project Deliverable D1.02, Grant Agreement No. 723242; CENTRELINE Consortium; European Commission Directorate-General for Research and Innovation: Brussels, Belgium, 2018.
45. Habermann, A.L.; Bijewitz, J.; Seitz, A.; Hornung, M. Performance bookkeeping for aircraft configurations with fuselage wake-filling propulsion integration. *CEAS Aeronaut. J.* **2019**, *11*, 529–551. [CrossRef]
46. Arntz, A. Civil Aircraft Aero-thermo-propulsive Performance Assessment by an Exergy Analysis of High-Fidelity CFD-RANS Flow Solutions. Ph.D. Thesis, Universite de Lille, Lille, France, 2014.
47. Della Corte, B.; van Sluis, M.; Gangoli Rao, A.; Veldhuis, L.L.M. Experimental Power Balance Analysis of an Axisymmetric Fuselage with an Integrated Boundary-Layer-Ingesting Fan. *AIAA J.* **2020**. in review.
48. Della Corte, B.; van Sluis, M.; Gangoli Rao, A. *Results of the Overall Configuration Wind Tunnel Testing*; Public Project Deliverable D3.02, Grant Agreement No. 723242; CENTRELINE Consortium; European Commission Directorate-General for Research and Innovation: Brussels, Belgium, 2020.
49. Graichen, J.; Gores, S.; Herold, A. *Überarbeitung des Emissionsinventars des Flugverkehrs*; ISSN-1862-4804; Umweltbundesamt: Dessau-Roßlau, Germany, 2010.
50. Seitz, A.; Habermann, A.L.; Van Sluis, M. Optimality considerations for propulsive fuselage power savings. *Proc. Inst. Mech. Eng. Part G J. Aerosp. Eng.* **2020**. [CrossRef]
51. Piegl, L. On NURBS: A Survey. *IEEE Comput. Graph. Appl.* **1991**, *11*, 55–71. [CrossRef]
52. Derksen, R.W.; Rogalsky, T. Optimum Airfoil Parametrization for Aerodynamic Design. *Comput. Aided Optim. Des. Eng.* **2009**, *106*. [CrossRef]
53. Van Sluis, M.; Della Corte, B.; Rao, A.G. Aerodynamic Design Space Exploration of a Fuselage Boundary Layer Ingesting Aircraft Concept. **2021**. to be published.
54. Hofer, D.; D'Errico, J.; Schwarz, M.D. Matlab to Ansys ICEM/Fluent and Spline Drawing Toolbox. Available online: http://mathworks.com/matlabcentral/fileexchange/66215 (accessed on 17 April 2019).
55. Menter, F.R. Two-equation eddy-viscosity turbulence models for engineering applications. *AIAA J.* **1994**, *32*, 1598–1605. [CrossRef]
56. Shields, M.D.; Zhang, J. The generalization of Latin hypercube sampling. *Reliab. Eng. Syst. Saf.* **2016**, *148*, 96–108. [CrossRef]
57. Chang, C.C.; Lin, C.J. LIBSVM: A Library for Support Vector Machines. *ACM Trans. Intell. Syst. Technol.* **2011**, *2*, 27. [CrossRef]
58. Van Sluis, M.; Della Corte, B. *Final PFC Aircraft Aerodynamic Design and Performance*; Public Project Deliverable D3.03, Grant Agreement No. 723242; CENTRELINE Consortium; European Commission Directorate-General for Research and Innovation: Brussels, Belgium, 2020.
59. Van Oudheusden, B.W.; Steenaert, C.B.; Boermans, L.M.M. Attachment-Line Approach for Design of a Wing-Body Leading-Edge Fairing. *J. Aircr.* **2004**, *41*, 238–246. [CrossRef]
60. Brandvik, T.; Pullan, G. An Accelerated 3D Navier–Stokes Solver for Flows in Turbomachines. *J. Turbomach.* **2010**, *133*, 021025. [CrossRef]
61. Spalart, P.; Allmaras, S. A one-equation turbulence model for aerodynamic flows. *Rech. Aerosp.* **1994**, *1*, 5–21.
62. Castillo Pardo, A.; Hall, C.A. Design of a Transonic Boundary Layer Ingesting Fuselage Fan. Paper ID: GPPS-CH-2020-0042, to be published. In Proceedings of the Global Power and Propulsion Society, GPPS, Chania, Greece, 7–9 September 2020.
63. Serpieri, J. Cross-Flow Instability: Flow Diagnostics and Control of Swept Wing Boundary Layers. Ph.D. Thesis, Delft University of Technology, Delft, The Netherlands, 2018.
64. Castillo Pardo, A.; Hall, C.A. Aerodynamics of Boundary Layer Ingesting Fuselage Fans. In Proceedings of the ISABE-2019-24162, 24th ISABE Conference, Canberra, Australia, 22–27 September 2019.
65. Taylor, J.V. Complete flow conditioning gauzes. *Exp. Fluids* **2019**, *60*, 35. [CrossRef]
66. Goraj, Z.; Kowalski, M.; Goliszek, B.; van Sluis, M. *Interim Report on Fuselage and Nacelle Aero-Structural Pre-Design*; Public Project Deliverable D2.06, Grant Agreement No. 723242; CENTRELINE Consortium; European Commission Directorate-General for Research and Innovation: Brussels, Belgium, 2018.
67. Goraj, Z.; Kowalski, M.; Goliszek, B. Optimisation of the loading structure for Propulsive Fuselage Concept. In Proceedings of the 24th ISABE Conference, International Society for Air Breathing Engines, Canberra, Australia, 22–27 September 2019.
68. Federal Aviation Administration. Design considerations for minimizing hazards caused by uncontained turbine engine and auxiliary power unit rotor failure. *FAA Advis. Circ.* **1997**.
69. Habermann, A.; Troeltsch, F.; Peter, F.; Maas, P.; van Sluis, M.; Kowalski, M.; Wortmann, G.; Bijewitz, B.; Seitz, A. *Summary Report on Multi-Disciplinary Design Results*; Public Project Deliverable D2.11, Grant Agreement No. 723242; CENTRELINE Consortium; European Commission Directorate General for Research and Innovation: Brussels, Belgium, 2020.
70. Seitz, A. Advanced Methods for Propulsion System Integration in Aircraft Conceptual Design. Ph.D. Thesis, Institut für Luft- und Raumfahrt, Technische Universität München, München, Germany, 2012.
71. Grieb, H. *Projektierung von Turboflugtriebwerken*; Birkhäuser Verlag: Basel, Switzerland; Boston, MA, USA; Berlin, Germany, 2004.
72. Kurzke, J. *GasTurb 11: Design and Off-Design Performance of Gas Turbines*; GasTurb GmbH: Aachen, Germany, 2007.
73. Kurzke, J. *Gasturb Details 6*; Software compiled with Delphi XE4, 2013.
74. Wittel, H.; Muhs, D.; Jannasch, D.; Voßiek, J. *Roloff/Matek Maschinenelemente, Normung, Berechnung, Gestaltung*, 20th ed.; Vieweg und Teubner Verlag: Wiesbaden, Germany, 2011; ISBN 978-3-8348-1454-8.

75. Stroh, A.; Wortmann, G.; Seitz, A. *Conceptual Sizing Methods for Power Gearboxes in Future Gas Turbine Engines*; Document ID 450100; Deutscher Luft- und Raumfahrtkongress: Garching, Germany, 2017.
76. European Aviation Safety Agency. *Certification Specifications and Acceptable Means of Compliance for Large Aeroplanes CS-25*, 23rd ed.; European Aviation Safety Agency: Cologne, Germany, 2019.
77. Stotler, C.L.; Coppa, A.P. *Containment of Compososite Fan Blades*; General Electric, NASA Contractor Report 159544; NASA Lewis Research Center: Cleveland, OH, USA, 1979.
78. Goraj, Z.; Kowalski, M.; Goliszek, B. Passenger Aircraft Composite Centre Wing Box Structure Optimization. In Proceedings of the Aerospace Europe Conference, Bordeaux, France, 25–28 February 2020.
79. Goraj, Z.; Kowalski, M.; Goliszek, B. The use of Finite Element Method and semi-empirical equations for weight estimation of the passenger aircraft utilizing innovative technological solutions. In Proceedings of the Research and Education in Aircraft Design (READ) Conference, Rzeszów, Poland, 21–23 October 2020.
80. Kowalski, M.; Goraj, Z.; Goliszek, B. The use of FEA and semi-empirical equations for weight estimation of a passenger aircraft. *Aircr. Eng. Aerosp. Technol. J.* **2020**. submitted.
81. Wortmann, G. *Electric Machinery Preliminary Design Report*; Public Project Deliverable D4.04, Grant Agreement No. 723242; CENTRELINE Consortium; European Commission Directorate-General for Research and Innovation: Brussels, Belgium, 2018.
82. Wortmann, G. *Electric Generator Final Design Report*; Confidential Project Deliverable D4.06, Grant Agreement No. 723242; CENTRELINE Consortium; European Commission Directorate-General for Research and Innovation: Brussels, Belgium, 2019.
83. Wortmann, G. *Final Design and Performance of FF Electric Drive Motor*; Confidential Project Deliverable D4.07, Grant Agreement No. 723242; CENTRELINE Consortium; European Commission Directorate-General for Research and Innovation: Brussels, Belgium, 2019.
84. Wortmann, G. *Final Design of the Electric Fuselage Drive Train System*; Confidential Project Deliverable D4.08, Grant Agreement No. 723242; CENTRELINE Consortium; European Commission Directorate-General for Research and Innovation: Brussels, Belgium, 2019.
85. Kellermann, H.; Habermann, A.; Vratny, P.; Hornung, M. Assessment of fuel as alternative heat sink for future aircraft. *Appl. Therm. Eng.* **2020**, *170*, 114985. [CrossRef]
86. Kellermann, H.; Habermann, A.L.; Hornung, M. Assessment of Aircraft Surface Heat Exchanger Potential. *Aerospace* **2019**, *7*, 1. [CrossRef]
87. Kurzke, J. *Compressor and Turbine Maps for Gas Turbine Performance Computer Programs–Issue 3*; GasTurb GmbH: Aachen, Germany, 2013.
88. Zhao, X. *Power Plant System Final Design and Performance Characteristics*; Public Project Deliverable D4.03, Grant Agreement No. 723242; CENTRELINE Consortium; European Commission Directorate-General for Research and Innovation: Brussels, Belgium, 2020.
89. Merkler, R. *Power Plant Integration Concept for Electric Generator*; Public Project Deliverable D4.05, Grant Agreement No. 723242; CENTRELINE Consortium; European Commission Directorate-General for Research and Innovation: Brussels, Belgium, 2020.
90. Campbell, C.E.; Welna, H.J. *Preliminary Evaluation of Turbine Performance with Variable-Area Turbine Nozzles in a Turbojet Engine*; National Advisory Committee for Aeronautics: Washington, DC, USA, 1960.
91. Schum, H.J.; Moffitt, T.P.; Behning, F.P. *Effect of Variable Stator Area on Performance of a Single-Stage Turbine Suitable for Air Cooling: III. Turbine Performance with 130 Percent Design Stator Area*; NASA: Cleveland, OH, USA, 1968.
92. Karstensen, K.W.; Wiggins, J.O. A Variable-Geometry Power Turbine for Marine Gas Turbines. *J. Turbomach.* **1990**, *112*, 165–174. [CrossRef]
93. Sobieszczanski-Sobieski, J.; Agte, J.; Sandusky, R. Bi-level integrated system synthesis (BLISS). AIAA 98-4916. In Proceedings of the 7th AIAA/USAF/NASA/ISSMO Symposium on Multidisciplinary Analysis and Optimization, St. Louis, MO, USA, 4 September 1998.
94. Sobieszczanski-Sobieski, J.; Emiley, M.; Agte, J.; Sandusky, R., Jr. *Advancement of Bi-Level Integrated System Synthesis (BLISS)*; NASA/TM-2000-210305; NASA Technical Memorandum, Langley Research Center: Hampton, VA, USA, 2000.
95. McKay, M.; Beckman, R.; Conover, W. A comparison of three methods for selecting values of input variables in the analysis of output from a computer code. *Technometrics* **1979**, *21*, 239–245.
96. Shamiyeh, M. *Definition of Multidisciplinary Interfacing Strategy*; Public Project Deliverable D2.01, Grant Agreement No. 723242; CENTRELINE Consortium; European Commission Directorate-General for Research and Innovation: Brussels, Belgium, 2018.
97. Troeltsch, F.; Bijewitz, J.; Seitz, A. Design Trade Studies for Turbo-electric Propulsive Fuselage Integration. In Proceedings of the 24th ISABE Conference, International Society for Air Breathing Engines, Canberra, Australia, 22–27 September 2019.
98. PACE Aerospace Engineering & Information Technology GmbH. Pacelab Aircraft Preliminary Design. Version 3.0.1. Software. 2012.
99. Torenbeek, E. *Synthesis of Subsonic Airplane Design: An Introduction to the Preliminary Design of Subsonic General Aviation and Transport Aircraft, with Emphasis on Layout, Aerodynamic Design, Propulsion and Performance*, 1st ed.; Kluwer: Dordrecht, The Netherlands, 1996.
100. Gologan, C. A Method for the Comparison of Transport Aircraft with Blown Flaps. Ph.D. Thesis, Technische Universität München, München, Germany, 2010.
101. Pornet, C. Conceptual Design Methods for Sizing and Performance of Hybrid-Electric Transport Aircraft. Ph.D. Thesis, Technische Universität München, München, Germany, 2018.

102. Vratny, P. Conceptual Design Methods of Electric Power Architectures for Hybrid Energy Aircraft. Ph.D. Thesis, Technische Universität München, München, Germany, 2019.
103. Society of Automotive Engineers. *ARP4761, Guidelines and Methods for Conducting the Safety Assessment Process on Civil Airborne Systems and Equipment*; SAE International: Warrendale, PA, USA, 1996.
104. Society of Automotive Engineers. *ARP4754 Certification Considerations for Highly-Integrated or Complex Aircraft Systems*; SAE International: Warrendale, PA, USA, 1996.
105. Kritzinger, D. *Aircraft System Safety. Assessments for Initial Airworthiness Certification*; Woodhead Publishing: Duxford, UK, 2016.
106. Castillo Pardo, A. *Results of the Fuselage-Fan Rig Testing*; Public Project Deliverable D3.05, Grant Agreement No. 723242; CENTRELINE Consortium; European Commission Directorate-General for Research and Innovation: Brussels, Belgium, 2020.
107. Gubisch, M. Why Airbus Foresees Laminar Wings on Next-Gen Aircraft. Flight Global. 2018. Available online: https://www.flightglobal.com/analysis/analysis-why-airbus-foresees-laminar-wings-on-next-gen-aircraft/128247.article?adredir=1 (accessed on 3 March 2020).
108. AIRBUS S.A.S. A340-500/-600 Aircraft Characteristics Airport and Maintenance Planning. AIRBUS S.A.S. Blagnac Cedex, France (Revision No. 19–Jun 01/20). Available online: https://www.airbus.com/aircraft/support-services/airport-operations-and-technical-data/aircraft-characteristics.html (accessed on 4 November 2020).
109. Peter, F.; Castillo Pardo, A.; Della Corte, B.; Goliszek, B.; Kowalski, M.; van Sluis, M.; Wortmann, G.; Zhao, X. *Propulsive Fuselage Technology Roadmap towards TRL6 in 2030*; Public Project Deliverable D5.06, Grant Agreement No. 723242; CENTRELINE Consortium; European Commission Directorate-General for Research and Innovation: Brussels, Belgium, 2021.
110. International Civil Aviation Organization. *Aerodrome Standards–Aerodrome Design and Operations. ICAO Annex 14*, 3rd ed.; International Civil Aviation Organization: Montreal, QC, Canada, 1999.
111. Seitz, A.; Engelmann, M. *Multi-Disciplinary Evaluation of the Optimized Propulsive Fuselage Aircraft*; Public Project Deliverable D2.03, Grant Agreement No. 723242; CENTRELINE Consortium; European Commission Directorate-General for Research and Innovation: Brussels, Belgium, 2020.
112. Green, J.E. Greener by Design—the Technology Challenge. *Aeronaut. J.* **2002**, *106*, 57–113. [CrossRef]
113. Tacina, K.M.; Wey, C. *NASA Glenn High Pressure Low NOx Emissions Research*; NASA/TM—2008-214974; National Aeronautics and Space Administration: Cleveland, OH, USA, 2008.
114. Grönstedt, T.; Dax, A.; Kyprianidis, K.; Ogaji, S. *Low-Pressure System Component Advancements and its Influence on Future Turbofan Engine Emissions*; GT2009-59950; ASME Turbo Expo: Orlando, FL, USA, 2009; pp. 505–516.
115. Depitre, A. *Noise Certification Workshop, Session 2: EPNdB Metric*; International Civil Aviation Organization: Montreal, QC, Canada, 2006.
116. Zhao, X. *Benchmark of the Propulsive Fuselage Concept against the Y2035 SRIA Targets*; Confidential Project Deliverable D1.04, Grant Agreement No. 723242; CENTRELINE Consortium; European Commission Directorate-General for Research and Innovation: Brussels, Belgium, 2021.
117. ICAO. *Aircraft Engine Emissions Databank*; International Civil Aviation Organization: Montreal, QC, Canada, 2020.
118. Clark, I.A.; Thomas, R.H.; Guo, Y. Noise Reduction Approaches for the NASA D8 Subsonic Transport Concept. In Proceedings of the 22nd Workshop of the Aeroacoustics Specialists Committee of the CEAS, Netherlands Aerospace Center, Amsterdam, The Netherlands, 6–7 September 2018.
119. Association of European Airlines. *Operating Economy of AEA Airlines*; Association of European Airlines: Brussels, Belgium, 2007.
120. Plötner, K.O.; Schmidt, M.; Baranowski, D.; Isikveren, A.T.; Hornung, M. Operating Cost Estimation for Electric-Powered Transport Aircraft. In Proceedings of the Aviation Technology, Integration, and Operations Conference, AIAA 2013-4281, Los Angeles, CA, USA, 12–14 August 2013.
121. Plötner, K.; Wesseler, P.; Phleps, P. Identification of key aircraft and operational parameters affecting airport charges. *Int. J. Aviat. Manag.* **2013**, *2*, 91–115. [CrossRef]
122. Gieras, J.F. Superconducting electrical machines–state of the art. *Prz. Elektrotechniczny* **2009**, *12*, 1–21.
123. Grilli, F.; Benkel, T.; Hänisch, J.; Lao, M.; Reis, T.; Berberich, E.; Wolfstädter, S.; Schneider, C.; Miller, P.; Palmer, C.; et al. Superconducting motors for aircraft propulsion: The Advanced Superconducting Motor Experimental Demonstrator project. *J. Phys. Conf. Ser.* **2020**, *1590*, 012051. [CrossRef]
124. Seitz, A.; Isikveren, A.T.; Bijewitz, J.; Mirzoyan, A.; Isyanov, A.; Godard, J.-L.; Stückl, S. Summary of Distributed Propulsion and Ultra-high By-pass Rotor Study at Aircraft Level. In Proceedings of the 7th European Aeronautics Days 2015, London, UK, 20–23 October 2015.

Article

Effect of Increased Cabin Recirculation Airflow Fraction on Relative Humidity, CO_2 and TVOC

Victor Norrefeldt [1,*], Florian Mayer [1], Britta Herbig [2], Ria Ströhlein [2], Pawel Wargocki [3] and Fang Lei [3]

1 Fraunhofer Institute for Building Physics IBP, 83626 Valley, Germany; florian.mayer@ibp.fraunhofer.de
2 Institute and Clinic for Occupational, Social and Environmental Medicine, LMU University Hospital Munich, 80336 Munich, Germany; Britta.Herbig@med.uni-muenchen.de (B.H.); Ria.Stroehlein@med.uni-muenchen.de (R.S.)
3 Department of Civil Engineering, Technical University of Denmark, 2800 Kgs. Lyngby, Denmark; paw@byg.dtu.dk (P.W.); fl@byg.dtu.dk (F.L.)
* Correspondence: victor.norrefeldt@ibp.fraunhofer.de

Abstract: In the CleanSky 2 ComAir study, subject tests were conducted in the Fraunhofer Flight Test Facility cabin mock-up. This mock-up consists of the front section of a former in-service A310 hosting up to 80 passengers. In 12 sessions the outdoor/recirculation airflow ratio was altered from today's typically applied fractions to up to 88% recirculation fraction. This leads to increased relative humidity, carbon dioxide (CO_2) and Total Volatile Organic Compounds (TVOC) levels in the cabin air, as the emissions by passengers become less diluted by outdoor, dry air. This paper describes the measured increase of relative humidity, CO_2 and TVOC level in the cabin air for the different test conditions.

Keywords: aircraft air quality; adaptive ECS; subject testing

Citation: Norrefeldt, V.; Mayer, F.; Herbig, B.; Ströhlein, R.; Wargocki, P.; Lei, F. Effect of Increased Cabin Recirculation Airflow Fraction on Relative Humidity, CO_2 and TVOC. *Aerospace* **2021**, *8*, 15. https://doi.org/10.3390/aerospace8010015

Received: 24 November 2020
Accepted: 23 December 2020
Published: 13 January 2021

1. Introduction

The environmental control system (ECS) is one of the major secondary energy consumers of an aircraft, consuming 3–5% of the power produced by the engines [1]. As outside conditions in cruise altitude are life threatening, the environmental control system fulfils a lifesaving function for the passengers by maintaining cabin pressurization and furthermore maintains a comfortable temperature and air quality.

In most commercial aircraft, the outdoor air used to ventilate the cabin is extracted from engine compressor stages upstream of the combustion chamber. Due to the compression, this air, called bleed air, is hot and its temperature exceeds 200 °C. An expansion to cabin pressure and cooling by ram air bring down temperature to an adequate level. In the mixing chamber, this outdoor air is mixed with HEPA (High Efficiency Particulate Arresting) filtered recirculation air. From there, the air supply passes through the riser ducts to the cabin air outlets, from where it enters the cabin [2].

Despite the low outdoor temperatures in cruise (below −50 °C) and the relatively thin layer of insulation between cabin and skin (order of 2–4 inches), the high passenger density usually leads to a cooling requirement for the cabin. Thus, the ECS primarily supplies air below cabin temperature. The moisture in the cabin is typically low, because the outside air is cold and thus dry. The major source for humidification are the passengers themselves, emitting water by breathing and sweating. Along with this, passengers exhale CO_2 and emit VOCs (volatile organic compounds) as a product of metabolism. Toiletry and cleaning agents are additional sources of VOCs in the cabin. These compounds are diluted by the outdoor air, while the recirculation airflow does not alter the composition. Aircraft ECS are required to comply with target values for both pressure and temperature. Typical values for the requirements are listed below [3].

- Minimum cabin air pressure limited to 750 hPa (equivalent to pressure at an altitude of 8.000 ft. or 2.400 m above sea level)
- Temperature range in the cabin between 18.3 and 23.9 °C (65 to 75 °F)
- Minimum outside airflow rate per passenger of 3.5 L/s (7.5 cfm)
- Recommendation of 9.4 L/s (20 cfm) total flow, to be met by outside and filtered recirculated air

The bleed air offtake reduces the efficiency of the engine because some compressed air is extracted rather than contributing to fuel burn and thrust production. Therefore, a reduction of bleed air is associated with lower fuel burn. In the building sector, demand controlled ventilation already today implements a similar philosophy based on reducing the supply airflow rate to that actually required, thereby avoiding excessive ventilation and thus energy use. Often, CO_2 concentration in the occupied space is used as a feedback signal for the control of the ventilation system as it is closely correlated with other human emissions and the occupation density [4]. DIN EN 15251 [5] gives quality criteria for the indoor air based on CO_2 level and states that concentrations below 800 ppm are optimal and concentrations above 1.200 ppm are not acceptable. Whether these design values can be directly applied to aircraft is not clear because buildings typically do not have HEPA filtration in the recirculation air, whereas this is standard for aircraft. A comparison of building and aircraft ventilation requirements is provided in [6].

A series of measurements taken during commercial flights by Giaconia et al. [7], found that average CO_2 concentrations varied between 925 and 1449 ppm. Unfortunately, the publication is not clear on whether the CO_2 readings were compensated for pressure or not. Typically, a pressure correction needs to include the factor $1.013/p_{cabin}$. [8]. In a similar approach on 179 domestic US flights Cao et al. [9] found an average CO_2 concentration of 1353 ± 290 ppm in the cabin. The researchers clearly state that sensor readings were pressure compensated. At the current stage, a review paper on cabin air quality is being prepared by the authors. This review confirms the magnitude of CO_2-concentrations reported.

In this context, the ComAir study was proposed to further investigate aircraft ventilation strategies and the impact of indoor air quality on passengers' comfort and well-being.

In today's aircraft, the adjustment of the ECS to the passenger count is limited to discrete, manual steps, like switching off an ECS pack or running it on low or high setting, however no controlled adaption of the flow is implemented. The research performed in the ComAir project shall give the basic design parameters for the possible future implementation of an adaptive ECS. This system is investigated in a CleanSky2 project [10]. For this, the impact of decreased outdoor air intake and increased recirculation flow rate is investigated in a subject study. Half full and fully booked conditions are investigated to assess the effect of passenger count.

The ComAir study was approved by the Ethics committee of the Faculty of Medicine, Ludwig-Maximilians-University, Munich (ID: 19-256) and written informed consent was obtained from all participants. The tests were performed in November 2019 and thus before the Sars-Cov-2 pandemic was an issue in Germany. This paper summarizes the resulting CO_2, humidity and TVOC levels measured in a cabin mock-up for the different outdoor air intakes.

2. Method

To investigate the effect of reduced outdoor air intake and increased recirculation airflow rate, a randomized controlled study with a total of 559 different participants was conducted. The study was hosted in the Flight Test Facility (FTF, https://www.hoki.ibp.fraunhofer.de/vr/virtual-tour_IBP/) located at the Fraunhofer-Institute for Building Physics IBP in Holzkirchen, Germany.

2.1. Flight Test Facility Test Setup

In the FTF, a former in-service front section of an A310 is located in a low pressure vessel (Figure 1a). The cabin (Figure 1b) hosts up to 80 passengers and is surrounded by crown, forward galley, cockpit, avionics, triangles, cargo and bilge compartments. The aircraft mock-up consists of the section from nose to front wing box wall (Figure 1c). The ECS is emulated by a building type air conditioning system with a sub freezer that cools and dehumidifies the supply air to a dew point of −20 °C. Thus, a similar low humidity level of supply air can be reached as in flight. Air was supplied through ceiling outlets above the stowage bin and lateral outlets that have been added to better emulate today's cabin design. The recirculation air is extracted from the cabin and HEPA filtered with original filters. A schematic view of the airflow path is shown in Figure 1d.

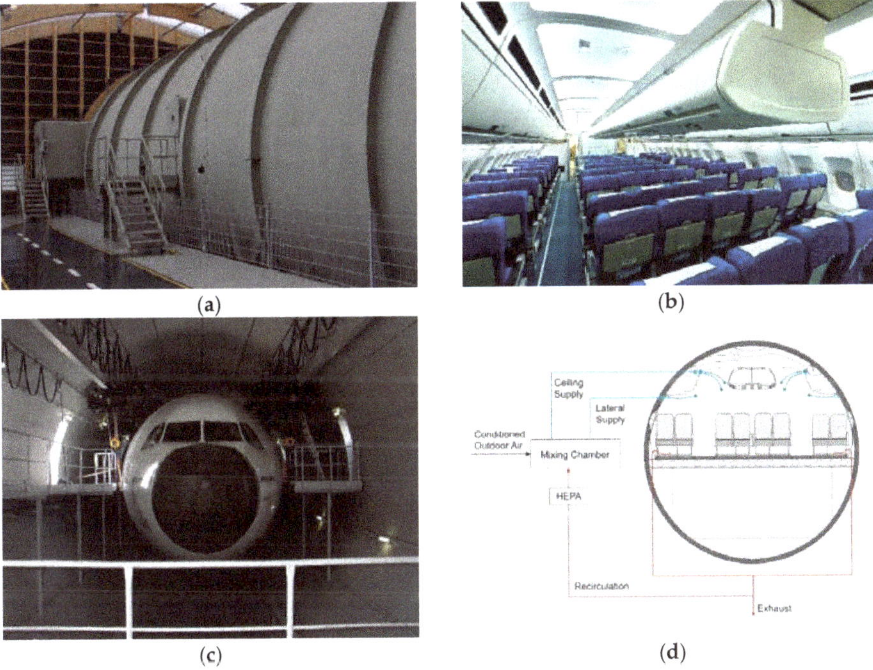

Figure 1. (**a**) Low pressure vessel of FTF, (**b**) Aircraft cabin mock-up, (**c**) Aircraft mock-up in the low-pressure vessel, (**d**) schematic ventilation pattern.

In the cabin, the temperature stratification was measured at eight locations at heights of 0.1, 0.6, 1.1, 1.7 and 2.2 m. CO_2 and relative humidity were measured at a height of 1.1 m at these locations (Figure 2). VOCs are pumped from the front left measurement location on sample tubes and later analyzed by GC-MS (gas chromatography-mass spectrometry) or by HPLC-DAD (high performance liquid chromatography with diode array detector) in the laboratory. CO_2 sensors were calibrated with 4,000 ppm calibration gas both at ground pressure (~940 hPa for Holzkirchen due to the place's elevation) and low pressure of 755 hPa in the vessel and a pressure correction was derived. The outdoor and recirculation airflow rates were measured with anemometers; a sufficient length of the straight pipe section to ensure accurate measurement was ensured. The following sensor types and specifications were used:

- Temperature: Four-wire PT100 thermocouples with an accuracy ±0.1 K @ 20 °C according to DIN EN 60751 [11] class A.
- Humidity: Rotronic HygroClip HC2-C05 sensor with ±1.5% RH [12]

- CO_2: Vaisala GMW20, range: 0–5.000 ppm, accuracy ±2%. Sensors were calibrated with 4.000 ppm calibration gas both at ground pressure (~940 hPa for Holzkirchen due to the place's elevation) and low pressure of 755 hPa in the vessel and a pressure correction was derived.
- Pumped tubes: samples drawn for 20 to 60 min with flow rates of 0.1 to 1.0 l/min (depending on target compounds) and analyzed by GC-MS (gas chromatography-mass spectrometry QP2010 SE, Shimadzu, Duisburg, Germany) or HPLC-DAD (high performance liquid chromatography with diode array detector, Agilent 1260 Infinity, Agilent Technologies, Waldbronn, Germany). VOCs were analyzed according to DIN ISO 16000-6 [13] and carbonyl compounds according to DIN ISO 16000-3 [14].
- Flow rate: Schmidt SS20.500 Sensors 0–35 m/s with an accuracy of ±3% [15].

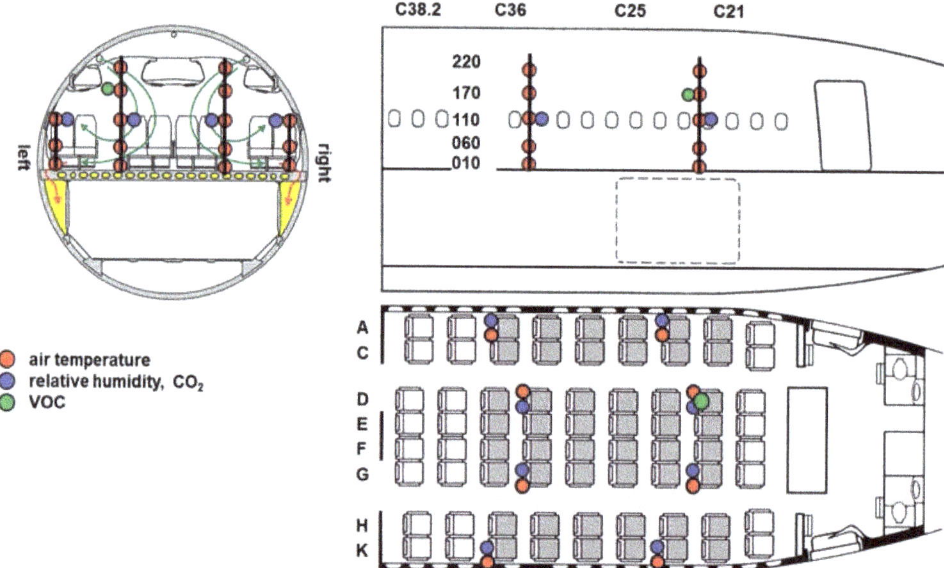

Figure 2. Cabin measurement locations.

2.2. Test Matrix and Sequence

The study investigated four outdoor airflow rates and two occupancy levels: uncongested/half and fully booked cabin (Table 1). In order to expose a similar number of subjects to each condition, the uncongested conditions were tested twice while the fully booked conditions were each tested once. Subjects were blinded to experimental condition and were only allowed to participate in one test, thus, nobody witnessed two conditions. The lower outdoor airflow rates were compensated with increased recirculation airflow rates in order to maintain a constant total flow rate per passenger of 9.4 L/s. The flow rates were adjusted for each test based on the actual count of subjects in the cabin. As a result, the fully booked conditions had a higher flow rate at the cabin air outlets than the uncongested conditions. Cabin temperature set-point was 23 °C in order to maintain a thermally comfortable situation. When setting up the test matrix, the following considerations were made:

- Baseline: Replication of today's typical CO_2 levels reported in aircraft
- ASHRAE: Replication of the minimum requirement for outdoor airflow rate (3.5 L/s/passenger) set out by ASHRAE 161

- ASHRAE half: Half the required outdoor airflow rate (1.8 L/s/passenger). Because the CO_2 level follows the inverse of the outdoor airflow rate, this point was chosen because it was pre-assessed to be in the middle between the ASHRAE and the Max. CO_2 condition.
- Max. CO_2: Lowest outdoor airflow rate designed to remain below 5.000 ppm limit [16].

Table 1. Test Matrix.

Conditions	Baseline	ASHRAE	ASHRAE Half	Max. CO_2
Outdoor airflow rate in L/s/passenger	5.2	3.5	1.8	1.1
Recirculation airflow rate in L/s/passenger	4.2	5.9	7.6	8.3
Total airflow rate in L/s/passenger	9.4	9.4	9.4	9.4
Fully booked (~70–80 PAX)	1 session	1 session	1 session	1 session
Uncongested (~35–40 PAX)	2 sessions	2 sessions	2 sessions	2 sessions

The following names will be used for the different test cases:

- Baseline–uncongested 1 (1st Session)
- Baseline–uncongested 2 (2nd Session)
- Baseline–fully booked
- ASHRAE–uncongested 1 (1st Session)
- ASHRAE–uncongested 2 (2nd Session)
- ASHRAE–fully booked
- ASHRAE half–uncongested 1 (1st Session)
- ASHRAE half–uncongested 2 (2nd Session)
- ASHRAE half–fully booked
- Max. CO_2–uncongested 1 (1st Session)
- Max. CO_2–uncongested 2 (2nd Session)
- Max. CO_2–fully booked

The test sequence is shown in Figure 3. Subject reception and medical pre-screening was 1h before the test started. After boarding, the pressure in the chamber was reduced (Figure 4). When flight altitude was reached, the desired ventilation regime was set. Subjects answered psychological and health questionnaires and were tested for performance in Batteries 1 and 2, and detailed questionnaires on comfort were distributed in the middle and at the end of the session (Comfort Battery and Battery 2). After test battery 2 the cabin was repressurized and subjects deboarded.

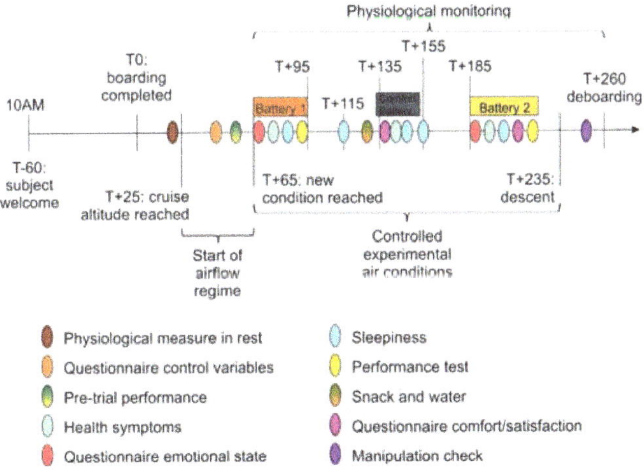

Figure 3. Test sequence.

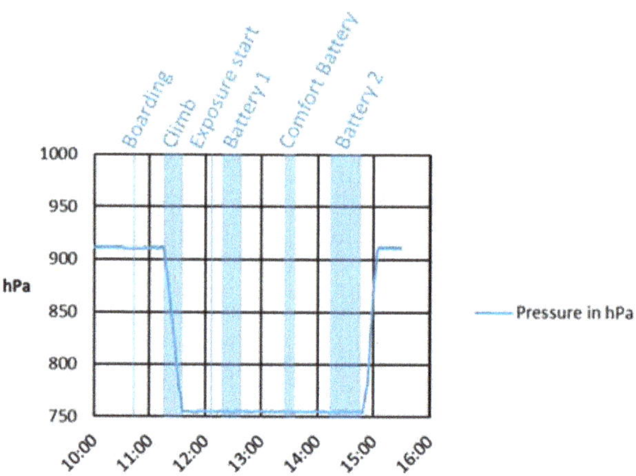

Figure 4. Example of Cabin pressure profile.

Only low emitting food (pretzels) and water were served to the subjects to not artificially alter the VOC composition in the cabin. In a real flight, coffee, tea, juices, alcoholic drinks and food with higher emission may be served as well. While for the high outdoor airflow rates, these events are expected to be flushed out quickly from the cabin, the low outdoor airflow rates show noticeable transient times and thus the effect of such events would persist longer. Furthermore, possible exhaust ingestion or engine oil (fume) events were disregarded in this study as it focused on emissions generated inside the cabin.

2.3. Assessment of Air Quality

The assessment of the indoor environment was performed by a trained sensory panel according to the requirements of ISO 16000-30 [17] and subjects' ratings via questionnaire at the end of the session.

For the first test of each condition, a sensory panel consisting of eight trained test persons entered the cabin and assessed the perceived intensity, the hedonic odor tone and the odor quality of the cabin air. Their evaluation of perceived intensity (pi) uses of a comparative scale, that assigns the odor intensity of air to a defined concentration of acetone (Figure 5). Hedonic odor tone is assessed on a 9 point category scale from −4 (very unpleasant), via 0 (neither pleasant nor unpleasant) to +4 (very pleasant).

Within the set of questionnaires at the end of exposure (i.e., before descent), all subjects were asked to:

- rate the smell in the cabin on a five point scale (How would you assess the odor intensity in this flight? no odor; slight odor; moderate odor; strong odor; overwhelming odor),
- evaluate the air quality with a five point Likert scale (How would you rate the air quality in this flight? very poor; poor; average; good; very good/excellent)
- assess the acceptability of the air quality with an approach used by [18,19].

Overall, 559 persons representative of flight passengers with regard to sex and age (283 men, 276 women; mean age 42.68 ± 15.85 years) participated in the study and rated these parameters.

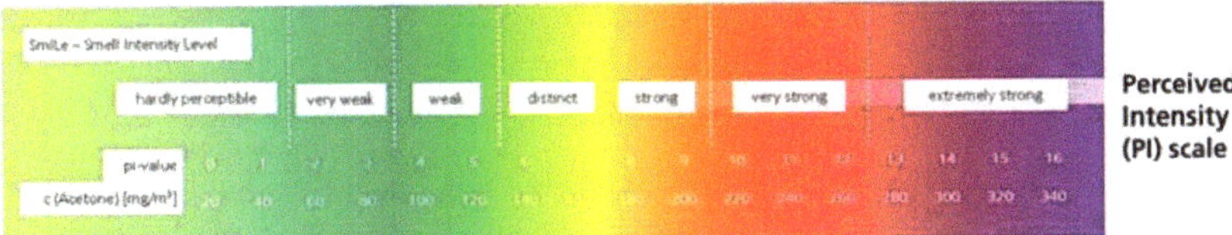

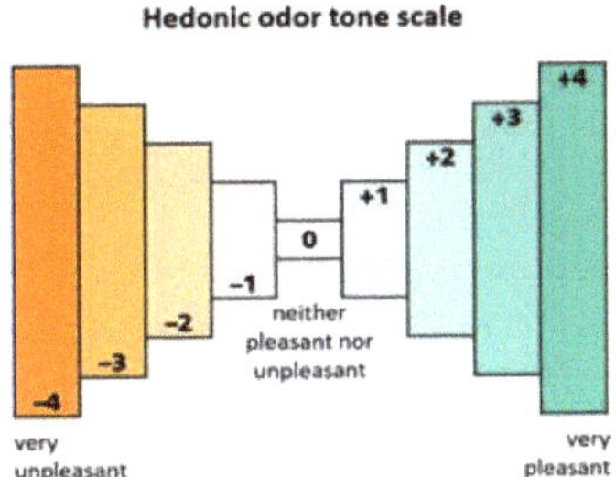

Figure 5. Top: Perceived intensity scale, **Bottom**: Hedonic odor tone scale.

2.4. *Test Preparations*

2.4.1. Considerations on Heat Balance

Based on the test matrix, some preliminary rule of thumb considerations were made on the heat balance of the cabin. As cabin cooling is provided by conditioned outdoor air, it was investigated to which extent the reduced flow of outdoor air is capable to fulfil this function. In this balance, the recirculation air is not considered, as from a thermal point of view it is extracted and re-injected into the cabin. For reasons of simplicity, it was assumed that each passenger brings in 100 W of heat by metabolism, electrical devices and in-flight entertainment. Neglecting heat losses through the envelope and heat gain from e.g., fan operation, the required outdoor air temperature entering the mixing box can be estimated by the equation below. Results of this assessment are summarized in Table 2. Especially for outdoor flow rates below the ASHRAE requirement of 3.5 L/s, the required temperature to maintain the cabin thermal balance becomes unrealistically low. Reaching such low temperatures would require a large ram air heat exchanger and would only be possible during cruise phase. Therefore, it was concluded that a recirculation heat exchanger connected to a mobile cooling machine will be needed for the experimental campaign. Based on the occupancy of 80 passengers, it was estimated that this cooling unit needs to extract up to 8 kW of heat assuming a close to zero cooling capacity of the lowest outdoor airflow rate of 1.1 L/s.

$$T_{supply} = T_{cabin} - \frac{\dot{Q}_{pax}}{\dot{V}_{outdoor} \cdot \rho \cdot c_p}$$

with: T_{supply}: Supply air temperature of outdoor air in °C, T_{cabin}: desired cabin air temperature (23 °C), Q_{pax}: Heat emission per passenger (100 W), $V_{outdoor}$: Outdoor airflow rate per passenger in L/s (norm conditions), ρ: air density (1.2 kg/m^3 at norm conditions) and c_p: 1004.5 J/(kg·K).

Table 2. Estimated requirement for outdoor air temperature and estimated cooling power required in the recirculation air path.

Conditions	Baseline	ASHRAE	ASHRAE Half	Max. CO_2
Outdoor airflow rate in L/s/PAX	5.2	3.5	1.8	1.1
Required outdoor air temperature	7 °C	−1 °C	−23 °C	−52 °C
Recirculation cooling power	n/a	n/a	4 kW	8 kW

2.4.2. Subject Safety

High priority was given to the safety of subjects in the cabin. Therefore, a detailed safety and hazard assessment was performed prior to the subject tests.

To ensure a safe conduction of the tests, a review of fire escape procedures was performed. Unnecessary equipment resulting in burn load was removed and smoke protection masks were purchased in sufficient number. It was ensured that fire extinguishers were in place and had a valid seal of approval. Escape maps were reviewed and updated. Prior to the test conduct, subjects were informed in the use of the smoke masks and the location of fire escape lanes, similar to the floating vest and exit instructions on a commercial flight.

It was ensured that at least one of the cabin crew had passed a recent first aid course. Furthermore, a medical doctor was onsite during the entire test carrying a medical emergency valet with him/her.

Prior to the test campaign, the test setup as well as the possibility to enter and leave the cabin through a pressure lock during low pressure operation was inspected by the local emergency rescue team.

The expected exposure in terms of CO_2 and VOCs was pre-assessed and it was assured that no regulatory limits were exceeded like e.g., FAR part 25, ref [16].

3. Results

3.1. Flow Rates

Figure 6 shows the flow rate per passenger. Obviously, the outdoor flow rate was maintained both for the uncongested and the fully booked cases. The total flow rate was close to 9.4 L/s/passenger for the tests. In the cases "ASHRAE uncongested 2" and "ASHRAE half uncongested 2" a slightly higher recirculation flow rate was applied because only a lower number of subjects participated (35 and 34 instead of 40). For the fully booked "Max. CO_2" case, the recirculation fan came to its limit resulting in slightly lower flow rate.

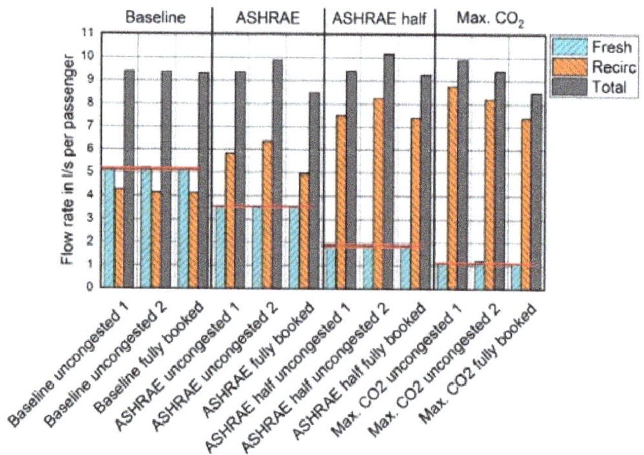

Figure 6. Passenger based airflow rates in the cabin.

Figure 7 shows the total airflow rates in the cabin and thereby proves the adaptive nature of the selected study approach. Due to the day by day adjustment of the airflow rate to the passenger count, each day results in different total flow rates and the fully booked cases have a noticeably higher total flow rate.

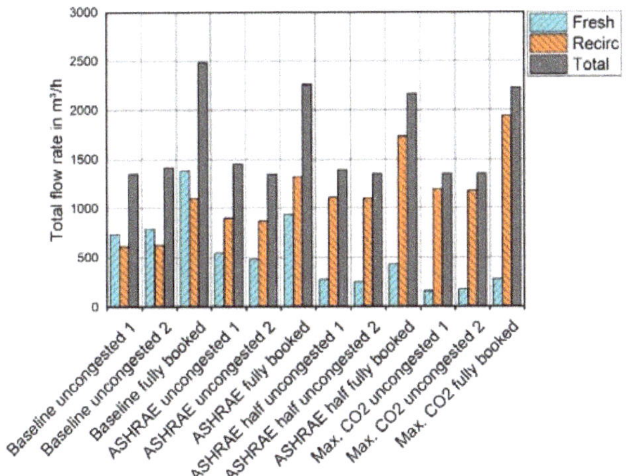

Figure 7. Total airflow rates in the cabin.

3.2. Temperatures

Temperatures were averaged for the entire exposure and for the timeslots the subjects filled out questionnaires (Figure 8). The cabin temperature and stratification usually were around ±1.5 K and thus below the limit of 2 K for a class A rating according to DIN EN ISO 7730 [20], therefore only the average temperature in the occupied zone (0–1.1 m height) is reported here. The temperature target of 23 °C was mostly maintained at ±1 K. It can be inferred by the measurements that the "Baseline fully booked" condition results in a colder cabin temperature because subjects were close to the control feedback sensor in the cabin and sufficient cooling power to react was available. For the "Max. CO_2 fully booked" condition, the available cooling power of the recirculation heat exchanger came to its limit and therefore temperature in the cabin was higher. For this case, a larger cooling system than the available one would have been necessary, possibly due to the additional heat injected by the recirculation fan operating at full speed.

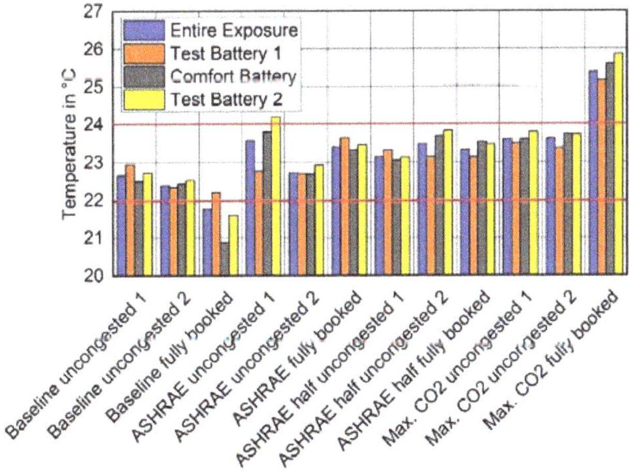

Figure 8. Average cabin temperature.

3.3. Cabin Humidity

A clear trend of increased cabin relative humidity with decreasing outside airflow rate is obvious (Figure 9). In the "Baseline" conditions, values around 15% RH were measured that are well in line with other publications [7,9]. For the "ASHRAE half" and "Max. CO_2" conditions, a gradual increase of relative humidity from test battery 1 over comfort battery to test battery 2 could be noticed. Due to the low outside airflow rate, the time constant of the cabin to reach steady-state was higher and thus the build-up phase time increased. Even at the end of the test, an increase in moisture was still obvious. For the fully booked case, the total flow rate was higher, and therefore humidity level was closer to convergence for the "Max. CO_2 fully booked" case, compared to the uncongested case at which a build-up could still be seen at the end of the test (Figure 10). Considering test battery 2 to be representative of the steady-state conditions reached at the end of the tests for all the experiments, a maximum value of 33.7% RH was measured in case of the most extreme condition ("Max. CO_2 fully booked") at a cabin temperature of 25.9 °C. This corresponds to a moisture content of 6.94 g/kg.

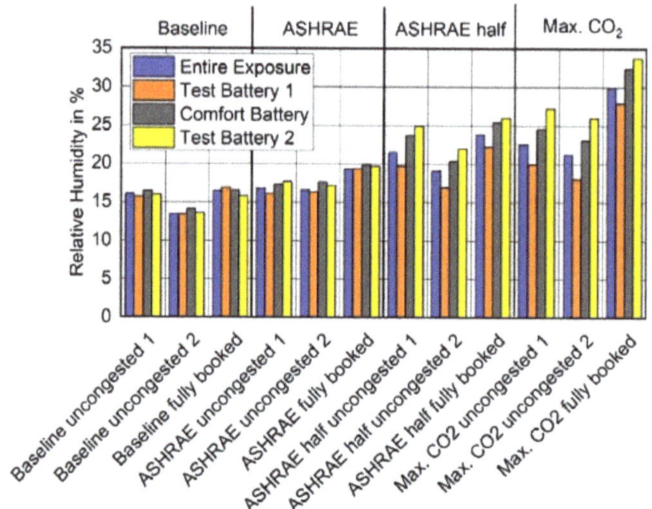

Figure 9. Measured average relative humidity in the cabin.

3.4. CO_2 Concentration

A clear trend of increased cabin CO_2 concentration with decreasing outside airflow rate is visible (Figure 11). For the baseline case, the content was 1660 to 1721 ppm (pressure corrected) at the end of the test. For the "ASHRAE half" and "Max. CO_2" cases, a gradual increase over the test time is obvious (Figure 12), similar to the findings for relative humidity and with a maximum CO_2 level in case of "Max. CO_2" between 3853 and 4520 ppm.

3.5. TVOCs

TVOC detection was performed by grab sampling and off-line analysis by GC-MS. Results are shown in Figure 13. The general tendency to increased TVOC levels with lower outdoor airflow rate is obvious, however less expressed than for humidity and CO_2. The peak in the "ASHRAE half–fully booked" is due to a peak in ethanol. Even though the subjects were told only to drink water and were observed by the cabin crew, it cannot be excluded that somebody brought alcoholic beverages on board. The peak in the "Max. CO_2–uncongested 2" condition is due to an event that made it necessary to disinfect the galley with isopropanol cleaning agent during the test conduction. Generally, VOCs are below or within the limits of UBA [4] level 2 (300–1.000 μg/m^3, "no relevant objections, increased ventilation recommended").

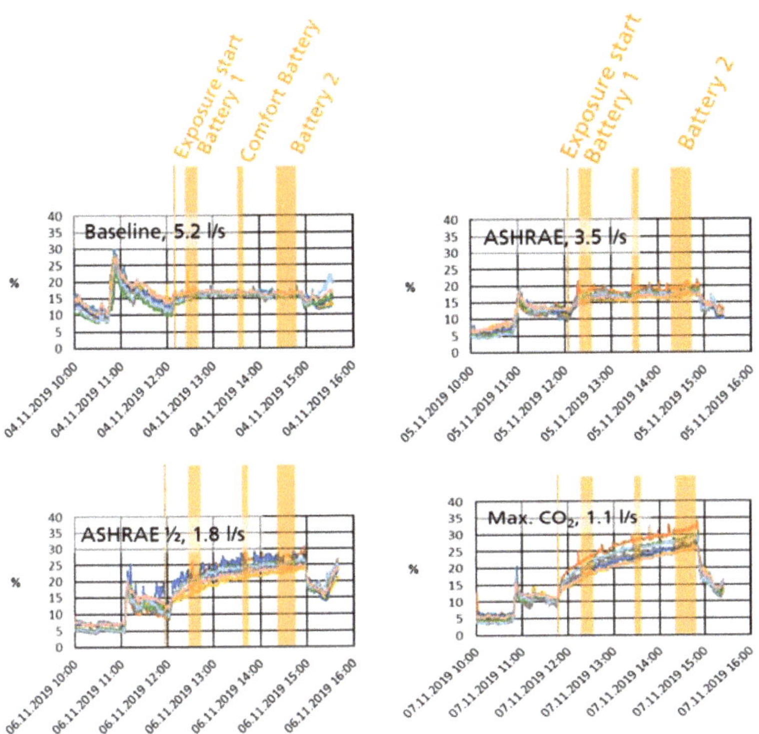

Figure 10. Transient evolution of relative humidity in the cabin for four different test days with different outside airflow rates.

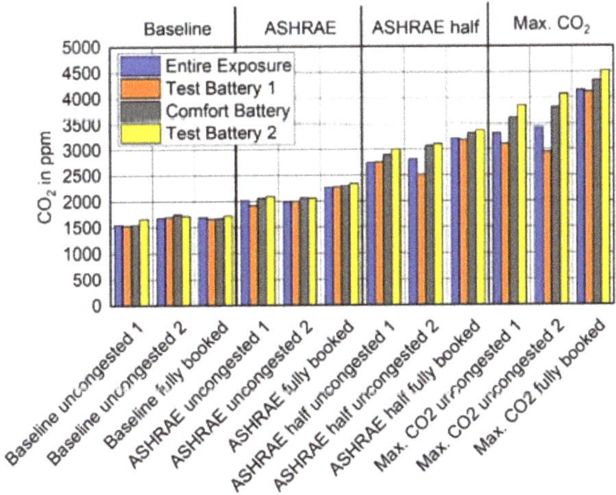

Figure 11. Measured average CO_2 concentration in the cabin (pressure corrected).

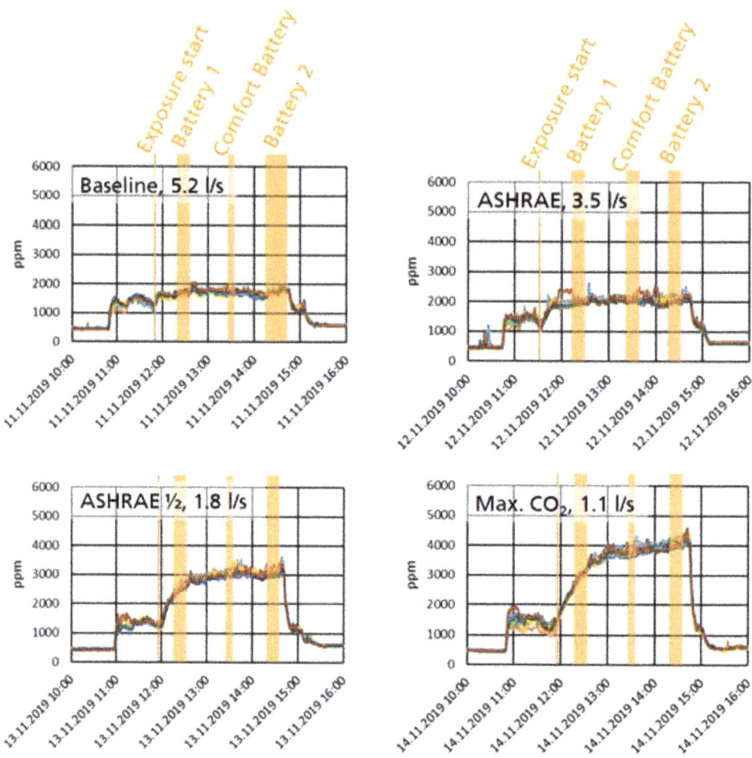

Figure 12. Transient evolution of CO_2 concentration (pressure corrected) in the cabin for four different test days with different outside airflow rates.

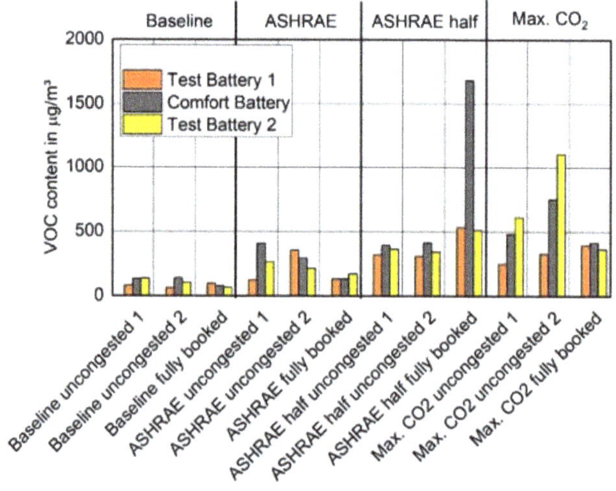

Figure 13. Measured TVOC concentration in the cabin air.

In order to assess whether VOCs could have been absorbed by condensed water in the recirculation heat exchanger, the drainage valve was opened after each test. Water was never observed and thus it was concluded that no major condensation occurred.

3.6. Trained Panel Votes

The panel of eight trained persons was used in the cases "Baseline uncongested 1", "ASHRAE uncongested 1", "ASHRAE half uncongested 1" and "Max. CO_2 uncongested 1". The panel entered the cabin after the comfort battery, slowly walked up and down the aisle and then left to independently give their votes. Through this short residence time, odor adaptation is avoided. The result shows no clear trend to increased perceived intensity or worse hedonic tone. The votes show a distinct intensity of smell and a slightly unpleasant tone (Figure 14).

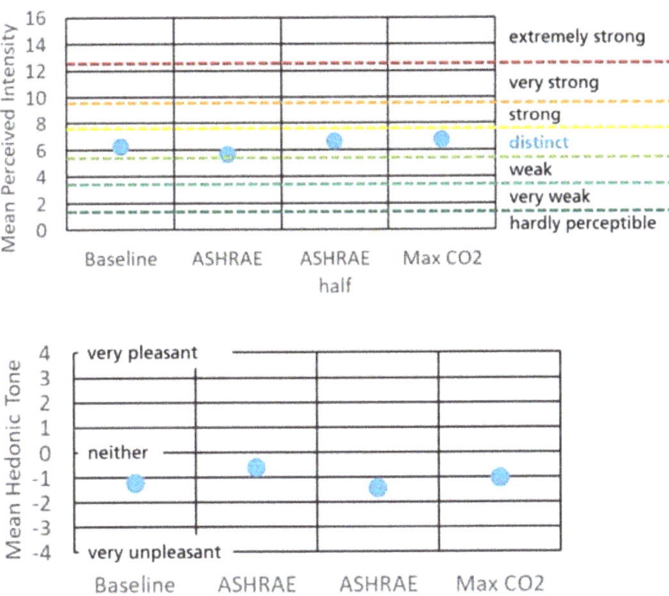

Figure 14. Assessment of air quality by trained odor panel.

3.7. Subject Votes

In all test cases, subjects were asked at the end of the exposure to assess the smell in the cabin and the acceptability of air quality. For the "Baseline" and "ASHRAE" conditions, the votes on smell are quite similar for the uncongested (Figure 15 left) and the fully booked case (Figure 15 right). For "ASHRAE half" and "Max. CO_2", the fully booked condition shows less votes for "no smell" and more votes for "slight" and "moderate" smell perception whereas for the uncongested condition, no systematic difference is found. Analyses of variance (ANOVA) confirm this: there is no main effect of ventilation regime but a small main effect of occupancy ($F = 2.99$, $p = 0.08$), interaction of both is not significant ($F = 1.72$, $p = 0.16$). That is, the slightly higher level of "slight" and "moderate" smell perceptions in the fully booked conditions with lower outdoor airflow rates are sufficient for an overall difference in perception between uncongested and fully booked sessions, although general levels of smell perception are very low. A similar pattern is found for the acceptability of air quality. However, analyses of variance for this variable show rather clear effects; a decrease of acceptability with decreasing outdoor airflow rate (main effect ventilation regime: $F = 3.36$, $p = 0.019$), lower acceptability in the fully booked condition (main effect occupancy: F = 5.68, p = 0.017) and an interaction effect between both (F = 3.97, p = 0.008), denoting that only the fully booked condition shows a linear decrease in the acceptability ratings. Figure 16 presents estimated marginal means from this analyses. Nevertheless, the overall acceptability is higher than 75% in all tests performed, and in the dichotomous approach by Wargocki [18,19] more than 93.4% of all participants rated the air quality as acceptable (4.1% inacceptable, 2.5 missing).

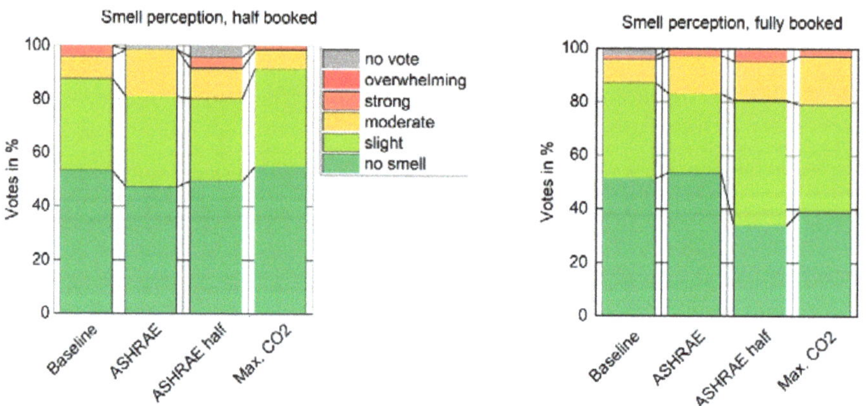

Figure 15. Smell perception votes for the different outdoor airflow rates. **Left**: uncongested, **Right**: Fully booked.

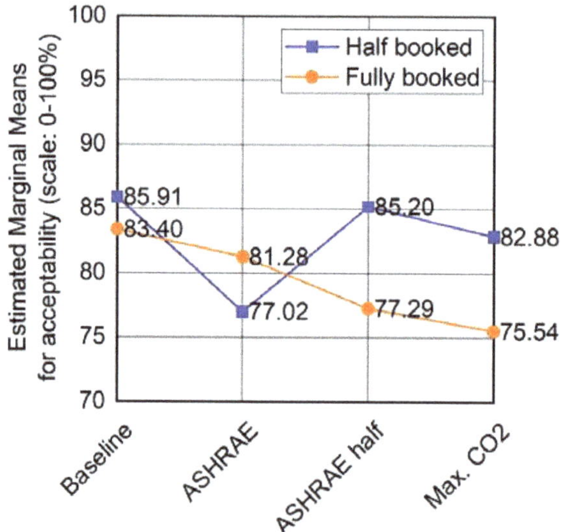

Figure 16. Acceptability ratings on a scale from 0–100 controlled for age, sex, smoking and self-reported multiple chemical sensitivity of participants.

4. Discussion

This paper shows the effect of lower outdoor airflow rate on cabin relative humidity, CO_2 concentration and TVOC level. The measurements were taken during a subject study with a randomized controlled design, blinded participants, representative of flight passengers with regard to age and sex. Thus, the emitted and measured species should be representative for the flying population.

The measured data for the "Baseline" condition show to realistically replicate humidity and CO_2 levels reported from commercial flights. The TVOC measurement may be impacted by the used mock-up and the strict behavior rules for the subjects. The votes on perceived intensity, hedonic tone, smell perception and acceptability thus must be critically considered with regard to generalizability to "normal" flights. Moreover, results can only be generalized to spaces where the occupants generate the major emission of VOCs (bioeffluents).

Nevertheless, the principle of an adaptive ECS operation was successfully proven in this study. Here, the airflow rate was adjusted to the number of passengers. Future research should be performed to identify markers in the cabin air that could be used as feedback signal for a controller. Such a marker could be CO_2, however it would not cover e.g., the event of increased TVOC found by cleaning requirement or the potential consumption of alcohol as detected in the cabin during this study.

Airflow rates were selected to generate different levels of CO_2. Whether these airflow rates are compatible with other requirements of the aircraft ventilation like e.g., exhaust airflow rates in the lavatories and galley, cabin pressurization, cooling of avionics, etc. has been disregarded.

The general trend, to have a higher satisfaction with the environment despite a similar exposure in the uncongested case compared to fully booked one was even reported for other comfort parameters in the "Baseline" case, whereas the other three airflow regimes are currently under evaluation [21].

5. Conclusions

This paper presents the indoor air measurements for different outdoor airflow rates in a realistic cabin mock-up with subjects in a simulated flight. The major results are:

- Relative humidity, CO_2 and TVOC clearly increase with decreasing outdoor airflow rate
- Singular effects like an ethanol or cleaning agent event showed higher impact on the TVOC levels than the airflow regime
- Neither a trained sensory panel nor subjects could differentiate smell or acceptability for the different airflow conditions. Only in a fully booked cabin slightly worse votes were given at lower outdoor air intake.
- Low outdoor airflow rates necessitate additional cooling capacity in the recirculation path. This would result in a possible need for redesign of the ECS compared to today's architecture.

Author Contributions: Conceptualization, V.N. and B.H.; methodology, V.N., F.M., B.H., F.L. and P.W.; validation, V.N., B.H. and P.W.; formal analysis, V.N., F.M., B.H. and R.S.; investigation, V.N., F.M. and R.S.; resources, V.N., F.M., B.H. and F.L.; data curation, V.N., B.H. and P.W.; writing—original draft preparation, V.N.; writing—review and editing, V.N., F.M., B.H. and P.W.; visualization, V.N.; supervision, B.H.; project administration, B.H.; funding acquisition, V.N. and B.H. All authors have read and agreed to the published version of the manuscript.

Funding: This study received funding from the Clean Sky 2 Joint Undertaking under the European Union's Horizon 2020 research and innovation programme under grant agreement No. 820872-ComAir-H2020-CS2-CFP07-2017-02.

Institutional Review Board Statement: The ComAir study was conducted according to the guidelines of the Declaration of Helsinki, and approved by the Ethics Committee at the Faculty of Medicine, Ludwig-Maximilians-University, Munich (ID: 19-256) on 17 June 2019.

Informed Consent Statement: Written informed consent was obtained from all subjects involved in this study.

Data Availability Statement: Some data presented in this study are available upon request from the corresponding author. Some data are not publicly available due to contractual restrictions and GDPR regulations regarding privacy for human subject studies.

Acknowledgments: We would like to thank Ivana Ivandic and Ines Englmann for their help with the subjects.

Conflicts of Interest: The authors are responsible for the content of this publication. The authors declare to have no conflict of interest.

References

1. Zavaglio, E.; Le Cam, M.; Thibaud, C.; Quartarone, G.; Zhu, Y.; Franzini, G.; Roux, P.; Dinca, M.; Walte, A.; Rothe, P. Innovative Environmental Control System for Aircraft. In Proceedings of the 49th International Conference on Environmental Systems ICES-2019-171, Boston, MA, USA, 7–11 July 2019.
2. Oehler, B. Modeling and Simulation of Global Thermal and Fluid Effects in an Aircraft Fuselage. In Proceedings of the 4th International Modelica Conference, Hamburg University of Technology, Hamburg-Harburg, Germany, 7–8 March 2005.
3. ASHRAE. *Standard 161-Air Quality Within Commercial Aircraft*; ASHRAE: Atlanta, GA, USA, 2007.
4. Umweltbundesamt: Beurteilung von Innenraumluftkontaminationen mittels Referenz- und Richtwerten. *Bundesgesundheitsbl Gesundheitsforsch Gesundheitsschutz* **2007**, *50*, 990–1005. [CrossRef] [PubMed]
5. *DIN EN 15251:2012-12: Indoor Environmental Input Parameters for Design and Assessment of Energy Performance of Buildings Addressing Indoor Air Quality, Thermal Environment, Lighting and Acoustics*; German Version EN 15251:2007; Beuth Verlag: Berlin, Germany, 2012.
6. Zavaglio, E.; Le Cam, M.; Quartarone, G.; Thibaud, C. An overview of indoor air quality and ventilation standards in commercial buildings and aircrafts. In Proceedings of the Indoor Air Conference, Philadelphia, PA, USA, 22–27 July 2018.
7. Giaconia, C.; Orioli, A.; Di Gangi, A. Air quality and relative humidity in commercial aircrafts: An experimental investigation on short-haul domestic flights. *Build. Environ.* **2013**, *67*, 69–81. [CrossRef]
8. Vaisala: Application Note-How to Measure Carbon Dioxide. 2019. Available online: https://www.vaisala.com/en/file/66231/download?token=k98ud14E (accessed on 1 October 2020).
9. Cao, X.; Zevitas, C.D.; Spengler, J.D.; Coull, B.; McNeely, E.; Jones, B.; Loo, S.M.; MacNaughton, P.; Allen, J.G. The on-board carbon dioxide concentrations and ventilation performance in passenger cabins of US domestic flights. *Indoor Built Environ.* **2019**, *28*, 761–771. [CrossRef]
10. EU Funding and Tenders. Available online: https://ec.europa.eu/info/funding-tenders/opportunities/portal/screen/opportunities/topic-details/jti-cs2-2015-cpw02-sys-02-02 (accessed on 19 June 2015).
11. *DIN EN 60751:2009-05: Industrial Platinum Resistance Thermometer Sensors*; German Version EN 60751:2008; Beuth Verlag: Berlin, Germany, 2009.
12. Rotronic: Gesamtkatalog 2009/10–Feuchte- und Temperaturmessung. 2009. Available online: https://www.rotronic.com/de-de/productattachments/index/download?id=419 (accessed on 1 October 2020).
13. *DIN ISO 16000-6:2012-11: Indoor Air—Part 6: Determination of Volatile Organic Compounds in Indoor and Test Chamber Air by Active Sampling on Tenax TA®Sorbent, Thermal Desorption and Gas Chromatography Using MS or MS-FID*, German version ISO 16000-6:2011; Beuth Verlag: Berlin, Germany, 2012.
14. *DIN ISO 16000-3:2013-01: Indoor air—Part 3: Determination of Formaldehyde and Other Carbonyl Compounds in Indoor Air and Test Chamber Air—Active Sampling Method*; German Version ISO 16000-3:2011; Beuth Verlag: Berlin, Germany, 2013.
15. Schmidt Technology: Flow Sensor SS20.500. 2020. Available online: https://schmidttechnology.de/wp-content/uploads/wpallimport/files/api_files/downloads/Instr_SS20.500_dt.pdf (accessed on 1 October 2020).
16. Federal Aviation Administration (FAA). *Airworthiness Standards: Transport Category Airplanes. Federal Aviation Regulation-Part 25*; FAA: Washington, DC, USA, 2005.
17. *DIN ISO 16000-30:2015-05: Indoor Air—Part 30: Sensory Testing of Indoor Air*; German Version ISO 16000-30:2014; Beuth Verlag: Berlin, Germany, 2015.
18. Wargocki, P. Measurements of the effects of air quality on sensory perception. *Chem. Sens.* **2001**, *26*, 345–348. [CrossRef] [PubMed]
19. Wargocki, P. Sensory pollution sources in buildings. *Indoor Air* **2004**, *14*, 82–91. [CrossRef] [PubMed]
20. *DIN EN ISO 7730:2006-05: Ergonomics of the Thermal Environment–Analytical Determination and Interpretation of Thermal Comfort Using Calculation of the PMV and PPD Indices and Local Thermal Comfort Criteria*; German Version ISO 7730:2005; Beuth Verlag: Berlin, Germany, 2006.
21. Herbig, B.; Ivandic, I.; Ströhlein, R.; Mayer, F.; Norrefeldt, V.; Lei, F.; Wargocki, P. Impact of different ventilation strategies on aircraft cabin air quality and passengers' comfort and wellbeing-the ComAir study. In Proceedings of the ICES-International Conference on Environmental Systems, Lisbon, Portugal, 12–16 July 2020.

MDPI

Article

Design and Optimization of Ram Air–Based Thermal Management Systems for Hybrid-Electric Aircraft

Hagen Kellermann [1,*], Michael Lüdemann [1], Markus Pohl [2] and Mirko Hornung [1]

1 Bauhaus Luftfahrt e. V., Willy-Messerschmitt Straße 1, 82024 Taufkirchen, Germany
2 Institute of Jet Propulsion and Turbomachinery, RWTH Aachen University, 52062 Aachen, Germany
* Correspondence: hagen.kellermann@bauhaus-luftfahrt.net

Abstract: Ram air–based thermal management systems (TMS) are investigated herein for the cooling of future hybrid-electric aircraft. The developed TMS model consists of all components required to estimate the impacts of mass, drag, and fuel burn on the aircraft, including heat exchangers, coldplates, ducts, pumps, and fans. To gain a better understanding of the TMS, one- and multi-dimensional system sensitivity analyses were conducted. The observations were used to aid with the numerical optimization of a ram air–based TMS towards the minimum fuel burn of a 180-passenger short-range partial-turboelectric aircraft with a power split of up to 30% electric power. The TMS was designed for the conditions at the top of the climb. For an aircraft with the maximum power split, the additional fuel burn caused by the TMS is 0.19%. Conditions occurring at a hot-day takeoff represent the most challenging off-design conditions for TMS. Steady-state cooling of all electric components with the designed TMS is possible during a hot-day takeoff if a small puller fan is utilized. Omitting the puller fan and instead oversizing the TMS is an alternative, but the fuel burn increase on aircraft level grows to 0.29%.

Keywords: thermal management; hybrid-electric aircraft; ram air–based cooling; compact heat exchangers; meredith effect

Citation: Kellermann, H.; Lüdemann, M.; Pohl, M.; Hornung, M. Design and Optimization of Ram Air–Based Thermal Management Systems for Hybrid-Electric Aircraft. *Aerospace* **2020**, *8*, 3. https://doi.org/10.3390/aerospace8010003

Received: 27 November 2020
Accepted: 15 December 2020
Published: 23 December 2020

1. Introduction

The introduction of (hybrid-)electric powertrains to future aircraft is one of the innovations that could help to achieve the ambitious goal of a 75% reduction in CO_2 emissions by the year 2050 set by the European Commission's Strategic Research and Innovation Agenda [1]. Thermal management is one of the key challenges for the successful realization of such powertrains [2].

Thermal management systems (TMS) were already part of early motorized aircraft, especially for the cooling of piston engines. When the engine power density increased, air cooling became insufficient and additional radiators were installed to reject heat from the oil system to ambiance. The Mustang P-51D and Messerschmitt Bf 109 are examples of aircraft which had these radiators installed inside a duct with a diffuser and a nozzle to reduce cooling air drag utilizing the so-called Meridith effect [3]. This principal architecture of a ram air–based cooling system is still present in modern aircraft systems, e.g., in the environmental control system [4].

With the introduction of gas turbines, and for turbofan engines especially, engine thermal management became a less critical issue for commercial aircraft because of the large, steady airflow that carries most of the engine's waste heat to ambiance. However, the continuous increase in turbine entry temperature and the introduction and further development of new technologies—for example, a gearbox for geared turbofan engines—have led to increased heat loads in modern aircraft engines. A summary of the development of engine waste heat and corresponding TMS developments can be found in [5].

Over the last two decades, research in (hybrid-)electric powertrains as an alternative to gas turbines has significantly increased. One of the key challenges for both realizing a theoretical benefit on aircraft level and successfully implementing first demonstrations is the thermal management of up to multi-megawatt electric powertrains [6,7]. Besides the high efficiency of electric components compared to gas turbines, they have no natural large heat rejection system such as the engine exhaust, so only small amounts of heat can be dissipated naturally via conduction through the structure. Therefore, the TMS has to manage their entire heat load. Additionally, electric components typically have low operating temperatures compared to combustion engines, which result in only small available temperature differences to ambient conditions for the TMS.

In recent research on hybrid-electric aircraft (HEA), the TMS is addressed more frequently and with increasing level of detail. For the NASA STARC-ABL concept, a specific power of 0.68 kW/kg of the TMS was assumed [8]. A hybrid version of the NASA N+4 Refined SUGAR research platform was designed with a dynamic model of a TMS for both the electric system and the engine oil system. The system was designed for conditions during a hot-day takeoff (HDTO), which, together with a low allowable battery temperature, resulted in a ram air cooler of about 150 kg. However, a 50% mass reduction was shown for an increase in battery temperature of 20 °F [9]. Further analysis of the concept, including various off-design points, showed an increase in design mission fuel burn (FB) of 3.4% due to TMS mass, power, and drag [10]. With additional optimization, such as decoupling the battery cooling loop, the FB increase was reduced to 0.75% [11]. In [12], a ram air–based TMS was designed for a vertical takeoff and landing (VTOL) vehicle with steady-state and transient methods. Sensitivities of key parameters of the developed compact heat exchanger (HEX) model were shown, as were Pareto fronts for a system optimization towards minimum system mass and power required by a puller fan. The final TMS of the VTOL had a mass of 171.63 kg and required 257.6 kW of power.

For HEA, the potential of using existing aircraft surfaces as alternative heat sinks was investigated, resulting in an indication that smaller aircraft can reject large parts of their heat load via the skin [13]. In a more detailed investigation, a TMS utilizing recirculating fuel underneath the wing surfaces for cooling of a 180-passenger short-range HEA was designed [14]. Despite the promising results, these surface cooling concepts have major disadvantages, such as the low available cooling power at low flight velocities and the low amount of coolant in case of fuel cooling towards the end of the mission. Therefore, a ram air–based TMS was considered for this study.

The research on ram air–based TMS has already developed some sensitivities and optimization for the compact HEX rather than solely solving the thermal management issue of one specific HEA. In this study, an even broader approach was chosen. The objective was threefold: Firstly, a static model of all necessary components for a ram air–based TMS was developed. Secondly, the overall system sensitivities were studied rather than just those of the compact HEX. Thirdly, different TMS architectures were optimized towards a weighted objective function derived from a 180-passenger short-range partial-turboelectric aircraft. The study will further improve knowledge of ram air–based TMS and their impacts on HEA. It will thereby enable future studies on HEA to assess their performances in more detail.

2. Models and Methods

The following section describes all required component models of the ram air–based TMS. At the end of this section, the partial-turboelectric aircraft and the derivation of its FB sensitivities for later use as an objective function are presented.

Figure 1 shows an exemplary centralized TMS architecture with all electric components being cooled in parallel. It requires the following components:

1. Coldplates to receive heat from the electric components and transfer it to the coolant.
2. A compact HEX to reject the collected heat to ambiance.
3. A diffuser to reduce cooling air speed and thereby the cold-side pressure loss of the compact HEX.
4. Optionally, a puller fan to increase cooling air flow.

5. A nozzle to recover some of the momentum of the cooling air and thereby reduce drag.
6. Pipes to transfer the coolant.
7. A pump to recover the pressure loss of the coolant.

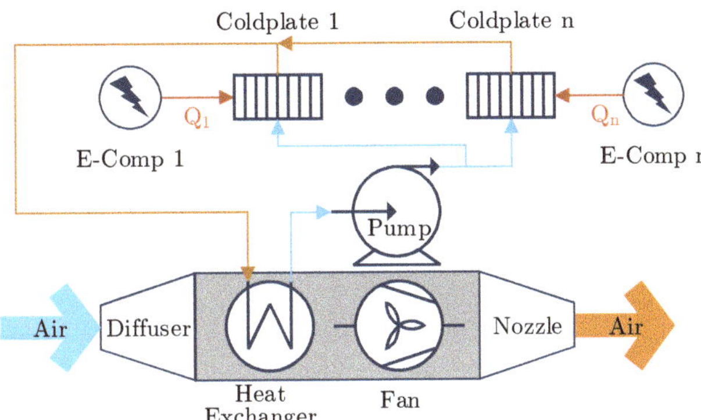

Figure 1. Centralized parallel thermal management system (TMS).

2.1. Coldplates

Coldplates are flat components with internal liquid flow to cool electronic devices, such as chips. Research trends towards lower thermal resistances (R_{th}) of future coldplates—for example, by decreasing the hydraulic diameters (d_H) of microchannels or by integrating the cooling channels closer to the working parts of the electronics [15]. Here, a simplified model of a coldplate is used not only for the cooling of the power electronics but also as a substitute for a model of the internal cooling of electrical machines. Despite the inlet properties (pressure (p), temperature (T), and heat load (Q)), the model only requires thermal insulance (r_{th}), maximum junction temperature (T_{cp}), area density (ρ_A), and design pressure loss (Δp_{des}) as inputs. These can be estimated from existing manufacturer data or research articles for future coldplate technology. The off-design performance is analytically derived, assuming straight parallel microchannels with laminar flow. A detailed explanation of the implemented coldplate model is provided in Appendix A.1. To validate the model, data from a numerical study of a microchannel coldplate is used [16]. The design point of the model was set to the highest Reynolds number (Re), and for the off-design performance, the mass flow rate (w) was subsequently decreased. All inputs to the design model are listed in Table A3 in Appendix A.2. The results of the validation are shown in Figure 2.

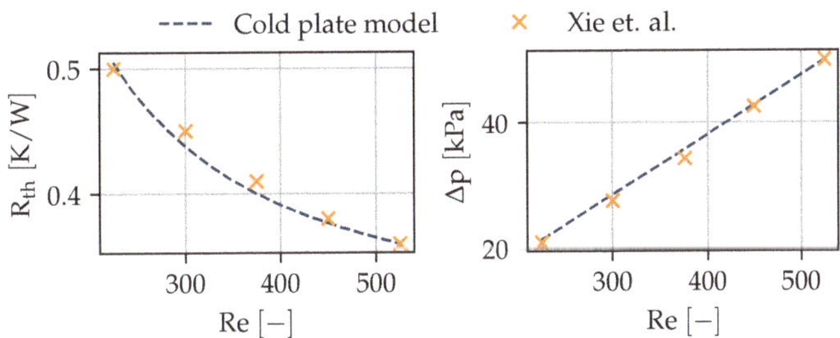

Figure 2. Coldplate model validation for thermal resistance (**left**) and pressure loss (**right**) with data from [16].

The predicted performances for both parameters (R_{th} and Δp) are within 2% of the validation data. The slight inaccuracy stems from the errors made in the visual acquisition of the data and the simplifications of the model. For the use in preliminary aircraft design, the accuracy is acceptable.

2.2. Compact Heat Exchanger

Heat exchangers can be built in many different architectures that have been described and categorized by different authors, e.g., [17,18]. Models attempting to cover all the different HEX types are therefore limited to a very low level of detail, which is not sufficient for the aim of this study to predict mass, dimensions, power, and drag of the TMS. However, due to the specific requirements of aircraft, only light, compact HEXs are considered. In [19], the most promising types of HEXs for aircraft applications are summarized as plate-fin heat exchangers (PFHE), printed circuit heat exchangers (PCHE), and in the future, microchannel heat exchangers.

There is no hard distinction between these three, as PFHE is a description of the overall architecture (plates and fins), PCHE is a description of the manufacturing technique (additive), and microchannel is a description of the layout on the microscopic level. Therefore, a HEX could match all three categories if it is an additively manufactured PFHE with very small channels. Thus, from a modeling perspective, it is only one type, which can be described as a single-phase, multi-pass, cross-flow HEX in overall counterflow arrangement. Both design and performance calculations were derived from the detailed procedures described in [17] for PFHE. Adaptions for the number of transfer units (NTU), the effectiveness (ϵ), and the dimensions of the HEX for multipass arrangements were implemented from [18]. The key equation for core mass velocity (cmv) from [17] then becomes:

$$cmv_{des} = \sqrt{2\Delta p_{des}} \cdot \left[\frac{f_{corr}}{j} \frac{ntu}{\eta_o} \cdot Pr^{\frac{2}{3}} \cdot \frac{1}{\rho_m} + 2 \cdot \left(\frac{1}{\rho_o} - \frac{1}{\rho_i} \right) + (1 - \sigma^2 + K_c) \cdot \frac{n_p}{\rho_i} \right.$$
$$\left. - (1 - \sigma^2 - K_e) \cdot \frac{n_p}{\rho_o} + (n_p - 1) \cdot K_{bt} \cdot \frac{\sigma^2}{\rho_m} \right]^{-0.5} \quad (1)$$

with corrected friction factor (f_{corr}), number of transfer units on one side (ntu), overall fin efficiency (η_o), Prandtl number (Pr), inlet, outlet, and mean density (ρ_i, ρ_o, and ρ_m), ratio of free flow to frontal area (σ), inlet, outlet, and bend loss coefficient (K_c, K_e, and K_{bt}), and number of passes (n_p).

The described algorithm can work with any HEX core as long as the parameters in Table 1 are given. The Colburn factor (j) and the Fanning friction factor (f) depend on Re, which means a correlation rather than one value has to be given. All other parameters are geometric and do not change in off-design operation. Three options for the HEX core are considered:

1. Rectangular microchannels.
2. Offset-strip fins.
3. Louvered fins.

A detailed explanation for the calculation of all parameters in Table 1 for all three types of HEX core can be found in Appendix B.

2.3. Diffuser, Nozzle, and Pipes

In many TMS models, e.g., the model presented in [12], the diffuser pressure loss is assumed to be constant. However, at low flight speeds, this simple assumption may overestimate the actual pressure loss and lead to the necessity of a puller fan. Its installation should be carefully considered because it usually is less efficient than the main propulsion devices. Therefore, in this study, a Mach number (Ma) dependent pressure loss model is used for the diffuser.

A drawing of the two-dimensional diffuser model is shown in Figure 3. It has a rectangular cross section, an opening angle (θ) in the z-direction, and a constant width (y-direction). Depending on the flight conditions, there is a pre-entry compression or expansion, i.e., $A_0 \neq A_1$. The changes in fluid properties between the flow cross sections A_0 and A_1 are calculated with the isentropic relations. Inside the diffuser, the ideal pressure recovery factor (c_p^*) can be obtained from correlations found in [20]:

$$c_p^* = g_1 \cdot g_2 \cdot \left\{ 1 - \frac{1.03 \cdot (1-B)^2}{\overline{A_R}^{2} \cdot \left[1 - 0.82 \cdot \overline{A_R}^{0.07} \cdot B^{1/(2 \cdot \overline{A_R} - 1)}\right]^2} \right\} \tag{2}$$

g_1 is a term depending on Ma and diffuser area ratio ($A_R = A_2/A_1$), and g_2 is a term depending on Re and relative inlet blockage (B). $\overline{A_R}$ is a corrected A_R to account for the influence of the aspect ratio of the inlet cross section. Using c_p^* implies a diffuser with optimal θ, which for the 2-D diffusers is around 8°. The outlet pressure is:

$$p_{2,s} = c_p^* \cdot \rho_1 \cdot v_1^2 + p_{1,s} \tag{3}$$

If $A_0 < A_1$, some air is spilled around the inlet and spillage drag occurs. It can be calculated according to [21,22]:

$$D_{spill} = K_{spill} \cdot [w_1 \cdot (v_1 - v_0) + A_1 \cdot (p_1 - p_0)] \tag{4}$$

K_{spill} is an empirical coefficient accounting for the lip suction effect. D_{spill} is added to the internal drag calculated from conservation of momentum equations over the entire system, i.e., from diffuser inlet to nozzle outlet.

Table 1. Required heat exchanger core parameters.

Name	Symbol	Unit
Colburn factor	j	–
Fanning friction factor	f	–
Hydraulic diameter	d_H	m
Plate space	b	m
Area density	β	m^2/m^3
Fin thickness	δ	m
Fin thermal conductivity	λ_f	W/(mK)
Ratio finned to total heat transfer area	A_f/A	–

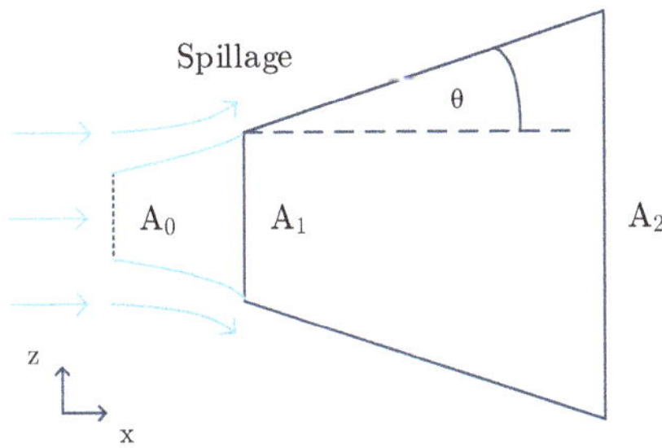

Figure 3. Diffuser model.

Since the nozzle has a negative static pressure gradient in the flow direction, its total pressure loss is less sensitive to shape and flow conditions than the diffuser. However, for the same reasons as mentioned above, it is important to have a pressure loss correlation sensitive to flow velocity rather than just a constant. It can be calculated according to [23]:

$$\Delta p_t = K_{loss} \cdot p_{1,t} \cdot \left[1 - \frac{p_{1,s}}{p_{1,t}}\right] \tag{5}$$

with shape-specific loss coefficient (K_{loss}) from [23]. Otherwise, the nozzle model uses area ratios to calculate outlet velocity and isentropic relations for the outlet fluid properties.

The pipe is modeled as a straight circular channel, and the well-known head loss formulas, e.g., from [24], are used to estimate pressure loss. For turbulent flow, the correlation from [25] is used to predict the friction factor.

All three models have simple geometric models to estimate their dry masses. In case of the pipe, a wet mass depending on the coolant is also available.

2.4. Pump and Fan

The puller fan is modeled as a repetition stage according to [22], i.e., the outlet velocity equals the inlet velocity. Isentropic relations are used to calculate the outlet fluid properties and compression work.

The pump model is simpler as the fluid is considered to be incompressible. Two efficiencies are implemented: The hydraulic efficiency (η_{hyd}) and the electric efficiency (η_{elec}). Mechanical power and outlet temperature are calculated as:

$$P_{mech} = \frac{\Delta p \cdot w}{\rho \cdot \eta_{hyd}} \tag{6}$$

$$T_2 = T_1 + P_{mech} \cdot \frac{1 - \eta_{hyd}}{c_v \cdot w} \tag{7}$$

2.5. Aircraft Fuel Burn Sensitivities

The aircraft used for the TMS design and optimization is designed to carry 180 passengers over a range of 1300 NM at a cruise speed of $Ma = 0.68$ (initial cruise altitude: 35,000 ft) and features a partial-turboelectric propulsion system.

The propulsion system is composed of advanced turboprop engines and turboelectrically driven wingtip propellers (WTPs). A key variable of this propulsion system architecture is the power split (S_P), which is defined as:

$$S_P = \frac{P_{WTP}}{P_{MP} + P_{WTP}} \tag{8}$$

where P_{WTP} is the shaft power of the WTP and P_{MP} is the shaft power of the turboprop engine's main propeller (MP). The design power of the electric system is determined by S_P and P_{MP} at the top of climb (TOC) of the aircraft design mission. This electric power remains constant unless the power of the gas turbine is lower than its TOC power of the design mission. In this case, P_{WTP} is lowered accordingly to match the desired S_P. Further details about the propulsion system and aircraft are provided in [26]. The aircraft investigated in [26] and this study is visualized in Figure 4.

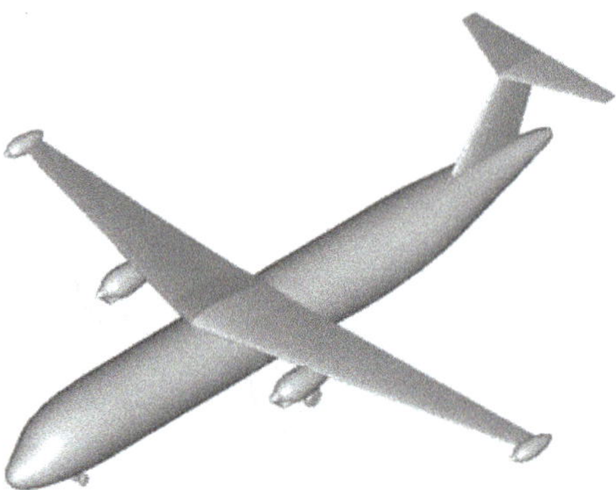

Figure 4. Aircraft design for $S_P = 30\%$.

To achieve an optimized TMS design on aircraft level, the impact of the variation of its most important parameters on an aircraft objective optimization variable is required. For this purpose, the sensitivity of the partial-turboelectric aircraft's FB (block fuel) to varying mass and drag increments due to the TMS integration was derived for three S_P values (10%, 20%, and 30%). Regarding the additional mass of a TMS (m_{TMS}), the operating empty mass (OEM) was gradually increased to include an assumed m_{TMS} of up to 1000 kg. In the same manner, the wing profile drag was increased to include an assumed TMS drag (D_{TMS}) of up to 1000 N since an integration into the wing was found to be reasonable. Consequently, every combination of m_{TMS} and D_{TMS} represents a new aircraft design. The resulting aircraft FB sensitivities for the three S_P variations are similar in their relative FB changes (ΔFB values) to the FB of the respective baseline aircraft design. An exemplary FB sensitivity is presented in Figure 5.

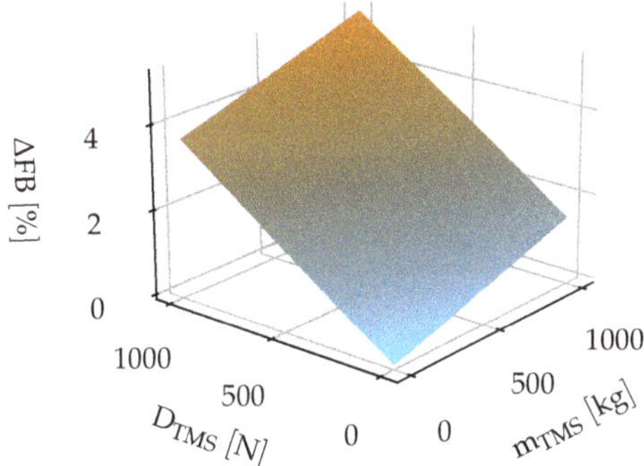

Figure 5. Aircraft FB sensitivity for $S_P = 30\%$.

Starting from the baseline aircraft design for $S_P = 30\%$, Figure 5 shows an increase in ΔFB of approximately 1.5% for a m_{TMS} increment of 1000 kg and approximately 3.6% if D_{TMS} is increased by 1000 N. These FB gradients of Δm_{TMS} and ΔD_{TMS} are almost independent of each other, which leads to a sensitivity plane with only minimal curvature.

3. System Sensitivity Analysis

The following section investigates an aircraft FB sensitivity to all relevant parameters of the system. It establishes a general understanding of the system and verifies the implementation of the models. Additionally, computational costs in the following optimization (see Section 4) are reduced when parameters with low sensitivity can be set to a constant value. The sensitivity analysis is conducted at TOC conditions. However, HDTO conditions are more challenging for the TMS and are considered later in Section 4.2. $S_P = 30\%$ is used for the sensitivity analysis. The trends shown in this section are also valid for the other S_P values. The heat loads of the design and the off-design point are shown in Figure 6.

Power electronics include inverters, rectifiers, and protection switches. The absolute values are rather close due to the aforementioned strategy of keeping the electric power near its maximum throughout the mission. In takeoff, the generator has a higher efficiency because of a better position in the operational characteristics and therefore less waste heat than in design. A 50%-water-glycol mixture is chosen as the coolant to cope with the low ambient temperatures at high altitudes.

3.1. One-Dimensional Sensitivities

The one-dimensional sensitivity analysis considers the sensitivity of each parameter isolated, i.e., only one parameter is varied at a time. In Section 3.2, some coupled or multi-dimensional sensitivities are discussed. The parameters considered for the one-dimensional analysis are summarized in Table 2 and the results are shown in Figure 7.

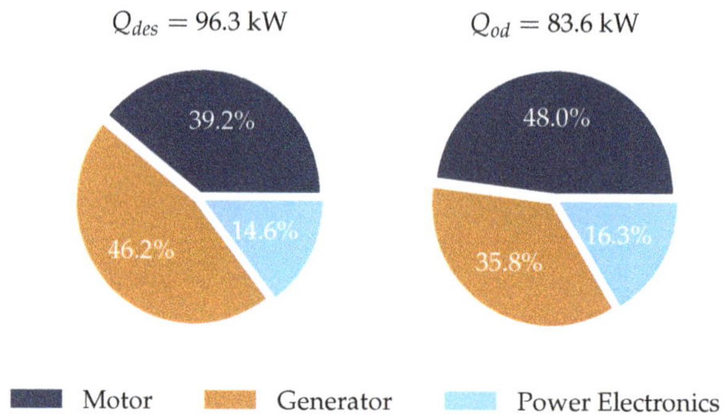

Figure 6. Design and off-design (HDTO) heat loads for $S_P = 30\%$ for one powertrain.

Table 2. Parameters considered in the one-dimensional sensitivity analysis.

Parameter	Symbol	Unit	Default Value
Coldplate surface temperature	T_{cp}	K	370
Heat capacity ratio HEX cold to hot side	C_R^*	–	1.0
Coldplate coolant inlet temperature	T_1	K	275
Pressure ratio HEX cold side	Π_c	–	0.95
Hydraulic diameter HEX cold side	$d_{H,c}$	mm	10.0
Coldplate effectiveness	ϵ_{cp}	–	0.4

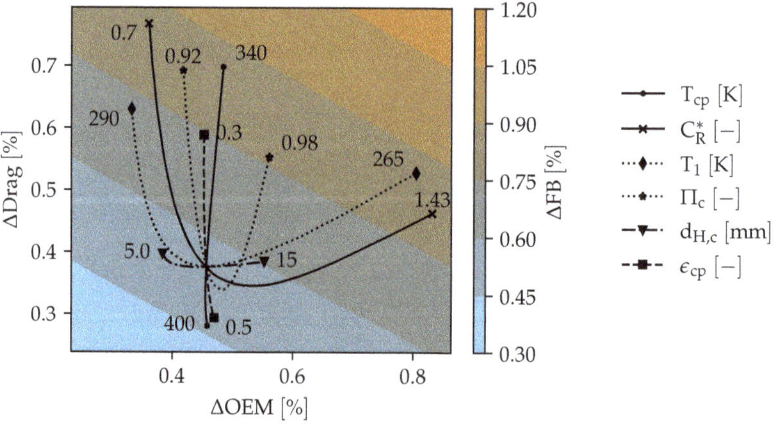

Figure 7. One-dimensional sensitivity analysis.

The default values from Table 2 are located at the intersection of all lines in Figure 7. The default values for each parameter are the median values of the respective parameter range. They are mostly not located at the middle of the resulting sensitivity line, indicating a higher sensitivity of the parameter to one end of the range. Increasing T_{cp} by 30 K from 370 K to 400 K, for example, results in roughly a 0.07% decrease in ΔFB, whereas decreasing it by 30 K to 340 K results in an approximate 0.3% increase in ΔFB.

T_{cp} and ϵ_{cp} have the highest proportionality with ΔFB. Increasing either one of them directly results in an increase of ΔT across the HEX, which leads to a decrease in HEX size. Both parameters cannot be freely chosen, but T_{cp} is constrained by the allowed operating temperature of the electric component and ϵ_{cp} by the possible size of the coldplate. High

ϵ_{cp} values require a longer length of stay of the cooling fluid inside the coldplate, which causes an increase in the size of the coldplate for constant heat loads.

All other parameters have an optimal value with a minimum in ΔFB inside the given range. Decreasing C_R^* to values lower than 1.0 is a direct increase of w_c. This improves the cold-side heat transfer coefficient (α_c), which results in a slightly smaller and lighter HEX; however, the corresponding increase in drag leads to an overall increased ΔFB. Increasing C_R^* past 1.0 has the opposite effect. The increased w_h causes an increased hot-side length (L_h) of the HEX to achieve the same T_1. This allows a shorter cold-side length (L_c) and thereby less drag. There is a limit to this effect—it will eventually result in an increase in drag again due to an unnecessarily large HEX area.

Π_c has a higher drag than OEM sensitivity. Low Π_c values directly result in more drag but also allow slightly lighter systems due to the increased cold-side flow velocity and thus higher α_c. The increased drag towards very high Π_c values originates in the diffuser. Very low face Ma are required for the HEX, leading to a large diffuser with larger internal losses and also larger spillage.

Decreasing T_1 further from the default value requires a more effective HEX, i.e., a larger HEX with increased L_h and L_c. Besides becoming heavier, the increased L_c also results in more drag for the system. At constant Π_c, an increased L_c requires a smaller face Ma with the above-described consequences for the diffuser. However, T_1 should not be infinitely increased either. Large T_1 values require large w_h values, a constant C_R^*, and large w_c values, resulting in a steep increase in drag.

$d_{H,c}$ is inversely proportional to mass because α_c increases with decreasing $d_{H,c}$. A very small $d_{H,c}$ leads to increased FB since for constant Π_c a very low face Ma is required, which again causes large diffuser losses and consequently drag, as mentioned above.

3.2. Multi-Dimensional Sensitivities

The results of Figure 7 may not be used to observe the optimal value for each parameter. This would only be possible if they were independent of each other. In reality, they are linked to each other via various interdependencies. Some of the more interesting ones are shown in Figure 8. Contrary to Figure 7, the lines in Figure 8 are lines of constant parameter values—e.g., along the dotted lines of the left image, Π_c has a constant value, which is indicated on the left side of each line.

The study settings are the same as in Table 2, except for the indicated parameters. On the left side $d_{H,h}$ and Π_c are varied. Varying Π_c shows the same curve shapes for each $d_{H,h}$ as in Figure 7. The minimum in ΔFB shifts. For $d_{H,h} = 4.5$ mm, the best Π_c would be about 0.94, whereas for smaller $d_{H,h}$, the ideal Π_c value increases slightly to about 0.96 for $d_{H,h} = 0.5$ mm. $d_{H,h}$ shows a rather clear trend indicating that lower $d_{H,h}$ values always result in less ΔFB. This statement is only valid as long as Π_c can be appropriately chosen. If, for example, Π_c is fixed at 0.9, the best $d_{H,h}$ value would be roughly 2 mm.

The main reason for the effects described above is the influence of $d_{H,h}$ on the HEX cold-side ratio of the free flow to the frontal area (σ_c). A larger $d_{H,h}$ increases the hot-side channel height (if the channel aspect ratio is not changed) and therefore decreases σ_c. If Π_c is left constant, the flow velocity in the cold-side channel is also about constant. However, due to the lower σ_c value, the face Ma must be smaller since a lower σ_c results in a higher difference between frontal and free flow velocity. Therefore, the diffuser must be larger, resulting in more drag. The system mass always decreases with decreasing $d_{H,h}$ because of an increased α_h and a more compact HEX.

On the right, the effects of varying d_H on both cold and hot sides of the HEX are shown. Again, reducing $d_{H,h}$ results in less ΔFB in every case. The reason is the same as described above. The optimal $d_{H,c}$ value depends heavily on the chosen $d_{H,h}$. For a large $d_{H,h}$, a larger $d_{H,c}$ should be chosen. A small $d_{H,c}$ with a large $d_{H,h}$ results in small σ_c values with its negative effects on drag as described above. For the lowest considered $d_{H,h}$ of 0.5 mm, the best $d_{H,c}$ value is about 5 mm. The factor between $d_{H,h}$ and the corresponding best $d_{H,c}$ value varies between 2 and 10. This large difference can be attributed to the different fluid properties, especially the large difference in thermal conductivity between water and air.

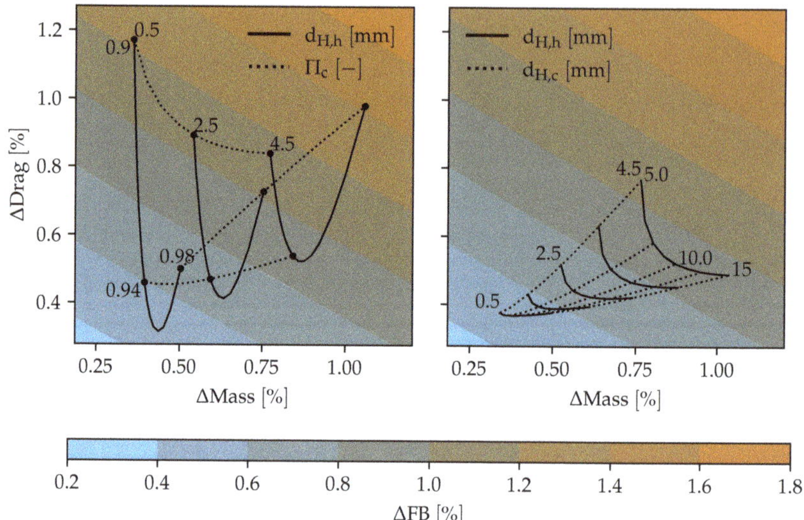

Figure 8. Two-dimensional sensitivity analysis of the hot-side hydraulic diameter with the cold-side pressure ratio (**left**) and the hot-side hydraulic diameter with the cold-side hydraulic diameter (**right**).

For TMS-equipped aircraft, a few interesting conclusions can be derived. The general trend in HEX design towards smaller d_H is only beneficial for the aircraft on the hot side if the drag is considered. Studies only focusing on HEX masses will still find smaller $d_{H,c}$ beneficial. For practical reasons, $d_{H,h}$ can be reduced far easier than $d_{H,c}$. The smaller the d_H, the higher the risk of congestion, and the greater the drop in performance for the HEX. The hot side is a closed loop, and therefore the fluid can be kept very pure through regular exchange and the incorporation of filter systems, thereby minimizing said risk. On the cold side, ambient air has to be used. The implementation of a filter would directly result in more drag and is therefore not a feasible option. With optimal $d_{H,c}$, values of more than 5 mm for maintenance are less of a problem than $d_{H,c}$ values of only a millimeter or less. Due to its obvious trends, $d_{H,h}$ does not need to be considered as a free variable but rather as direct input constrained mainly by manufacturing techniques for the optimization studies in Section 4 if mass, drag, and *FB* are the only relevant metrics.

3.3. Heat Exchanger Size

While mass, drag, and *FB* are the most relevant metrics for the aircraft performance, the system size cannot be neglected since the TMS has to be integrated into the aircraft. The influences of $d_{H,h}$ and $d_{H,c}$ on the three HEX dimensions L_h, L_c, and stack height (H_{stack}) are shown in Figure 9. The study settings are equal to those in Figure 8, except for a smaller range of considered values for both d_H.

Clearly, d_H on both sides has a direct influence on overall HEX dimensions. In any size constrained optimization problem, $d_{H,h}$ should therefore be considered as a free variable as well. Increasing $d_{H,h}$ results in an increase of L_h because Π_h is kept constant. To have the same pressure drop for a lower f_h, L_h needs to be higher. As a consequence of the increased L_h, H_{stack} is reduced because Q is also constant. Without a reduction in H_{stack}, the total heat exchange area would be larger, and therefore Q would be higher than actually required. Increasing $d_{H,c}$ shows an analogue trend.

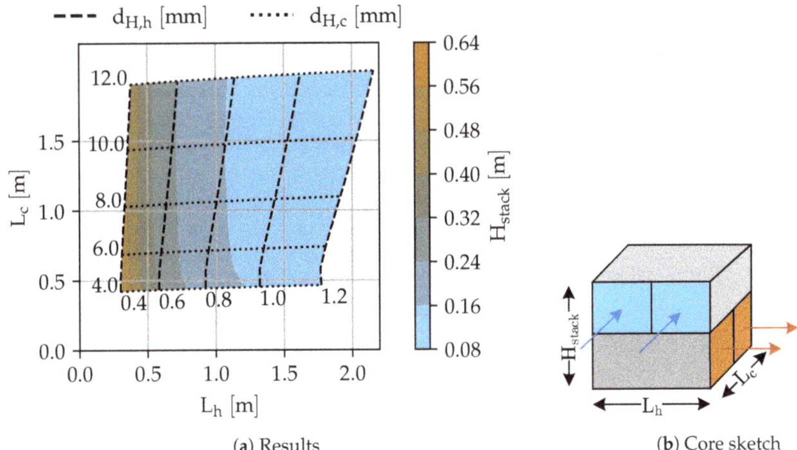

(**a**) Results (**b**) Core sketch

Figure 9. Heat exchanger size sensitivity in three dimensions: hot-side length, cold-side length, and stack height over hot- and cold-side hydraulic diameter.

4. Design and Off-Design Optimization for the Application Case

This section uses the previously gathered knowledge to design and optimize TMS for the application case of a HEA (here, "hybrid" refers to power hybridization) described in Section 2.5. The section is divided into design, off-design, and multi-point design.

4.1. Design Point Optimization

The settings of the study have already been described in the previous sections. The design point of choice is the TOC, which is also the design point of the gas turbine. The aircraft has been designed with three different S_P values, so a TMS was designed for each of them. Free variables for the optimization were $d_{H,h}$, $d_{H,c}$, Π_h, Π_c, C_R, $(A_0/A_1)_{diff}$, and T_1. The cumulative optimization results of two identical TMS (one for each powertrain) are shown in Figure 10.

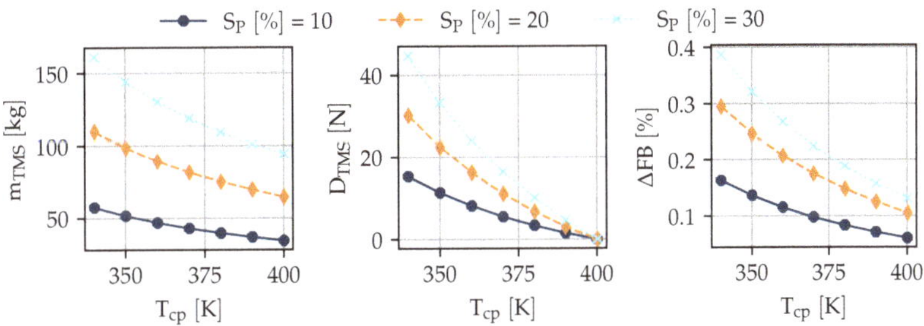

Figure 10. Design optimization results for different S_P variations.

For each S_P, multiple designs for different T_{cp} were made, as T_{cp} is subject to electric component technology and therefore not certainly known. S_P values were imposed by the aircraft studies [26], and the given range of T_{cp} was chosen to include current electric component technology. Results are shown for m_{TMS}, D_{TMS}, and ΔFB, which was the objective function of the optimization. As expected, all three parameters grow with increasing S_P and decreasing T_{cp}. The exponential behavior towards decreasing T_{cp} was also anticipated from the results shown in Figure 7. D_{TMS} gets reduced to almost 0 N when increasing T_{cp} to 400 K, due to the Meridith effect. The heat rejected by the HEX is recovered as thrust and compensates for the pressure loss of the TMS. If even higher T_{cp} values are possible, the aircraft FB sensitivities have to be extended towards negative drags, i.e., thrust from the TMS.

With state-of-the-art electric components, i.e., motors, generators, and power electronics, a T_{cp} of 380 K is realistic. For the three different S_P values, ΔFB is 0.09%, 0.15%, and 0.19%, respectively. There are several reasons for these very low values. Firstly, S_P is not very large, and therefore Q stays relatively low (see Figure 6). Secondly, the partial-turboelectric architecture only includes electric components with comparably high maximum operating temperatures. If a large battery or fuel cell is included in the powertrain, the TMS design becomes more complex and will likely have a higher impact on ΔFB. Thirdly, the currently implemented system mass estimations have to be refined in a more detailed analysis. So far, redundancy is not considered. Additionally, the technology assumptions for the HEX have been rather optimistic, with wall thicknesses for the plates assumed at 0.5 mm and for the fins at 0.1 mm.

Fourthly, integration of the TMS has not been considered in the design yet. For T_{cp} = 380 K and S_P = 30%, the HEX would measure $L_c \times L_h \times H_{stack}$ = 0.48 m × 0.73 m × 0.18 m. The diffuser and nozzle would be 2.2 m and 0.9 m long, respectively, resulting in an overall cold-side system length of 3.5 m. If needed, the diffuser could be shortened, trading efficiency. The current model (see Section 2.3) only allows diffusers with $\theta = 8°$. In this case, a fuselage integration seems feasible, but cargo space would be reduced. Another option could be the installation on top of the wing near the root, but it would possibly require additional cowlings, resulting in additional mass and drag.

It is worth noting that the numeric optimization resulted in ΔFB values of less than 0.2% for $S_P = 30\%$ and $T_{cp} = 380$ K, whereas even the best values in the sensitivity studies (see Figures 7 and 8) were above 0.4%. While the difference in percentage points is not of large relevance to the aircraft in this case, the relative difference achieved through numeric optimization is remarkable, i.e., a reduction of more than 50%.

4.2. Off-Design Point Optimization

An exemplary off-design optimization was conducted for $T_{cp,des} = 380$ K and S_P = 30%. The objective function was the electric power required to drive the TMS (P_{TMS}), which includes the power for the hydraulic pump and the fan. The efficiencies of the pump η_{hyd} and η_{elec} were assumed to be 0.75 and 0.95, respectively, and the fan efficiency (η_{fan}) was set to 0.50. In a more detailed study, proper maps should be implemented for pump and fan efficiency to accurately predict their behavior with changing operating conditions. Variables of the study were the international standard atmosphere (ISA) temperature deviation (ΔT_{ISA}) and the differences between cooling fluid outlet and inlet temperatures of the electric components (ΔT_{cp}). The results are shown in Figure 11.

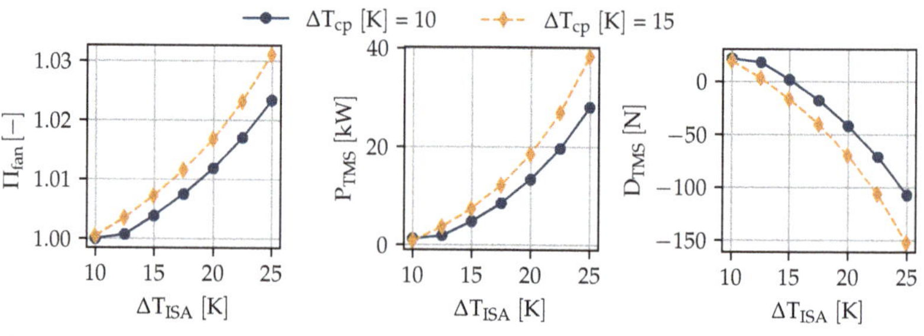

Figure 11. Off-design optimization at takeoff for a TMS designed for $T_{cp,des}$ = 380 K and S_P = 30%.

Hot days are a particular challenge for the TMS because the available ΔT between cooling fluid and ambient is smaller. Raising ΔT_{ISA} results in an exponential increase in required fan pressure ratio (Π_{fan}). ΔT_{cp} is an operational parameter that can be controlled via P_{pump}. A lower P_{pump} results in a smaller w_h, and thereby a higher ΔT_{cp}. A higher ΔT_{cp}

value does require a larger Π_{fan} because the ΔT between hot-side HEX inlet to outlet is larger, and therefore a higher α_c is needed. P_{TMS} follows Π_{fan} almost directly because P_{pump} is at a different order of magnitude, i.e., only 1.1 kW and 0.5 kW for $\Delta T_{cp} = 10$ K and 15 K, respectively. The large difference between P_{pump} and P_{fan} is due to the fact that the pump compresses an incompressible fluid, and the compressor a compressible one. About 25% P_{TMS} can be saved on a hot day by choosing the lower ΔT_{cp} value.

P_{TMS} has not been considered in the aircraft *FB* sensitivities. During the majority of the mission, the fan is not required and could either be removed from the flow path or set to idle. The takeoff segment is rather short compared to the overall mission length, and even if the maximum load of 60 kW is required, the impact on the powertrain is negligible. The generators have a combined power of more than 2 MW, and some of the P_{TMS} is actually converted to useful thrust, as seen by the negative drag values of up to −150 N.

4.3. Multi-Point Optimization

From the previous section, the question arises of whether it is possible to design a TMS without the additional puller fan. Though its impact on ΔFB is negligible, it is still an additional component with costs and requirements for certification and maintenance. To answer the question, a multi-point study was conducted that combined the previous design point with an additional constraint to achieve the required cooling power in off-design as well. The objective function was again ΔFB—the same as in Section 4.1. The results are shown in Figure 12.

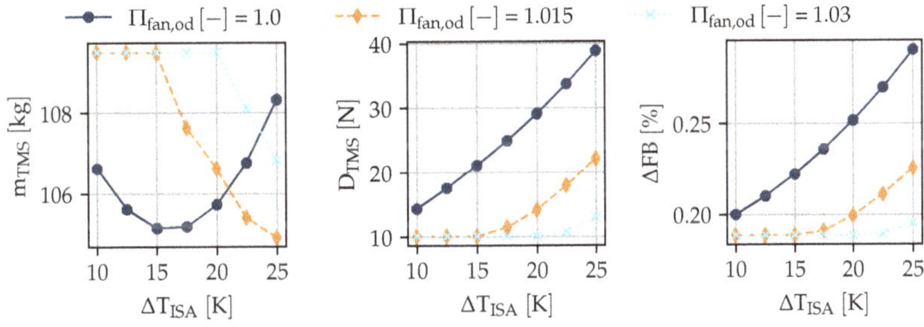

Figure 12. Multi-point optimization for a TMS for $T_{cp,des} = 380$ K and $S_P = 30\%$.

Three different off-design fan pressure ratios ($\Pi_{fan,od}$) were investigated. If $\Pi_{fan,od}$ is 1.0, no fan installation is required. For the larger values of $\Pi_{fan,od}$, all results form horizontal lines for lower ΔT_{ISA}. This implies that the optimal design is only dependent on the design point, and the additional off-design constraint is met because $\Pi_{fan,od}$ is oversized. Only when ΔT_{ISA} increases beyond a certain threshold, the off-design constraint becomes relevant.

If no fan is installed ($\Pi_{fan,od} = 1.0$), the constraint is relevant even at low ΔT_{ISA}, immediately resulting in a larger TMS with increased ΔFB. ΔFB grows exponentially with ΔT_{ISA}. It is certainly possible to design the TMS without the puller fan, however, assuming a maximum ΔT_{ISA} of 25 K, ΔFB would increase from 0.19% to 0.29%. In absolute numbers, this difference is negligible, but for a TMS with a larger *FB* impact, it could be better to install the fan. Using a puller fan also has the advantage of an additional degree of freedom for the system that can help to better adapt to operational changes.

5. Conclusions and Outlook

Ram air–based thermal management systems (TMS) were investigated regarding their overall impacts on an aircraft's fuel burn. Fuel burn sensitivities were derived from a 180-passenger short-range partial-turboelectric aircraft equipped with wingtip propellers by adding an assumed TMS design drag and mass to it.

A TMS model consisting of coldplates for heat acquisition, pipes and pumps for hot-side heat transfer, a two-pass cross-flow plate-fin heat exchanger for heat rejection, and a diffuser and a nozzle for cold-side flow velocity control was developed. Variations of one- and multi-dimensional parameter sensitivities were used to gain an understanding of the system. The system reacted very sensitively to seven parameters that were selected as free variables for a numeric optimization.

Alternating the hydraulic diameter of the main heat exchanger on both sides was shown to be one of the most effective ways to control the overall system dimensions and therefore manage the integration problem.

TMS optimization studies were conducted. It was found that increasing electric component junction temperature to about 400 K could eliminate parasitic drag from the TMS in cruise entirely. For a more realistic temperature of 380 K, additional fuel burn for an aircraft with 30% power split was 0.19%. The system could withstand hot-day takeoff conditions with the help of a small puller fan installed behind the main heat exchanger. Alternatively, oversizing the TMS removed the need for a puller fan but increased additional fuel burn to 0.29%.

In the future, the mass of the system should be re-investigated. Redundancy considerations are most likely going to cause an increase in system mass of up to 100%. In this study, only rectangular channels were considered for the heat exchanger core. Other options, such as offset-strip fins and louvered fins should be considered in the future. Additionally, integration of the TMS, including secondary mass and drag increases, will be discussed in the future. The integration of the TMS seems to be one of the largest challenges. In concrete aircraft applications, this problem should be addressed and possibly solved in a synergistic manner—e.g., by installing the ram-air inlets behind an open rotor. Additionally, adaptive nozzle geometries are an idea to better adapt TMS performance in different operating conditions.

Author Contributions: Conceptualization, methodology, simulation, analysis, and writing of all aspects of the research except for Section 2.5, H.K.; conceptualization, methodology, simulation, analysis, and writing of all aspects of the research of Section 2.5, M.L.; conceptualization, methodology, simulation, and analysis of the powertrain, M.P.; supervision, M.H. All authors have read and agreed to the published version of the manuscript.

Funding: This research received funding as part of the IVeA project, a research project supported by the Federal Ministry for Economic Affairs and Energy in the national LuFo program.

Data Availability Statement: Not applicable.

Acknowledgments: We would like to thank everyone involved in the IVeA project for their dedication to its success. Additionally, we thank Arne Seitz for his continued support and fruitful discussions.

Conflicts of Interest: The authors declare no conflict of interest. The funders had no role in the design of the study; in the collection, analyses, or interpretation of data; in the writing of the manuscript, or in the decision to publish the results.

Abbreviations

HDTO	Hot-day takeoff
HEA	Hybrid-electric aircraft
HEX	Heat exchanger
ISA	International standard atmosphere
MP	Main propeller
NASA	National Aeronautics and Space Administration
OEM	Operating empty mass
PCHE	Printed circuit heat exchanger
PFHE	Plate fin heat exchanger
TMS	Thermal management system

TOC	Top of climb	
VTOL	Vertical takeoff and landing	
WTP	Wingtip propeller	

Roman Symbols

A	Area	m^2
A_R	Diffuser area ratio	–
$\overline{A_R}$	Corrected diffuser area ratio	–
b	Heat exchanger plate space	m
B	Diffuser inlet blockage	–
c_p	Specific heat capacity at constant pressure	J/(kgK)
c_p^*	Ideal diffuser pressure recovery factor	–
c_v	Specific heat capacity at constant volume	J/(kgK)
C	Absolute heat capacity	W/K
C_R	Heat capacity ratio (C_{min}/C_{max})	–
C_R^*	Side-specific heat capacity ratio (C_h/C_c)	–
cmv	Core mass velocity	$kg/(m^2s)$
d_H	Hydraulic diameter	m
D	Drag	N
f	Fanning friction factor	–
FB	Fuel burn	kg
g	Diffuser pressure recovery geometry factor	–
j	Colburn factor	–
K_{bt}	Bend loss coefficient	–
K_c	Inlet loss coefficient	–
K_e	Outlet loss coefficient	–
K_{loss}	Nozzle pressure loss coefficient	–
K_{spill}	Spillage coefficient	–
L	Length	m
m	Mass	kg
Ma	Mach number	–
n_p	Number of passes	–
ntu	Number of transfer units on one side	–
NTU	Number of transfer units	–
p	Pressure	Pa
P	Power	W
Pr	Prandtl number	–
q	Area-specific heat flow rate	W/m^2
Q	Heat flow rate	W
r_{th}	Thermal insulance	m^2K/W
R_{th}	Thermal resistance	K/W
Re	Reynolds number	–
S_P	Power split	%
t	Channel width	m
T	Temperature	K
U	Overall heat transfer coefficient	$W/(m^2K)$
v	Velocity	m/s
V	Volume	m^3
w	Mass flow rate	kg/s

Greek Symbols

α	Heat transfer coefficient	$W/(m^2K)$
δ	Fin thickness	m
Δ	Difference	–

ϵ	Heat exchanger effectiveness	–
η_o	Overall fin efficiency	–
Φ	Aspect ratio	–
Π	Pressure ratio	–
ρ	Density	kg/m^3
ρ_A	Area density	kg/m^2
σ	Heat exchanger ratio of free flow to frontal area	–
θ	Diffuser opening angle	deg

Subscripts

c	Cold
$cond$	Conductive
$conv$	Convective
$corr$	Corrected
cp	Coldplate
cs	Cross section
des	Design
f	Finned
h	Hot
i	Inlet
m	Mean
o	Outlet
od	Off-design
s	Static
$spill$	Spillage
tot	Total

Appendix A. Coldplate Model

Appendix A.1. Model Description

For the coldplate design model, all input and output parameters are listed in Table A1. The input parameters have to be estimated or obtained from manufacturer data.

The area-specific heat load (q_{des}) is calculated from the thermal insulance ($r_{th,des}$) and the coldplate surface temperature ($T_{cp,des}$) [27,28].

$$q_{des} = (T_{cp,des} - T_{i,des})/r_{th} \tag{A1}$$

The outlet temperature ($T_{o,des}$) can be obtained from the effectiveness (ϵ_{des}). The evaluation of fluid properties inside a heat exchanging device is conducted at an average temperature (T_m):

$$T_{o,des} = (T_{cp,des} - T_{i,des}) \cdot \epsilon_{des} + T_{i,des} \tag{A2}$$

$$T_m = (T_i + T_o)/2 \tag{A3}$$

The specific heat capacity of the cooling fluid (c_v) is a function of T_m and p_i (the pressure drop is neglected here as c_v has a much larger temperature than pressure sensitivity) and is evaluated from the CoolProp fluid database [29]. The required mass flow (w_{des}) can be calculated from Q_{des} and the area of the coldplate (A_{cp}) from q_{des}:

$$w_{des} = Q_{des}/(c_v \cdot (T_{o,des} - T_{i,des})) \tag{A4}$$

$$A_{cp} = Q_{des}/q_{des} \tag{A5}$$

The dry mass is then calculated from the area density (ρ_A):

$$m_{dry} = A_{cp} \cdot \rho_A \tag{A6}$$

The product of the overall heat transfer coefficient and heat exchange area ($(UA)_{des}$) is required for later off-design calculations (note: $A_{des} \neq A_{cp}$ since A_{cp} is the coldplate base area and A_{des} the inner channel surface area). It is calculated from the number of transfer units (NTU). The NTU–ϵ relation for heat exchanging devices with a heat capacity ratio of $C_r = 0$ is found in many thermodynamic textbooks, e.g., [24].

$$NTU_{des} = -\ln(1 - \epsilon_{des}) \tag{A7}$$

$$UA_{des} = NTU_{des} \cdot c_p \cdot w_{des} / A_{cp} \tag{A8}$$

Finally, the outflow pressure (p_o) is calculated:

$$p_{o,des} = p_{i,des} - \Delta p_{des} \tag{A9}$$

Table A1. Design parameters for the coldplate model.

Parameter	Symbol	Unit
	Inputs	
Inlet pressure	$p_{i,des}$	Pa
Inlet temperature	$T_{i,des}$	K
Effectiveness	ϵ_{des}	–
Heat load	Q_{des}	W
Coldplate surface temperature	$T_{cp,des}$	K
Thermal insulance	$r_{th,des}$	m^2K/W
Area density	ρ_A	kg/m^2
Pressure drop	Δp_{des}	Pa
	Outputs	
Design mass flow	w_{des}	kg/s
Outlet pressure	$p_{o,des}$	Pa
Outlet temperature	$T_{o,des}$	K
Area-specific heat load	q_{des}	W/m^2
Coldplate area	A_{cp}	m^2
Dry mass	m_{dry}	kg
Number of transfer units	NTU_{des}	–
U-A product	$(UA)_{des}$	W/K

In off-design calculations, the dimensions of the coldplate are fixed. Only fluid inlet conditions (T_i, p_i, w_{od}) vary, as does the off-design heat load (Q_{od}). All input and output parameters of the off-design model are listed in Table A2.

Since A_{cp} has been defined in the design model, the off-design area-specific heat flow (q_{od}) can be calculated:

$$q_{od} = Q_{od} / A_{cp} \tag{A10}$$

T_m is calculated from (A3), and c_p is obtained from tabulated data. The off-design mass flow (w_{od}) is determined from (A4) with off-design inputs. The off-design coldplate temperature ($T_{cp,od}$) can be determined from the off-design effectiveness (ϵ_{od}).

$$NTU_{od} = (UA)_{od} / (c_p \cdot w_{od}) \tag{A11}$$

$$\epsilon_{od} = 1 - e^{-NTU_{od}} \tag{A12}$$

$$T_{cp,od} = q_{od} / h_{od} + T_m \tag{A13}$$

With $(UA)_{od} = (UA)_{des}$. This will be proven in the following paragraph.

Table A2. Off-design parameters for the coldplate model.

Parameter	Symbol	Unit
	Inputs	
Inlet pressure	p_i	Pa
Inlet temperature	T_i	K
Outlet temperature	T_o	K
Heat load	Q_{od}	W
	Outputs	
Off-design mass flow	w_{od}	kg/s
Outlet pressure	p_o	Pa
Coldplate temperature	$T_{cp,od}$	K
Area-specific thermal resistance	$r_{th,od}$	m^2K/W
Effectiveness	ϵ_{od}	–

The area is constant as no geometries are changed. For a coldplate, U is comprised of conductive (α_{cond}) and convective (α_{conv}) heat transfer coefficients. α_{cond} does not change in off-design situations because material and thickness are constant. The change in thermal conductivity of the material (λ) is neglected because the mean material temperature is not expected to differ greatly between design and off-design. α_{conv} can be calculated from:

$$\alpha_{conv} = Nu \cdot \lambda / d_H \tag{A14}$$

with Nusselt number (Nu) and hydraulic diameter (d_H) [30]. Microchannels provide compact, light-weight coldplates with the ability to absorb very high q_{des} as required by modern chip generations. The flow in such small channels is typically laminar due to the very small d_H [31]. In laminar flow, Nu is constant regardless of the flow velocity [24]. For this model, laminar flow is assumed in all operating points. To ensure this assumption is true, the coldplate should always be designed for maximum mass flow, and off-design operating points should have smaller mass flows ($w_{des} > w_{od}$). d_H is also constant as it is a fixed geometry. Neglecting the T-p dependency of λ the equity of both α_{conv} follows:

$$\alpha_{conv,des} = \alpha_{conv,od} \tag{A15}$$

U_{des} is known from (A7). If the temperature differences between design and off-design are large, the λ-T-p sensitivity can be accounted for by means of a ratio $\lambda_{od}/\lambda_{des}$. The off-design thermal insulance ($r_{th,od}$) is:

$$r_{th,od} = (T_{cp,od} - T_i)/q_{od} \tag{A16}$$

Since no exact geometry is known from the design model, the off-design pressure loss (Δp_{od}) has to be derived from its design counterpart and the design/off-design w-ratio:

$$\Delta p_{od} = f(\Delta p_{des}, w_{des}/w_{od}) \tag{A17}$$

In general, Δp can be calculated from [24]:

$$\Delta p = h \cdot \rho \cdot g \tag{A18}$$

with head loss (h) and gravitational constant (g). The head loss is [32]:

$$h = f \cdot L \cdot u^2/(d_H \cdot 2 \cdot g) \tag{A19}$$

with friction factor (f), flow length (L), and flow velocity (u). In laminar flow, f is a function of Re and a channel geometry depending constant (c_{geom}) [24]:

$$f = c_{geom}/Re \tag{A20}$$

$$Re = u \cdot d_H/\nu \tag{A21}$$

with kinematic viscosity (ν). Combining (A18)–(A21) results in:

$$\Delta p = c_{geom} \cdot L \cdot u \cdot \rho \cdot \nu/(2 \cdot d_H^2) \tag{A22}$$

c_{geom}, L, and d_H do not change from design to off-design conditions; the difference in ρ and ν is neglected so that:

$$\Delta p_{od}/\Delta p_{des} = u_{od}/u_{des} \tag{A23}$$

$$u = w/(\rho \cdot A_{cs}) \tag{A24}$$

with flow cross section area (A_{cs}). Again, A_{cs} stays constant, and the difference in ρ is neglected, finally resulting in:

$$\Delta p_{od} = \Delta p_{des} \cdot w_{od}/w_{des} \tag{A25}$$

$p_{o,od}$ can now be calculated via (A9).

Appendix A.2. Coldplate Validation Design Inputs

Table A3. Design inputs for coldplate validation.

Parameter	Unit	Value
$T_{1,des}$	K	294
ϵ_{des}	–	0.47
Q_{des}	W	100
$T_{cp,des}$	K	330
$r_{th,des}$	m^2K/W	2.88×10^{-5}
Δp_{des}	Pa	50×10^3

Appendix B. Compact Heat Exchanger Core Model

This section describes how the core geometry parameters and Colburn factor (j) and Fanning friction factor (f) for the different core surfaces of a compact heat exchanger are calculated.

1. **Rectangular microchannels**. j and f are calculated according to the methods described for rectangular channels in [24]. Of the parameters in Table 1, d_H and δ are used as known inputs, and the other parameters are calculated. The aspect ratio of the channels is also an input and defined as:

$$\Phi = b/t \tag{A26}$$

with channel width (t). Starting from (A26) and the definition of d_H:

$$d_H = \frac{4A_{cs}}{P} \tag{A27}$$

with channel cross section area A_{cs} and perimeter P, rearrangement leads to:

$$b = d_H \cdot \frac{1+\Phi}{2} \tag{A28}$$

In a similar fashion, using basic geometry and regarding the sidewalls of the channels as fins results in:

$$A_f/A = \frac{\Phi}{\Phi + 1} \quad \text{(A29)}$$

with finned area A_f and total heat exchange area A. The area density is defined as:

$$\beta = \frac{A}{V} \quad \text{(A30)}$$

with core volume V. Combining (A26), (A27), and (A30) concludes after some rearrangements in:

$$\beta = \frac{4 \cdot (1 + \Phi)}{d_H \cdot (1 + \Phi) + 2 \cdot \Phi \cdot \delta} \quad \text{(A31)}$$

2. **Offset-strip fins**. The model for this core is entirely based on [33]. j and f correlations were directly adapted and used within the given limits. For offset-strip fins, the fin length (L_f) is required as an additional input parameter. The missing geometries were derived from Figure 1 in [33]. If offset-strip fins could be realized without additional material on the top or bottom b, A_f/A, and β could be calculated from (A28), (A29), and (A30), respectively. With enhanced manufacturing techniques, it may become possible. Hence, for this model, the additional material thickness on the top and bottom is neglected.
3. **Louvered fins**. The correlation for j was directly implemented from [34] and for f from [35]. b is used as a direct input for this model. A_f/A and β were calculated with (8.76–8.84) from [17]. Additional input parameters to be considered here are louver angle, louver pitch, and louver cut length, which should be selected carefully within the valid ranges given in [34,35].

References

1. Advisory Council for Aviation Research and Innovation in Europe. *Strategic Research and Innovation Agenda: Volume 1: 2017 Update*; Advisory Council for Aviation Research and Innovation in Europe: Brussels, Belgium, 2017.
2. Singh, R.; Freeman, J.; Osterkamp, P.; Green, M.; Gibson, A.; Schiltgen, B. Challenges and opportunities for electric aircraft thermal management. *Aircr. Eng. Aerosp. Technol.* **2014**, *86*, 519–524.
3. Piancastelli, L.; Frizziero, L.; Donnici, G. The Meredith Ramjet: An Efficient Way To Recover The Heat Wasted in Piston Engine Cooling. *J. Eng. Appl. Sci.* **2015**, *10*, 5327–5333.
4. Pérez-Grande, I.; Leo, T.J. Optimization of a commercial aircraft environmental control system. *Appl. Therm. Eng.* **2002**, *22*, 1885–1904. [CrossRef]
5. Jafari, S.; Nikolaidis, T. Thermal Management Systems for Civil Aircraft Engines: Review, Challenges and Exploring the Future. *Appl. Sci.* **2018**, *8*, 2044. [CrossRef]
6. Brelje, B.J.; Martins, J.R. Electric, hybrid, and turboelectric fixed-wing aircraft: A review of concepts, models, and design approaches. *Prog. Aerosp. Sci.* **2018**, *104*, 1–19. [CrossRef]
7. National Acadamy of Sciences. *Commercial Aircraft Propulsion and Energy Systems Research*; National Academies Press: Washington, DC, USA, 2016.
8. Jansen, R.; Bowman, C.; Jankovsky, A. Sizing Power Components of an Electrically Driven Tail Cone Thruster and a Range Extender. In Proceedings of the 16th AIAA Aviation Technology, Integration, and Operations Conference, Washington, DC, USA, 13–17 June 2016.
9. Lents, C.E.; Hardin, L.W.; Rheaume, J.; Kohlman, L. Parallel Hybrid Gas-Electric Geared Turbofan Engine Conceptual Design and Benefits Analysis. In Proceedings of the 52nd AIAA/SAE/ASEE Joint Propulsion Conference, Salt Lake City, UT, USA, 25–27 July 2016.
10. Rheaume, J.; Lents, C.E. Design and Simulation of a Commercial Hybrid Electric Aircraft Thermal Management System. In Proceedings of the 2018 AIAA/IEEE Electric Aircraft Technologies Symposium, Cincinnati, OH, USA, 9–11 July 2018.
11. Rheaume, J.M.; MacDonald, M.; Lents, C.E. Commercial Hybrid Electric Aircraft Thermal Management System Design, Simulation, and Operation Improvements. In Proceedings of the 2019 AIAA/IEEE Electric Aircraft Technologies, Indianapolis, IN, USA, 19–22 August 2019.

12. Chapman, J.W.; Schnulo, S.L. Development of a Thermal Management System for Electrified Aircraft. In Proceedings of the AIAA Scitech 2020 Forum, Orlando, FL, USA, 6–10 January 2020; p. 2273.
13. Kellermann, H.; Habermann, A.L.; Hornung, M. Assessment of Aircraft Surface Heat Exchanger Potential. *Aerospace* **2020**, *7*, 1. [CrossRef]
14. Kellermann, H.; Habermann, A.L.; Vratny, P.C.; Hornung, M. Assessment of fuel as alternative heat sink for future aircraft. *Appl. Therm. Eng.* **2020**, *170*, 114985. [CrossRef]
15. Wei, T. All-in-one design integrates microfluidic cooling into electronic chips. *Nature* **2020**, *585*, 188–189. [CrossRef] [PubMed]
16. Xie, G.; Li, S.; Sunden, B.; Zhang, W.; Li, H. A Numerical Study of the Thermal Performance of Microchannel Heat Sinks with Multiple Length Bifurcation in Laminar Liquid Flow. *Numer. Heat Transf. A Appl.* **2014**, *65*, 107–126. [CrossRef]
17. Shah, R.K.; Sekulić, D.P. *Fundamentals of Heat Exchanger Design*; Wiley-Interscience: Hoboken, NJ, USA, 2003.
18. Kays, W.M.; London, A.L. *Compact Heat Exchangers*, 3rd ed.; Krieger: Malabar, FL, USA, 1998.
19. Sundén, B.; Fu, J. *Heat Transfer in Aerospace Applications*; Academic Press: London, UK, 2017.
20. Pittaluga, F. A set of correlations proposed for diffuser performance prediction. *Mech. Res. Commun.* **1981**, *8*, 161–168. [CrossRef]
21. Malan, P.; Brown, E.F. Inlet drag prediction for aircraft conceptual design. *J. Aircr.* **1994**, *31*, 616–622. [CrossRef]
22. Bräunling, W.J.G. *Flugzeugtriebwerke: Grundlagen, Aero-Thermodynamik, Ideale und Reale Kreisprozesse, Thermische Turbomaschinen, Komponenten, Emissionen und Systeme*, 4th ed.; VDI-Buch; Springer: Berlin, Germany, 2015.
23. Walsh, P.P.; Fletcher, P. *Gas Turbine Performance*, 2nd ed.; Blackwell Science: Oxford, UK, 2008.
24. Incropera, F.P.; DeWitt, D.P.; Bergman, T.L.; Lavine, A.S. *Principles of Heat and Mass Transfer*, 7th ed.; International Student Version ed.; Wiley: Singapore, 2013.
25. Haaland, S.E. Simple and Explicit Formulas for the Friction Factor in Turbulent Pipe Flow. *J. Fluids Eng.* **1983**, *105*, 89. [CrossRef]
26. Pohl, M.; Köhler, J.; Jeschke, P.; Kellermann, H.; Lüdemann, M.; Hornung, M.; Weintraub, D. Integrated Preliminary Design of Turboelectric Aircraft Propulsion Systems. 2021. Manuscript in preparation.
27. Matsuda, M.; Mashiko, K.; Saito, Y.; Nguyen, T.; Nguyen, T. *Mico-Channel Cold Plate Units for Cooling Super Computer*; Fujikura: Tokyo, Japan, 2015.
28. Advanced Thermal Solutions. The Thermal Resistance of Microchannel Cold Plates. *Qpedia Therm. Emagazine* **2012**, *2012*, 6–9.
29. Bell, I.H.; Wronski, J.; Quoilin, S.; Lemort, V. Pure and Pseudo-pure Fluid Thermophysical Property Evaluation and the Open-Source Thermophysical Property Library CoolProp. *Ind. Eng. Chem. Res.* **2014**, *53*, 2498–2508. [CrossRef] [PubMed]
30. VDI-Wärmeatlas. *Mit 320 Tabellen*, 11th ed.; VDI-Buch; Springer: Berlin, Germany, 2013.
31. Denkenberger, D.C.; Brandemuehl, M.J.; Pearce, J.M.; Zhai, J. Expanded microchannel heat exchanger: Design, fabrication, and preliminary experimental test. *Proc. Inst. Mech. Eng. A J. Power Energy* **2012**, *226*, 532–544. [CrossRef]
32. Nakayama, Y.; Boucher, R.F. *Introduction to Fluid Mechanics*; Butterworth Heinemann: Oxford, UK; Boston, MA, USA, 1999.
33. Wieting, A.R. Empirical Correlations for Heat Transfer and Flow Friction Characteristics of Rectangular Offset-Fin Plate-Fin Heat Exchangers. *J. Heat Transf.* **1975**, *97*, 488–490. [CrossRef]
34. Chang, Y.J.; Wang, C.C. A generalized heat transfer correlation for Iouver fin geometry. *Int. J. Heat Mass Transf.* **1997**, *40*, 533–544. [CrossRef]
35. Chang, Y.J.; Hsu, K.C.; Lin, Y.T.; Wang, C.C. A generalized friction correlation for louver fin geometry. *Int. J. Heat Mass Transf.* **2000**, *43*, 2237–2243. [CrossRef]

 aerospace

MDPI

Article

Advanced Materials and Technologies for Compressor Blades of Small Turbofan Engines†

Dmytro Pavlenko [1,*], Yaroslav Dvirnyk [1,2] and Radoslaw Przysowa [3]

1 Mechanical Engineering Department, National University "Zaporizhzhia Polytechnic", 64 Zhukovskogo st, 69063 Zaporizhzhia, Ukraine; dvirnyk@gmail.com
2 Motor Sich JSC, 15 Motorostroiteley Ave., 69068 Zaporizhzhia, Ukraine
3 Instytut Techniczny Wojsk Lotniczych, ul. Księcia Bolesława 6, 01-494 Warsaw, Poland; radoslaw.przysowa@itwl.pl
* Correspondence: dvp1977dvp@gmail.com

† This paper is an extended version of our paper published in 10th EASN International Conference on Innovation in Aviation & Space to the Satisfaction of the European Citizens.

Abstract: Manufacturing costs, along with operational performance, are among the major factors determining the selection of the propulsion system for unmanned aerial vehicles (UAVs), especially for aerial targets and cruise missiles. In this paper, the design requirements and operating parameters of small turbofan engines for single-use and reusable UAVs are analysed to introduce alternative materials and technologies for manufacturing their compressor blades, such as sintered titanium, a new generation of aluminium alloys and titanium aluminides. To assess the influence of severe plastic deformation (SPD) on the hardening efficiency of the proposed materials, the alloys with the coarse-grained and submicrocrystalline structure were studied. Changes in the physical and mechanical properties of materials were taken into account. The thermodynamic analysis of the compressor was performed in a finite element analysis system (ANSYS) to determine the impact of gas pressure and temperature on the aerodynamic surfaces of compressor blades of all stages. Based on thermal and structural analysis, the stress and temperature maps on compressor blades and vanes were obtained, taking into account the physical and mechanical properties of advanced materials and technologies of their processing. The safety factors of the components were established based on the assessment of their stress-strength characteristics. Thanks to nomograms, the possibility of using the new materials in five compressor stages was confirmed in view of the permissible operating temperature and safety factor. The proposed alternative materials for compressor blades and vanes meet the design requirements of the turbofan at lower manufacturing costs.

Keywords: turbofan; unmanned aerial vehicles; cruise missile; aerial target; axial compressor; blade; titanium alloy; aluminium alloy; titanium aluminide; safety factor

Citation: Pavlenko, D.; Dvirnyk, Y.; Przysowa, R. Advanced Materials and Technologies for Compressor Blades of Small Turbofan Engines. *Aerospace* **2021**, *8*, 1. https://dx.doi.org/10.3390/aerospace8010001

Received: 5 November 2020
Accepted: 16 December 2020
Published: 22 December 2020

1. Introduction

Currently, one of the most promising areas in the aerospace and defence industry is the development of unmanned aerial systems for various purposes. They are based on unmanned aerial vehicles (UAVs) of both reusable and single use. Ukrainian [1] and global manufacturers offer gas-turbine engines for UAVs of various types [2,3]. While full-scale turboprops and turbofans, as a rule, are based on engines designed for manned aircraft [4], small turbofans are custom-made [5,6]. They typically have a compact and simplified single-shaft structure, determined by the tactical and technical characteristics of the platform [7].

Small turbofan engines (Table 1) are designed for target drones (Streaker, Lakshya) and cruise missiles such as R-360 Neptune, Kite, Kh-55, Tomahawk and Harpoon. Their main performance characteristics include a short life cycle (if used as weapons), small size and weight and, as a result, high thrust-to-weight ratio. Also, operation on an unmanned platform contributes to the fact that they are not subject to the aviation safety regulations. Such engines are produced by JSC Motor Sich, SE Ivchenko Progress and a number of foreign firms. Engines of this class have thrust in the range of 1.9–4 kN, a low bypass ratio and a small dry mass not exceeding 60–85 kg. At the same time, to ensure high efficiency, such turbofan engines rotate at several tens of thousands of revolutions per minute, which imposes special requirements on the design of their components and selection of materials. First of all, they should exhibit high specific strength under static loads and a relatively low manufacturing cost. At the same time, their durability, due to the short life cycle and lack of pilot, is not of prime importance.

Table 1. Small gas-turbine engines. Data from minijets.org, uasresearch.org, wikipedia.org and [2,3,8].

Producer	Model	Thrust kN	Weight kg	Thrust /Weight	Length mm	Diameter mm	Platform
Turbomeca	Arbizon IIIB2	4.02	115	3.56	1361	421	Otomat missile
Microturbo	TRI 60-30	5.70	61	9.53	841	343	Apache missile
Teledyne CAE	J402-CA-702	4.20	63	6.85	762	317	MQM-107D Streaker
HAL	PTAE-7	3.72	65	5.83	1270	330	Lakshya PTA drone
Mitsubishi	TJM4	2.84	56	5.19	1092	355	Subaru drone
Williams Int.	F107WR402	3.11	66	4.60	1262	305	BGM-109 Tomahawk
Motor Sich	MS-400	3.92	85	4.70	850	320	R-360 Neptune missile
Ivchenko Progress	AI-305	3.04	61	5.08	650	232	Ultra light aircraft
Soyuz	R95-300	3.55	100	3.62	850	315	Kh-55 missile
Saturn	36MT	4.54	100	4.63	850	330	Kh-59 missile
Price Induction	DGEN 380	2.55	85	3.06	1126	469	Personal Light Jet

In small turbofan engines both radial and axial compressors are used. Currently, various types of titanium alloys are successfully used for manufacturing the blades and vanes of axial compressors [9,10]. The most common are VT6 (Ti-6Al-4V), VT3-1 (Ti-6.7Al-2.5Mo-1.8Cr-0.5Fe-0.25Si) and VT8 (Ti-6.8Al-3.5Mo-0.32Si). For compressor stages with increased air temperature along the gas path, heat-resistant titanium alloys of the VT25 (Ti-6.8Al-2.0Mo-2.0Zr-2.0Sn-1.0W-0.3Si) type are used [11,12]. For the last stages of the compressor, taking into account the temperature level, heat-resistant nickel-based alloys such as Inconel 718 (EP718-ID) and similar are used. A common drawback of these materials, along with the high cost and energy costs of production, is their poor machinability. Having a combination of the properties necessary for the compressor blades of a manned aircraft's engine, they are redundant when used on UAVs. This leads to the increased cost of engines and UAVs in general. To meet the requirements for UAV power plants, it is necessary to introduce new materials and technologies, which reduce their manufacturing cost.

There are several modern technologies which can be used for manufacturing gas turbines for UAVs [13]. With regard to compressor blades, a number of candidate materials is considered, for example, sintered powder alloys; rare earth aluminium alloys; alloys based on titanium aluminides and others [14,15].

At present, only surfaces of compressor blades are hardened [16], primarily by laser shock peening [17]. However, surface hardening does not modify the inner structure of the alloy. Therefore, to significantly increase the strength and ductility of aircraft materials, severe plastic deformation (SPD) technologies are used [18,19] but the size of produced ingots is still limited. What is more, for each compressor stage, there are limitations both in the operating temperature and mechanical properties which have to be met by the introduced materials.

In this work, to reduce the manufacturing cost of a selected small turbofan engine, alternative materials and technologies for producing compressor aerofoils are introduced and evaluated. To ensure structural integrity, the static safety factor is assessed for the blades and vanes of individual stages, taking into account their operating temperature. The key objectives of this work include material selection, strength testing, airflow simulation of the compressor to obtain pressure fields on aerofoil surfaces of all stages as well as gas temperature and finally the structural analysis of components which evaluates their stress and static safety factor.

2. Materials and Methods

2.1. Twist Extrusion

Various SPD methods [20,21] are introduced to improve mechanical, physical, and functional properties of metals and alloys by forming their submicrocrystalline structure [22,23]. Twist extrusion (TE) is a variant of the simple shear deformation process that was introduced by Beygelzimer [24]. Under TE processing, a prismatic billet is extruded through a twist die.

In this work, a number of standard and powder metal alloys (Table 2) sourced from various contractors were processed with TE. The titanium billets were made from annealed VT8 rods of increased quality, 32 mm in diameter (GOST 26492-85), produced by VSMPO-AVISMA Corporation. Sintered titanium was synthesised in laboratory by pressing and subsequent vacuum sintering of powder mixtures based on PT5 titanium powder (TU U14-10-026-98) produced by the Zaporizhzhia Titanium-Magnesium Plant. The grain size was 160–500 μm [25].

Table 2. Analysed alloys, their composition and related publications.

VT8 (Ti-6.8Al-3.5Mo-0.32Si), OST 190013-81, GOST 26492-85, [25–31]											
Composition % mass						Impurities % max					
Ti	Al	Mo	Sn	Si	C	Fe	Zr	O	N	H	
base	5.8–7.0	2.8–3.8	≥0.4	0.2–0.4	0.1	0.3	0.5	0.15	0.05	0.015	
γ-TiAl (Ti-46Al-5Nb-2W), [32–34]											
Ti	Al	Nb	W								
Base	44–47	4.2–5.5	1.5–2.5								
7055+Sc (Al-Zn-Mg-Cu-Sc), OST 190013–81, [35–40]											
Al	Zn	Mg	Cu	Zr	Sc	Fe	Mn	Si	Ti	Cr	Ni
base	6.8–8.4	1.5–2.5	1.6–2.9	0.1–0.5	0.1–0.25	0.13	0.01	0.03	0.01	0.01	0.01

The billet (Figure 1) was 70 mm long with the cross-section of 18 × 28 mm. It was placed in a matrix with a helical channel of rectangular cross section with an angle of the helix inclination to the TE axis. The extrusion pressure was FP = 1600 MPa for all the studied alloys. To increase their plasticity, back pressure BP = 200 MPa was applied to the front end of the billet. To transmit back pressure, a deformable medium was used, which was either a mixture based on the low-melting glass or a copper billet [41].

There are different approaches to modelling and optimising the TE process [42,43], usually based on FE methods. Calculations and experiments are aimed to obtain high plastic strain and uniform ultrafine grains [21]. In this work the Beygelzimer's approach [41,44] is followed. The total relative shear deformation Λ per pass was calculated as follows [41]:

$$\Lambda = \frac{2}{\sqrt{3}} \tan \gamma_{max}, \tag{1}$$

where γ_{max} is the maximum inclination angle between the twist line and the extrusion axis. As the deflection angle of the helical channel was 45° for all investigated materials, the total shear deformation per pass was approximately 1.15. Five TE passes were carried out, so the total relative shear deformation of the billet was 5.77.

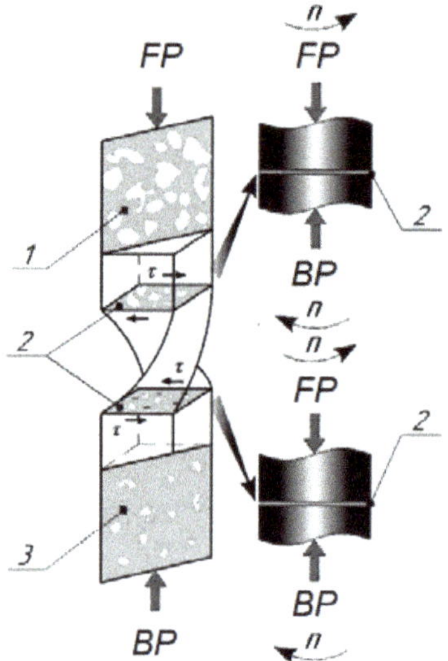

Figure 1. Deformation of a porous billet by twist extrusion: 1—before TE; 2—deformation zone; 3—after TE, *FP*—forward pressure, *BP*—back pressure.

The porosity of the specimens was measured by the hydrostatic weighing method (GOST 18847-84) and by analysing the micrographs of metallographic specimens (GOST 9391-80) [45]. In the first case, the specimens were submerged into distilled water, whose temperature was measured by a mercury thermometer. There was no porosity in specimens made of VT8 titanium alloy. After TE, a slight (within a few percent) increase in porosity was observed, which could be associated with an increase in the number of crystal lattice defects.

For the investigation of structure and the fractographic analysis of fracture surfaces, a NEOPHOT light microscope and a JEOL scanning electron microscope was used [46]. The average grain size in the samples after five TE passes was in the range of 200–500 nm for titanium alloys [45]. The grain size in the original material was 150–300 µm.

2.2. Sintered Titanium

One of the well-known methods of reducing the manufacturing cost of the axial compressor is using sintered titanium alloys [47,48] but their residual porosity and low ductility are the reasons for which up to now they are used in aircraft engines for a narrow circle of lightly loaded, non-critical components. Therefore, powder materials need consolidation and grain refinement, which can be effectively achieved by the SPD process of high-pressure torsion (HPT) [49,50]. However, HPT can produce only very small samples which cannot be used for manufacturing compressor blades. Therefore, our recent paper [51] uses the physical similarity of the processes occurring in a thin layer of material during HPT and TE to simulate twist extrusion with the available HPT data.

In this work, among others, alloys synthesised from a mixture of selected powder components [25,52] were evaluated. Doped elements (pure Al, Mo and Si metals) were mixed with the matrix titanium powder in a mixer drum at 60–80 rpm to ensure the required chemical composition of the test alloy after sintering. The powders were subjected to single-action compaction in rigid dies at room temperature. The compaction force was 730–760 MPa. The compacts were sintered in vacuum in the range of 1250–1270 °C with an isothermal holding time of 2.5–3 h and cooled down in the furnace in vacuum.

Our previous paper [46] showed that the characteristics of sintered titanium alloys subjected to SPD in some indicators exceed similar values for regular alloys in cast and deformed states. The preliminary structural analysis of blades made of sintered titanium with subsequent SPD confirmed that their safety margin meets the operating conditions [53].

However, an important factor that limits the use of alloys in the compressor design is the elevated temperature caused by air compression in the gas path. Also, given the high rotational speed of the engine, close to 40–50 thousand revolutions per minute, stress analysis results depend heavily on the calculated pressure field. As the information on the operating temperature and pressure in the compressor stages of small turbofan engines is very limited [54,55], air flow and thermal analysis is performed in this work.

2.3. Aluminium-Based Alloys

Aluminium alloys with lithium and scandium are well suited to be used in turbofan engines, given their high specific strength which exceeds those of titanium alloys [35–37]. SPD effectively hardens the cast structure of aluminium, so it could be used instead of homogenization annealing [40,56]. However, it is necessary to take into account the operating temperature of components since the heat resistance of aluminium alloys is significantly lower than that of titanium and nickel ones.

Intermetallic Fe_3Al-based alloys could potentially substitute more expensive superalloys and creep-resistant steels. They are characterized by a combination of interesting functional characteristics such as excellent resistance to oxidation, sulfidation and carburizing, good resistance to seawater corrosion, wear, erosion, or cavitation, and high strength to weight ratio [57,58].

In this work, a variant of the standard aerospace aluminium alloy 7055+Sc was used (Al-Zn-Mg-Cu-Sc). It was obtained in laboratory by melting with the additive of scandium. Its initial porosity was 3–4% and it reduced to less than 1.5% after TE.

Lightweight, heat-resistant and weldable alloys based on titanium aluminides [34] make it possible to design more efficient compressors. These materials offer a number of unique properties—low density, relatively high melting point, high modulus of elasticity, resistance to oxidation and fire, high specific heat resistance, and so forth. They are well suited for the last stages of compressor blades but their effectiveness is controversial. On one hand, due to the combination of specific strength and heat resistance, they can replace traditional nickel-based alloys [33,59]. On the other hand, the technology of their manufacturing and processing is quite energy-intensive, which makes them cost-ineffective in the case of small turbofan engines. While heat explosion is a significantly cheaper technology to synthesise such materials [60], their mechanical properties are not satisfactory for aircraft components, in particular for aero-engines.

In this case, a promising, cost-saving technology for the preparation of semi-finished intermetallic γ-TiAl alloys for aircraft, in particular compressor blades, was self-propagating high-temperature synthesis and subsequent TE of the initial ingots [32]. The initial porosity of the γ-TiAl alloy was 35–40% and it decreased to 4–5% after TE.

Taking into account that this technology not only reduces the cost of manufacturing compressor blades, but also increases the level of their mechanical characteristics, assessing the possibility of their use in the design of engines for UAVs is important.

2.4. Strength Testing

To determine the mechanical properties of alloys, 11 mm $\times$ 11 mm $\times$ 56 mm billets were used to produce standard tensile samples in accordance with GOST 1497-84. Strength testing was carried out on the INSTRON 8802 servohydraulic machine under programmed loading at room and elevated temperature. Five reference samples, mass-produced from VT8 alloy bars, were measured to validate the test procedure. The extensometer span was 25 mm. The specimen test portion strain was controlled with an accuracy of 1 µm. The accuracy of stress measurements in the specimen cross-section was $\pm$3 MPa. Extensometer and spring dynamometer readings were ADC-processed and sampled with a rate of $\Delta t = 0.01$ s [25,61]. The actual tensile testing covered more than three specimens for each case.

Table 3 presents the physical and mechanical properties of considered blade materials. The last column shows that materials subjected to TE become less heat-resistant because intensive grain grow begins at a lower temperature. The ratio of Young's modulus and material ultimate strength to density characterises the specific stiffness and specific strength of the material. From the point of view of strength, for the production of aircraft engine components, the most promising materials are those with the maximum values of the specified characteristics. This makes it possible to ensure not only a high level of their strength reliability (safety factor), but also a decrease in the mass. It is known that reducing the rotor mass is one of the best ways for improving the design of a gas turbine, since it effectively reduces the level of dynamic loads and vibration [62].

Taking into account that the analysed technologies for obtaining ingots for compressor components (powder metallurgy and severe plastic deformation) lead to a change in the indicated characteristics of materials at the level of 10%, they were not considered as a criterion for choosing a production technology. At the same time, when choosing a material, preference was given to that material, the specific stiffness and strength of which is higher while ensuring equal safety margins.

Table 3. Mechanical and physical properties of the alloys considered for compressor aerofoils.

Material	E MPa	ρ kg/m^3	UTS MPa	$\sigma_{0.2}$ MPa	ν	E/ρ Nm/kg	UTS/ρ Nm/kg	T_{max} °C
VT8	(1.20 ± 0.05) e5	4520 ± 198	980 ± 42	850 ± 38	0.30	26.5 e6	0.22 e6	500^{+20}
VT8_spd	(1.08 ± 0.04) e5	4400 ± 201	1250 ± 34	1150 ± 44	0.38	24.5 e6	0.28 e6	460^{+20}
VT8_spk	(0.95 ± 0.04) e5	4000 ± 226	700 ± 40	450 ± 42	0.10	23.8 e6	0.18 e6	500^{+20}
VT8_spk_spd	(1.10 ± 0.05) e5	4400 ± 180	1040 ± 35	960 ± 36	0.32	25.0 e6	0.21 e6	460^{+10}
γ-TiAL	(9.50 ± 0.43) e4	4200 ± 189	720 ± 32	650 ± 29	0.30	22.6 e6	0.17 e6	750^{+20}
γ-TiAL_spd	(8.50 ± 0.38) e4	4100 ± 166	920 ± 30	880 ± 36	0.34	20.7 e6	0.22 e6	680^{+10}
7055+Sc	(6.90 ± 0.30) e3	2700 ± 121	75 ± 3	60 ± 3	0.33	2.6 e6	0.03 e6	120^{+20}
7055+Sc_spd	(6.20 ± 0.30) e3	2680 ± 114	203 ± 7	180 ± 7	0.35	2.3 e6	0.08 e6	100^{+10}

UTS—ultimate tensile strength, SPD—alloy of a submicrocrystalline structure formed by TE SPK (sintered metal powder)—alloy obtained by powder metallurgy methods.

2.5. Modelling the Compressor

The effectiveness of the use of candidate materials for manufacturing blades and vanes was evaluated for an axial compressor with the geometry representative of small turbofan engines. The stress-strain state of compressor components was estimated by a coupled Finite Element (FE) Analysis which included a flow calculation and stress analysis. The obtained pressure and temperature fields were applied directly to aerofoil surfaces to determine the stresses and strains in components [63–65]. The analysis was performed for a 6-stage axial compressor (Figure 2). The fan is not considered in this paper as its blades are too large for SPD technology and also sintered alloys do not provide the necessary level of strength.

The profile section of the first compressor stage is shown in Figure 3. The geometry of the compressor blades corresponds to the standard aerodynamic profile of NACA 7404-7405 AIRFOIL. The total number of blades in the compressor stages is given in Table 4.

Using the Unigraphics NX system, models of blades and vanes (one pair per each stage) were built. To develop the aerofoil profile, the surface modelling method was used, while for roots, the method based on Boolean operations with geometric primitives (Figure 4). To create finite element models, an ICEM CFD grid generator was used. The mesh models of the blades consisted of 15,000–18,000 hexagonal SOLID 186 elements. ANSYS Workbench version 2019 R3 was used for the calculations. Blades were fixed at the root plane.

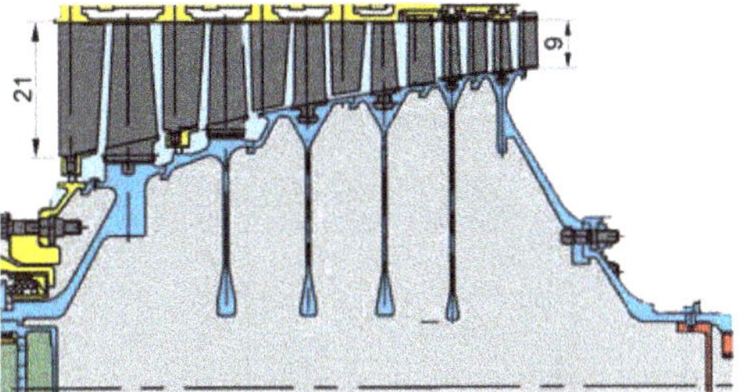

Figure 2. Axial compressor.

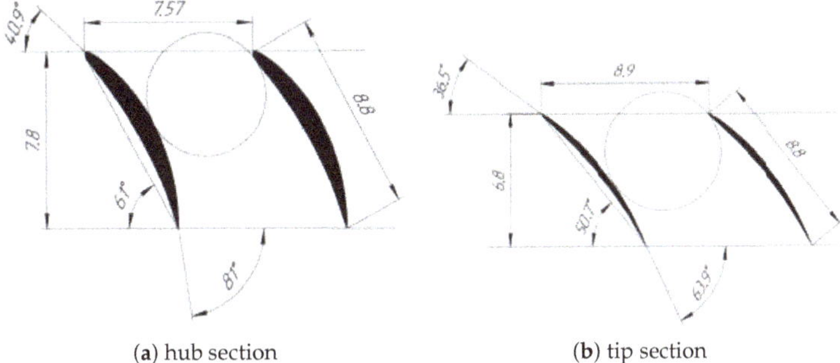

(**a**) hub section (**b**) tip section

Figure 3. Profile of the first stage blade.

(**a**) compressor aerofoils (**b**) 1st stage blade root

Figure 4. Structural model.

Table 4. Number of compressor blades.

Compressor Stage	**R1**	**R2**	**R3**	**R4**	**R5**	**R6**
Number of blades	37	43	59	67	73	81

2.6. CFD Model

Temperature along the compressor gas path and pressure on the aerodynamic surfaces of the blades was determined by flow calculation in Ansys CFX with the finite element method. The CFD model of the compressor inter-blade channel was obtained by arranging the domains of each compressor stage in the axial and radial directions. To build a mesh of the compressor flow, the TurboGrid grid generator was used (Figure 5). Volumetric finite elements intended for CFD calculations were used. To reduce the required computing power, one blade was modelled for each compressor stage with the cyclic symmetry along the boundaries of the domain (Figure 5c). The boundary conditions were set in the form of total inlet pressure, mass flow at the compressor outlet, and rotational speed (Figure 5d).

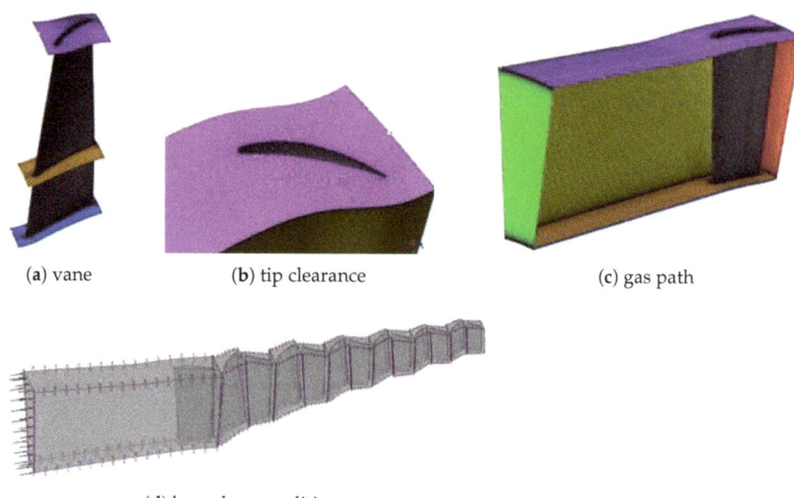

(**a**) vane (**b**) tip clearance (**c**) gas path

(**d**) boundary conditions

Figure 5. Airflow model of the compressor.

An interface between stationary and rotating regions (Stage Mizing-Plane) was defined on the mating boundaries of regions belonging to different steps, which allows for the interpolation between mating grids. A satisfactory criterion for the convergence of the calculation was the value of the mean square residual at the level of 10^{-6}. This convergence was achieved at 1200–1400 iterations. We used the SST (Menter's Shear Stress Transport) $k - \omega$ model of turbulence [66,67], as the most accurate and reliable for flows with a positive pressure gradient when flowing around profiles. At the inlet and outlet of the compressor, the mass flow rate and temperature corresponding to the engine emergency operation were set. The the simulation results were validated according to the methodology described in Reference [68].

2.7. Thermal Structural Analysis

To assess the stress-strain state of the components and temperature distribution, the results of the flow calculation were used. The aerodynamic surfaces of the blades (pressure and suction sides) were loaded with the pressure and temperature fields obtained as a result of preliminary flow calculation.

Typically, both static and fatigue strength are evaluated for new components [69,70] which requires reliable material data to check the safety factor. It includes the endurance limit of laboratory samples at operating temperature, the amplitude of alternating stress at the time of failure, as well as the effective coefficient of stress concentration and the magnitude of their variation. Considering that when analysing the suitability of new materials, these data were not available, the static safety factor (SF) was evaluated with the following formula [71]:

$$SF = \frac{\sigma_{0.2}}{\sigma_{Mises}}, \tag{2}$$

where $\sigma_{0.2}$—conditional yield strength of the blade material, σ_{Mises}—maximum value of the von Mises stress in the compressor blades.

3. Results and Discussion

Figure 6 shows the calculated pressure fields on blade surfaces, and the flow temperature. The flow temperature was used as the initial data for thermal analysis as the boundary condition of the third kind to calculate the surface temperature of the blades (Figure 7). The obtained operating temperature of the compressor blades makes it possible to evaluate the suitability of the considered materials. Given that the blade has a relatively small profile thickness, the temperature distribution over the cross-section was considered uniform.

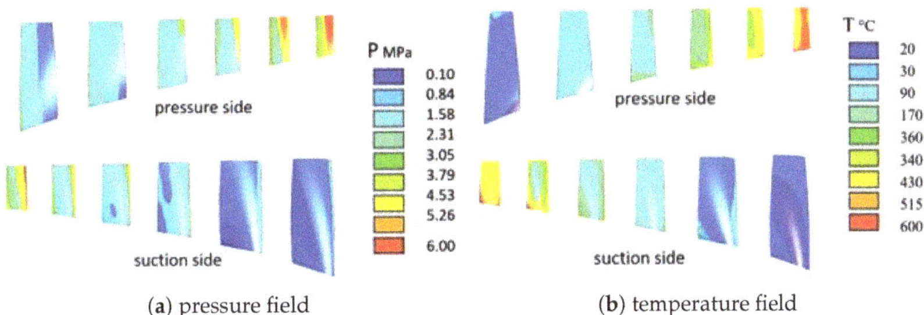

(**a**) pressure field (**b**) temperature field

Figure 6. Pressure and temperature field on blade surfaces.

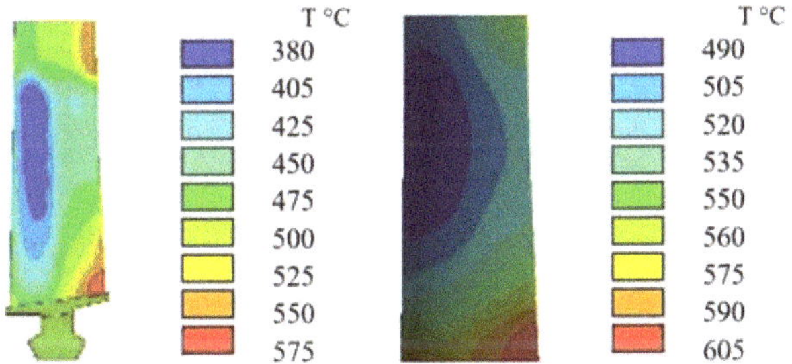

Figure 7. Temperature field for blades and vanes of stage 6.

The calculated stress distribution in the aerofoils (Figure 8) made it possible to evaluate candidate materials and processing technologies in view of their structural integrity. Values of maximum equivalent stress and static safety factor of blades and vanes made from advanced materials and technologies are given in Tables 5 and 6. Materials with safety factor less than the threshold of 1.1 cannot be used in the particular stage [72]. This value was selected by the manufacturer on the basis of industrial experience and reliability data. Under certain conditions, such a low SF threshold is acceptable in aircraft components, especially for unmanned and single-use platforms.

Analysing the obtained data, we can conclude that the candidate materials and processing technologies can be used for manufacturing compressor components. Considering that material selection by the temperature and strength criteria is complicated due to the variety of limiting factors, nomograms were developed for this purpose (Figures 9 and 10). It can be noted that VT8 alloy is limited to rotor stages 1–2 in terms of its strength reliability. The use of SPD methods expands the scope of its application up to the 7th stage; however, in terms of the temperature limit, VT8 usage is limited to blades of the first five stages.

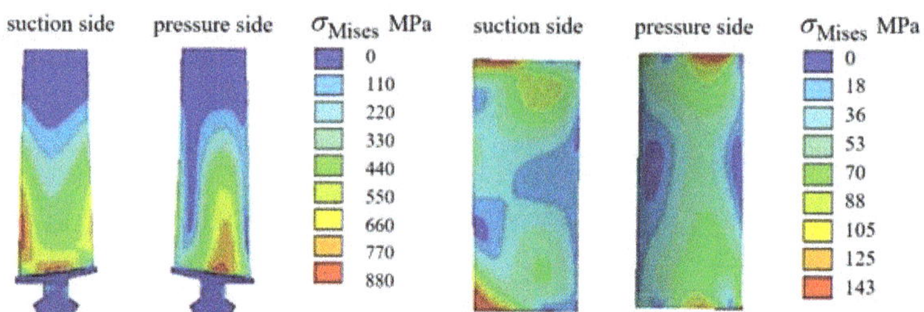

Figure 8. Von Misses stress in blades and vanes made from VT8_spk_spd in engine emergency mode.

Table 5. Equivalent stress and static safety factor of blades made from candidate materials.

Rotor Stage	**R1**		**R2**		**R3**		**R4**		**R5**		**R6**	
Alloy/Process	σ_{max} **MPa**	*SF*	σ_{max} **MPa**	*SF*	σ_{max} **MPa**	*SF*	σ_{max} **MPa**	*SF*	σ_{max} **MPa**	*SF*	σ_{max} **MPa**	*SF*
VT8	480.2	1.77	481.1	1.77	805.8	1.05	893.3	0.95	717.6	1.19	864.5	0.98
VT8_spd	481.9	2.39	451.6	2.25	802.4	1.43	859.4	1.34	718.4	1.60	889.4	1.29
VT8_spk	477.2	0.94	517.3	0.87	801.2	0.56	938.6	0.48	719.9	0.63	882.2	0.51
VT8_spk_spd	481.7	1.99	474.2	2.02	804.3	1.91	886.0	1.08	717.4	1.33	872.2	1.10
γTiAl	483.6	1.34	473.6	1.37	803.1	0.81	892.3	0.73	717.2	0.91	873.5	0.74
γTiAl_spd	483.6	1.82	462.9	1.90	801.3	1.10	877.3	1.00	717.5	1.23	885.5	0.99
7055+Sc	460.0	0.13	451.1	0.13	789.8	0.08	877.0	0.07	717.0	0.08	922.4	0.07
7055+Sc_spd	460.0	0.39	450.9	0.40	789.0	0.23	868.7	0.21	717.3	0.25	928.3	0.19

Table 6. Equivalent stress and static safety factor of vanes made from candidate materials.

Stator Stage	**S1**		**S2**		**S3**		**S4**		**S5**		**S6**	
Alloy/Process	σ_{max} **MPa**	*SF*	σ_{max} **MPa**	*SF*	σ_{max} **MPa**	*SF*	σ_{max} **MPa**	*SF*	σ_{max} **MPa**	*SF*	σ_{max} **MPa**	*SF*
VT8	8.8	96.2	45.5	18.7	54.0	15.8	133.5	6.4	140.6	6.1	142.4	6.0
VT8_spd	8.9	129.7	43.6	26.4	53.8	21.4	128.4	9.0	140.8	8.2	146.5	7.9
VT8_spk	8.8	51.3	48.9	9.2	53.7	8.4	140.2	3.2	141.1	3.2	145.3	3.1
VT8_spk_spd	8.8	84.6	44.8	16.7	53.9	13.9	132.4	5.7	140.6	5.3	143.6	5.2
γ-TiAl	8.9	73.0	44.8	14.5	53.8	12.1	133.3	4.9	140.5	4.6	143.8	4.5
γ-TiAl_spd	8.9	98.9	43.7	20.1	53.7	16.4	131.1	6.7	140.6	6.3	145.8	6.0
7055+Sc	8.5	7.1	42.6	1.4	52.9	1.1	131.0	0.5	140.5	0.4	151.9	0.4
7055+Sc_spd	8.5	21.3	42.6	4.2	52.9	3.4	129.8	1.4	140.6	1.3	152.9	1.2

It can be inferred that the strength of the blades of all compressor stages made of sintered titanium, is below the acceptable threshold (SF = 1.1). Therefore, they cannot be used, despite the significantly lower manufacturing cost in comparison with an alloy in a deformed state. However, the use of SPD methods, due to the elimination of porosity, the formation of a submicrocrystalline structure in the entire cross-section and the homogenization of alloying elements, contributes to a significant increase in strength and, as a consequence, the expansion of their application to all stages.

At the same time, the operating temperature of the submicrocrystalline alloy is lower than one with the standard structure which does not allow for their use in 6th stage blades (Figure 9). Considering that the compressor vanes experience a load only from the flow, the field of application of the VT8 alloy is limited only by its operating temperature, regardless of the technology of production and processing. Despite the great strength, the sintered titanium processed with TE, in comparison with the regular sintered alloy, has the operating temperature lower by 40 °C, which may limit its use. Taking into account the lower cost of obtaining sintered titanium alloys, their use is the most rational in the blades of the first five stages.

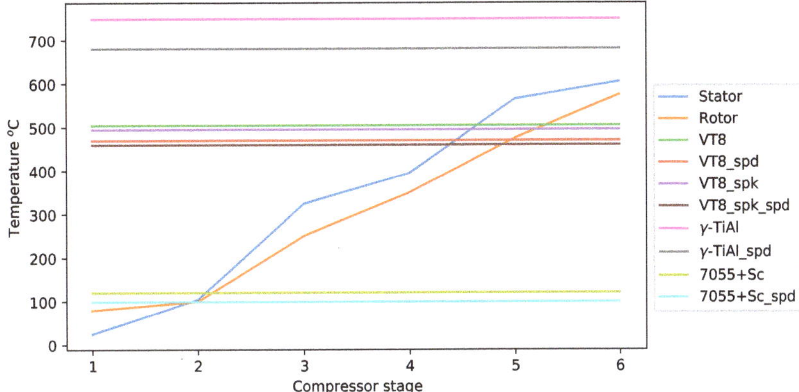

Figure 9. Material maximum temperature vs the operating temperature of compressor stages.

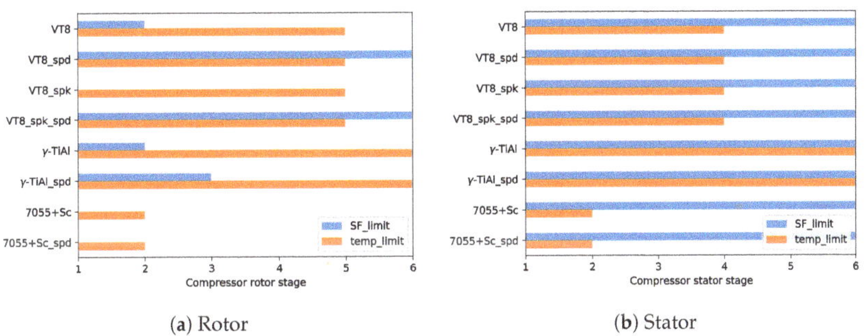

(**a**) Rotor (**b**) Stator

Figure 10. Rotor and stator stages for which the strength and temperature limits of materials are met.

Aluminium alloys with a coarse-grained and submicrocrystalline structure according to the thermal criterion can be applied only to blades of the first and second stages. However, a safety factor assessment indicates that their application is limited to stator vanes. At the same time, modern aluminium alloys can be used to make vanes without SPD processing, which reduces the manufacturing cost. Given the low weight and cost of aluminium vanes compared to titanium ones, the replacement of the material is justified. Moreover, the well-known problems of aluminium alloys, such as low hardness and resistance to sand erosion, are an uncritical factor for UAV engines.

Alloys based on titanium aluminides are the most heat resistant of the considered ones, which predetermines their use for manufacturing blades of the last compressor stages. From the point of view of the permissible operating temperature, this alloy can be applied to blades of all stages regardless of their structural state (Table 3). At the same time, from the point of view of strength reliability for blades, their use is allowed up to stage 2 without additional strain hardening and up to 3rd stage with TE processing (Table 5).

For all stator stages, the safety factor of vanes made from titanium aluminides is higher than the threshold (Table 6). Thus, this alloy can be used for manufacturing vanes of stages 5 and 6, for which, due to temperature limitations, lighter titanium alloys may not be applicable. Nevertheless, the replacement of more heat-resistant Inconel 718 alloys with titanium aluminides would reduce the weight of gas turbine engines.

It should be noted that the considered temperature limitations of submicrocrystalline alloys are associated with the onset of recrystallization processes. Considering that this processes take a relatively long time, exceeding the mission time of single-use UAVs (cruise missiles, disposable reconnaissance vehicles, aerial targets, etc.), this restriction can be removed for such turbofan engines. In this case, their maximum allowable temperature will be similar to alloys in a coarse-crystalline state. The calculated values of the safety factors for compressor components made from considered alloys and technologies let us propose their field of application (Figure 11).

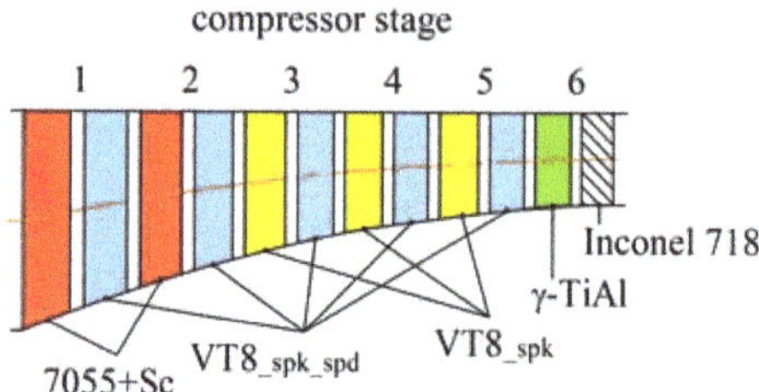

Figure 11. Materials recommended for individual compressor stages.

4. Conclusions

The analysis of the thermal and stress-strain state of the compressor blades and vanes, in combination with the tensile testing of the candidate alloys, made it possible to develop recommendations for their use:

1. It was found that the vanes of the first fifth stator stages can be made of sintered VT8 titanium alloy without strain hardening. Respectively, the blades of the first fifth rotor stages can be made of sintered VT8 titanium alloy, subjected to SPD processing.
2. 7055+Sc aluminium alloy, regardless of the use of TE, can be used to make vanes of the first two stages.
3. Titanium aluminides (γ-TiAl) processed with TE can be used for the blades of stages 1–3 and all stator stages. Considering the lower cost of sintered titanium compared to γ-TiAl alloy, it is reasonable to use it only for the 6th-stage vanes.
4. None of the candidate materials are suitable for making 6th-stage blades, so a superalloy such as Inconel 718 has to be used instead.

The thermal and structural analysis of this high-speed axial compressor shows that its blades are extremely loaded up to the strength and temperature limits of the available alloys. Taking into account that the change in the physical and mechanical properties of materials can affect not only the stress-strain state of the blades but also their dynamic characteristics, the natural frequencies of blades need to be evaluated in the next stage of research. For the compressor under study, Campbell diagrams and the surge margin will be calculated. Also, the damping properties of alloys in various conditions should be analysed.

Author Contributions: D.P. and Y.D. conceived and designed the research; D.P synthesised and processed the alloys; Y.D. and D.P developed FEM and CFD models and performed structural analysis; D.P. and R.P. verified and evaluated the results. D.P., Y.D. and R.P. drew conclusions and produced the paper. All authors have read and agreed to the published version of the manuscript.

Funding: This research received no external funding.

Acknowledgments: We would like to thank Wieslaw Beres and Sylwester Klysz for their comments on an earlier version of the manuscript, although any errors are our own and should not tarnish the reputations of these esteemed persons.

Conflicts of Interest: The authors declare no conflict of interest. Motor Sich JSC had no role in the design, execution, interpretation, or writing the study. The views, information, or opinions expressed herein are solely those of the authors and do not necessarily represent the position of any organization.

Abbreviations

The following abbreviations and symbols are used in this manuscript:

ν	Poisson's ratio
ρ	density
$\sigma_{0.2}$	conditional yield strength
σ_{Mises}	von Mises stress
BP	back pressure
E	Young's modulus
FP	forward pressure
p	pressure
SF	safety factor
T	temperature
CFD	computational fluid dynamics
FE	finite element
HPT	high-pressure torsion
IGV	inlet guide vanes
JSC	join-stock company
rpm	revolutions per minute
SE	state enterprise
SPD	severe plastic deformation
SPK	sintered metal powder
SST	Menter's Shear Stress Transport model of turbulence
TE	twist extrusion
UAV	unmanned aerial vehicle
UTS	ultimate tensile strength
VT8	Titanium wrought alloy

References

1. Telesyk. Motor Sich Engines for UAVs (Dvigateli "Motor Sich" Dlja BPLA). Available online: https://telesyk.livejournal.com/146218.html (accessed on 2 November 2020).
2. Brooks, V.E. Small Turbine Engine Evolution. *SAE Int. J. Aerosp.* **2008**, *1*, 2008-01-2874. [CrossRef]
3. Costa, F.P.; Henrique, L.; Whitacker, L.; Bringhenti, C.; Tomita, J.T. An Overview of Small Gas Turbine Engines. In Proceedings of the 24th ISABE conference, Canberra, Australia, 22–27 September 2019; ISABE 2019, ISABE-2019-24387.
4. Weinberg, M.; Wyzykowski, J. Development and Testing of a Commercial Turbofan Engine for High Altitude UAV Applications. *SAE Tech. Pap.* **2001**. Available online: https://saemobilus.sae.org/content/2001-01-2972/ (accessed on 8 May 2020). [CrossRef]
5. Rodgers, C. Affordable Smaller Turbofans. Volume 1: Turbo Expo 2005. *ASMEDC* **2005**, *1*, 1–10. [CrossRef]
6. Large, J.; Pesyridis, A. Investigation of micro gas turbine systems for high speed long loiter tactical unmanned air systems. *Aerospace* **2019**, *6*, 55. [CrossRef]
7. Nelson, J.R.; Dix, D.M. *Development of Engines for Unmanned Air Vehicles: Some Factors to Be Considered*; Technical Report; Institute for Defense Analyses: Alexandria, VA, USA, 2003. [CrossRef]
8. Razinsky, E.; Cae, T. The J402-CA-702-A Modern 1000 Lb. Thrust RPV Engine. In Proceedings of the AIAA/ASME/SAE/ASEE 24th Joint Propulsion Conference & Exhibit, Boston, MA, USA, 11–13 July 1988.
9. Jackson, M. Titanium—21st century. *Mater. World* **2007**, *15*, 33–34.
10. Leyens, C.; Peters, M. (Eds.) *Titanium and Titanium Alloys*; Wiley-VCH Verlag GmbH & Co. KGaA: Weinheim, Germany, 2003. [CrossRef]
11. Kashapov, O.; Novak, A.; Nochovnaya, N.; Pavlova, T. Sostojanie, problemy i perspektivy sozdanija zharoprochnyh titanovyh splavov dlja detalej GTD (State, problems and prospects of heat-resistant titanium alloys for GTE parts). *Proc. VIAM* **2013**, *3*, 1–12.
12. Whittaker, M. Titanium in the Gas Turbine Engine. In *Advances in Gas Turbine Technology*; Benini, E., Ed.; InTech: Rijeka, Croatia, 2011; Volume 4. [CrossRef]
13. Moustapha, H. Future Technology Challenges for Small Gas Turbines. In *AIAA International Air and Space Symposium and Exposition: The Next 100 Years*; American Institute of Aeronautics and Astronautics: Reston, VA, USA, 2003; pp. 1–11. [CrossRef]
14. Liu, S.; Shin, Y.C. Additive manufacturing of Ti6Al4V alloy: A review. *Mater. Des.* **2019**, *164*, 107552. doi:10.1016/j.matdes.2018. 107552 [CrossRef]
15. Salvati, E.; Lunt, A.J.G.; Ying, S.; Sui, T.; Zhang, H.J.; Heason, C.; Baxter, G.; Korsunsky, A.M. Eigenstrain reconstruction of residual strains in an additively manufactured and shot peened nickel superalloy compressor blade. *Comput. Methods Appl. Mech. Eng.* **2017**, *320*, 335–351. [CrossRef]
16. Boguslaev, V.A.; Pukhal'Skaya, G.V.; Koval', A.D.; Stepanova, L.P.; Tkachenko, V.V. The effect of methods for hardening finish treatment of blades made of titanium alloys on the state of their surface layer. *Met. Sci. Heat Treat.* **2008**, *50*, 18–24. [CrossRef]
17. Zou, S.; Wu, J.; Zhang, Y.; Gong, S.; Sun, G.; Ni, Z.; Cao, Z.; Che, Z.; Feng, A. Surface integrity and fatigue lives of Ti17 compressor blades subjected to laser shock peening with square spots. *Surf. Coat. Technol.* **2018**, *347*, 398–406. [CrossRef]

18. Azushima, A.; Kopp, R.; Korhonen, A.; Yang, D.Y.; Micari, F.; Lahoti, G.D.; Groche, P.; Yanagimoto, J.; Tsuji, N.; Rosochowski, A.; et al. Severe plastic deformation (SPD) processes for metals. *CIRP Ann. Manuf. Technol.* **2008**, *57*, 716–735. [CrossRef]
19. Segal, V. Review: Modes and Processes of Severe Plastic. *Materials* **2018**, *11*, 1175. [CrossRef] [PubMed]
20. Valiev, R.Z.; Estrin, Y.; Horita, Z.; Langdon, T.G.; Zehetbauer, M.J.; Zhu, Y. Producing Bulk Ultrafine-Grained Materials by Severe Plastic Deformation: Ten Years Later. *JOM* **2016**, *68*, 1216–1226. [CrossRef]
21. Husaain, Z.; Ahmed, A.; Irfan, O.M.; Al-Mufadi, F. Severe Plastic Deformation and Its Application on Processing Titanium: A Review. *Int. J. Eng. Technol.* **2017**, *9*, 426. [CrossRef]
22. Pavlenko, D.V.; Beygelzimer, Y.E. Vortices in Noncompact Blanks During Twist Extrusion. *Powder Metall. Met. Ceram.* **2016**, *54*, 517–524. [CrossRef]
23. Estrin, Y.; Vinogradov, A. Extreme grain refinement by severe plastic deformation: A wealth of challenging science. *Acta Mater.* **2013**, *61*, 782–817. [CrossRef]
24. Beygelzimer, Y.E.; Orlov, D.; Korshunov, A.; Synkov, S.; Varyukhin, V.; Vedernikova, I.; Reshetov, A.; Synkov, A.; Polyakov, L.; Korotchenkova, I. Features of twist extrusion: Method, structures & material properties. *Solid State Phenom.* **2006**, *114*, 69–78. [CrossRef]
25. Bykov, I.O.; Ovchinnikov, A.V.; Pavlenko, D.V.; Lechovitzer, Z.V. Composition, Structure, and Properties of Sintered Silicon-Containing Titanium Alloys. *Powder Metall. Met. Ceram.* **2020**, *58*, 613–621. [CrossRef]
26. Moiseev, V.N. Titanium in Russia. *Met. Sci. Heat Treat.* **2005**, *47*, 371–376. [CrossRef]
27. Moiseyev, V.N. *Titanium Alloys. Russian Aircraft and Aerospace Applications*; CRC Press: Boca Raton, FL, USA, 2005. doi:10.1201/9781420037678. [CrossRef]
28. Ermachenko, A.G.; Lutfullin, R.Y.; Mulyukov, R.R. Advanced technologies of processing titanium alloys and their applications in industry. *Rev. Adv. Mater. Sci.* **2011**, *29*, 68–82.
29. Pavlova, T.; Kashapov, O.; Nochovnaja, N. Titanovye splavy dlja gazoturbinnyh dvigatelej (Titanium alloys for gas turbine engines). *Proc. VIAM* **2012**, *5*, 8–14.
30. Kommel, L. Microstructure evolution in titanium alloys enforced by joule heating and severe plastic deformation concurrently. *J. Manuf. Technol. Res.* **2010**, *2*, 59–75.
31. Semenova, I.P.; Raab, G.I.; Valiev, R.Z. Nanostructured titanium alloys: New developments and application prospects. *Nanotechnol. Russ.* **2014**, *9*, 311–324. [CrossRef]
32. Pavlenko, D.V.; Belokon', Y.; Tkach, D.V. Resource-Saving Technology of Manufacturing of Semifinished Products from Intermetallic γ-TiAl Alloys Intended for Aviation Engineering. *Mater. Sci.* **2020**, *55*, 118–124. [CrossRef]
33. Nochovnaya, N.A.; Panin, P.V.; Kochetkov, A.S.; Bokov, K.A. Modern Refractory Alloys Based on Titanium Gamma-Aluminide: Prospects of Development and Application. *Met. Sci. Heat Treat.* **2014**, *56*, 364–367. [CrossRef]
34. Appel, F.; Paul, J.D.H.; Oehring, M. *Gamma Titanium Aluminide Alloys*; Wiley-VCH Verlag GmbH & Co. KGaA: Weinheim, Germany, 2011.
35. Zakharov, V. Effect of Scandium on the Structure and Properties of Aluminum Alloys. *Met. Sci. Heat Treat.* **2003**, *45*, 246–253. [CrossRef]
36. Røyset, J.; Ryum, N. Scandium in aluminium alloys. *Int. Mater. Rev.* **2005**, *50*, 19–44. doi:10.1179/174328005X14311. [CrossRef]
37. Ahmad, Z. The properties and application of scandium-reinforced aluminum. *JOM* **2003**, *55*, 35–39. doi:10.1007/s11837-003-0224-6. [CrossRef]
38. Liddicoat, P.V.; Liao, X.Z.; Zhao, Y.; Zhu, Y.; Murashkin, M.Y.; Lavernia, E.J.; Valiev, R.Z.; Ringer, S.P. Nanostructural hierarchy increases the strength of aluminium alloys. *Nat. Commun.* **2010**. doi:10.1038/ncomms1062. [CrossRef]
39. Bahadori, S.R.; Mousavi, S.A.A.A.; Shahab, A.R. Sequence effects of twist extrusion and rolling on microstructure and mechanical properties of aluminum alloy 8112. *J. Phys. Conf. Ser.* **2010**, *240*, 012132. doi:10.1088/1742-6596/240/1/012132. [CrossRef]
40. Seikh, A.H.; Baig, M.; Ur Rehman, A. Effect of Severe Plastic Deformation, through Equal-Channel Angular Press Processing, on the Electrochemical Behavior of Al5083 Alloy. *Appl. Sci.* **2020**, *10*, 7776. [CrossRef]
41. Beygelzimer, Y.; Kulagin, R.; Estrin, Y.; Toth, L.S.; Kim, H.S.; Latypov, M.I. Twist Extrusion as a Potent Tool for Obtaining Advanced Engineering Materials: A Review. *Adv. Eng. Mater.* **2017**, *19*. [CrossRef]
42. Yalçinkaya, T.; Şimşek, Ü.; Miyamoto, H.; Yuasa, M. Numerical Analysis of a New Nonlinear Twist Extrusion Process. *Metals* **2019**, *9*, 513. [CrossRef]
43. Joudaki, J.; Safari, M.; Alhosseini, S.M. Hollow Twist Extrusion: Introduction, Strain Distribution, and Process Parameters Investigation. *Met. Mater. Int.* **2019**, *25*, 1593–1602. [CrossRef]
44. Latypov, M.I.; Alexandrov, I.V.; Beygelzimer, Y.E.; Lee, S.; Kim, H.S. Finite element analysis of plastic deformation in twist extrusion. *Comput. Mater. Sci.* **2012**, *60*, 194–200. [CrossRef]
45. Pavlenko, D.V. Effect of Porosity Parameters on the Strength of Gas Turbine Compressor Blades Made of Titanium Alloys. *Strength Mater.* **2019**, *51*, 887–899. [CrossRef]
46. Pavlenko, D.V.; Ovchinnikov, A.V. Effect of Deformation by the Method of Screw Extrusion on the Structure and Properties of VT1-0 Alloy in Different States. *Mater. Sci.* **2015**, *51*, 52–60. [CrossRef]
47. Ivasishin, O.M.; Anokhin, V.M.; Demidik, A.N.; Sawakin, D.G. Cost-effective blended elemental powder metallurgy of titanium alloys for transportation application. *Key Eng. Mater.* **2000**, *188*, 55–62. [CrossRef]

48. Fang, Z.Z.; Sun, P. Pathways to optimize performance/cost ratio of powder metallurgy titanium— A perspective. *Key Eng. Mater.* **2012**, *520*, 15–23. [CrossRef]
49. Pavlenko, D. Povyshenie tehnologicheskoj plastichnosti spechennyh titanovyh splavov (Improving the technological plasticity of sintered titanium alloys). *Process. Mech. Process. Mach. Build.* **2015**, *15*, 102–112.
50. Kulagin, R.; Zhao, Y.; Beygelzimer, Y.; Toth, S.L.; Shtern, M. Modeling strain and density distributions during high-pressure torsion of pre-compacted powder materials. *Mater. Res. Lett.* **2017**, *5*, 179–186. [CrossRef]
51. Pavlenko, D. Structural and chemical inhomogeneities in the sintered titanium alloys after severe plastic deformation. *Metalozn. Obrobka Met.* **2020**, *95*, 37–45. [CrossRef]
52. Beygelzimer, Y.E.; Pavlenko, D.V.; Synkov, O.S.; Davydenko, O.O. The Efficiency of Twist Extrusion for Compaction of Powder Materials. *Powder Metall. Met. Ceram.* **2019**, *58*, 7–12. [CrossRef]
53. Pavlenko, D.V.; Pribora, T.I.; Kocjuba, V.J.; Paholka, S.N. Perspektivnye materialy i tehnologii dlja detalej rotora kompressora GTD (Promising materials and technologies for the rotating components of axial compressor). *Aerosp. Sci. Technol.* **2016**, *8*, 128–138.
54. Lu, W.; Huang, G.; Xiang, X.; Wang, J.; Yang, Y. Thermodynamic and aerodynamic analysis of an air-driven fan system in low-cost high-bypass-ratio turbofan engine. *Energies* **2019**, *12*, 1917. [CrossRef]
55. Chivukula, V.; Mohla, R.; Srinivas, G. The flow visualization of small-scale aircraft engine axial flow turbine rotor using numerical technique. *Int. J. Mech. Prod. Eng. Res. Dev.* **2019**, *9*, 777–784. [CrossRef]
56. Beygelzimer, Y.; Kulagin, R.; Raspornya, D.; Varukhin, D. Deformation homogenization of aluminum alloys through twist extrusion. In Proceedings of the 10th International Conference on Technology of Plasticity (ICTP 2011), Aachen, Germany, 25–30 September 2011; pp. 241–243.
57. Łyszkowski, R.; Czujko, T.; Varin, R.A. Multi-axial forging of Fe3Al-base intermetallic alloy and its mechanical properties. *J. Mater. Sci.* **2017**, *52*, 2902–2914. [CrossRef]
58. Łyszkowski, R.; Polkowski, W.; Czujko, T. Severe plastic deformation of Fe-22Al-5Cr alloy by cross-channel extrusion with back pressure. *Materials* **2018**, *11*, 1–17. [CrossRef]
59. Imayev, V.; Imayev, R.; Gaisin, R.; Nazarova, T.; Shagiev, M.; Mulyukov, R. Heat-resistant intermetallic alloys and composites based on titanium: microstructure, mechanical properties and possible application. *Mater. Phys. Mech.* **2017**, *33*, 80–96. [CrossRef]
60. Belokon, K.; Belokon, Y. The Usage of Heat Explosion to Synthesize Intermetallic Compounds and Alloys. In *Processing, Properties, and Design of Advanced Ceramics and Composites II: Ceramic Transactions*; The American Ceramic Society: Columbus, OH, USA, 2018; Volume 261, pp. 109–115. [CrossRef]
61. Karpinos, B.S.; Pavlenko, D.V.; Kachan, O.Y. Deformation of a submicrocrystalline VT1-0 titanium alloy under static loading. *Strength Mater.* **2012**, *44*, 100–107. [CrossRef]
62. Przysowa, R.; Russhard, P. Non-Contact Measurement of Blade Vibration in an Axial Compressor. *Sensors* **2020**, *20*, 68. [CrossRef] [PubMed]
63. Masud, J.; Ahmed, S. Design Refinement and Performance Analysis of Two-Stage Fan for Small Turbofan Engines. In *Proceedings of the 45th AIAA Aerospace Sciences Meeting and Exhibit*; American Institute of Aeronautics and Astronautics: Reston, VA, USA, 2007; Volume 1, pp. 161–168. [CrossRef]
64. Patel, K.S.; Ranjan, R.; Maruthi, N.H.; Deshpande, S.M.; Narasimha, R. Predictions of aero-thermal loading of an HPT stator blade of a typical small turbofan engine Turbomachinery Flows. In Proceedings of the 19th AeSI Annual CFD Symposium, Bengaluru, India, 10–11 August 2017.
65. Rehman, M.; Afzal, R. Design and analysis of a 11:1 centrifugal compressor for a small turbofan engine. In Proceedings of the 2019 16th International Bhurban Conference on Applied Sciences and Technology (IBCAST), Islamabad, Pakistan, 8–12 January 2019; pp. 189–196. [CrossRef]
66. Evans, S.; Lardeau, S. Validation of a turbulence methodology using the SST k-ω model for adjoint calculation. In Proceedings of the 54th AIAA Aerospace Sciences Meeting, San Diego, CA, USA, 4–8 January 2016; American Institute of Aeronautics and Astronautics: Reston, VA, USA, 2016. [CrossRef]
67. Piovesan, T.; Magrini, A.; Benini, E. Accurate 2-D modelling of transonic compressor cascade aerodynamics. *Aerospace* **2019**, *6*, 1–19. [CrossRef]
68. Dvirnyk, Y.; Pavlenko, D.; Przysowa, R. Determination of Serviceability Limits of a Turboshaft Engine by the Criterion of Blade Natural Frequency and Stall Margin. *Aerospace* **2019**, *6*, 132. [CrossRef]
69. Będkowski, W. Assessment of the fatigue life of machine components under service loading—A review of selected problems. *J. Theor. Appl. Mech.* **2014**, *52*, 443–458.

70. Mehdizadeh, O.; Zhang, C.; Shi, F. Flow-Induced Vibratory Stress Prediction on Small Turbofan Engine Compressor Vanes Using Fluid-Structure Interaction Analysis. In Proceedings of the 44th AIAA/ASME/SAE/ASEE Joint Propulsion Conference & Exhibit, Hartford, CT, USA, 21–23 July 2008; American Institute of Aeronautics and Astronautics: Reston, Virigina, 2008; pp. 1–8. [CrossRef]
71. Bhandari, V. *Design of Machine Elements*; Tata McGraw-Hill Education: New Delhi, India, 2010.
72. Mulville, D.R. *Structural design and Test Factors of Safety for Spaceflight Hardware*; Technical Report; Technical Report NASA-STD-5001; NASA: Washington, DC, USA, 1996.

Article

Advanced Passenger Movement Model Depending On the Aircraft Cabin Geometry

Marc Engelmann *, Tim Kleinheinz and Mirko Hornung

Bauhaus Luftfahrt e.V., Willy-Messerschmitt-Str. 1, 82024 Taufkirchen, Germany; tim.kleinheinz@tum.de (T.K.); mirko.hornung@bauhaus-luftfahrt.net (M.H.)

* Correspondence: marc.engelmann@bauhaus-luftfahrt.net; Tel.: +49-89-3074-849-55

Received: 12 November 2020; Accepted: 16 December 2020; Published: 20 December 2020

Abstract: The aircraft cabin and boarding procedures are steadily increasing focus points for both aircraft manufacturers and airlines, as they play a key part in the customer experience. In the German research project AVACON (AdVAnced Aircraft CONcepts), the boarding procedure is assessed using the PAXelerate boarding simulation. As the project demands an increased level of detail concerning the passenger movement model, this publication introduces an improved methodology. Additions to the model include the development of a method capable of describing the passenger walking speed in dependence of the surrounding objects, their proximity as well as the location of other passengers within the cabin. The validation of the model is performed using the AVACON research baseline and an Airbus A320. The model is then applied to an altered version of the Airbus A320 with an extended aisle and to a COVID-19 safe distance scenario. Regarding the results, an extended aisle width delivers boarding times reduced by up to 3%, whereas the COVID-19 assessment delivers a 67% increase in boarding times. Concluding, the integration of the newly developed model empowers PAXelerate to simulate a more detailed boarding process and enables a better understanding of the influence of cabin layout changes to an aircraft's boarding performance.

Keywords: boarding; simulation; cabin; aircraft; passenger; movement; Covid-19

---

1. Motivation

The aircraft cabin is steadily becoming more important for both aircraft manufacturers and airlines, as it is the most prominent aspect of an aircraft that passengers are in contact with during a commercial flight. In regards to this, the boarding performance of an aircraft plays a key part in the customer experience and enables cost saving potential by reducing the critical path and thus turnaround time at the airport [1,2].

An important aspect of today's aircraft operational research is optimizing the boarding procedure and uncovering options for a reduction in boarding time while increasing passenger comfort throughout the process [2–4]. In alignment with this focus, the boarding process is assessed during the course of the German research project "AdVanced Aircraft CONcepts" (AVACON). The project's target is to develop a mid-range aircraft concept with a projected entry into service in the year 2028 [5]. A secondary target is to strengthen the interdisciplinary connections within the German aerospace industry.

Throughout the course of the project, multiple design iterations of the aircraft concept are performed. For each of these concepts, the boarding process performance of the corresponding cabin layout has to be assessed. To do so, Bauhaus Luftfahrt is applying the PAXelerate boarding simulation tool to the different cabin concepts. As the project demands a greater level of detail for PAXelerate regarding the passenger movement, a novel methodology was both created and implemented into PAXelerate and is now introduced in the scope of this publication.

In order to assess different cabin layout modifications and changes to the passenger behavior throughout the project, it is required to model the passenger movement in more detail than previously.

Especially the location of cabin monuments (e.g., lavatories, galleys, or cabin dividers) and their respective impact as well as other passengers within the cabin are of increased interest. The target is thus to develop a model capable of describing the passenger walking speed depending on the surrounding objects and their proximity. Other passengers walking ahead or already seated and their influence on moving passengers are also taken into consideration. The following Figure 1 depicts the modelling approach for this problem. It is based on the fundamental assumption that every object in proximity of the passenger, its dimensions as well as its location in relation to a walking human affects the resulting walking speed. The challenges of this modelling approach are, as depicted in Figure 1, the determination and modelling of the influence of all objects such as arm rests of seats, overhead bins, the ceiling as well as other passengers already seated or preceding the walking passenger. Specifically, the strength and range of these different influences as well as their interaction with each other has to be determined and properly dimensioned.

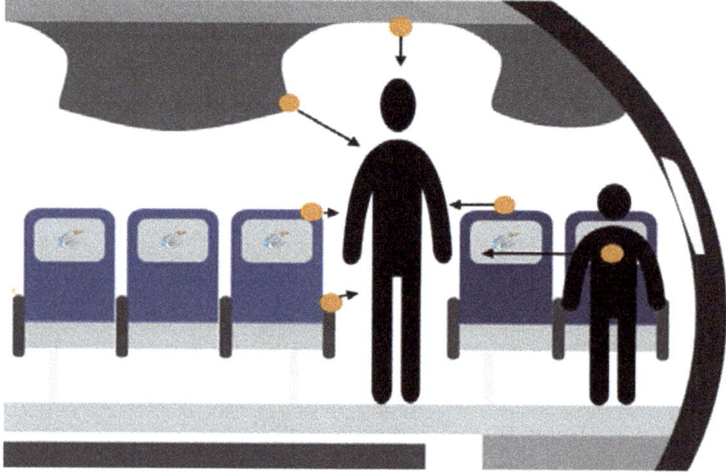

Figure 1. Schematic highlighting the influence of all surrounding objects and passengers on a walking passenger as the basic concept behind the modelling approach.

In recent developments, the global COVID-19 pandemic has also put the spotlight on the aircraft cabin and the health risks its passengers are exposed to. These risks are especially high during the boarding and de-boarding procedures where moving passengers walk within close proximity of each other. This is why airlines have introduced several measures that aim to reduce these risks by requiring masks at all times [6], reducing on-board catering options, and restricting passenger movement during flight [7]. During the boarding specifically, passengers are asked to keep a safe distance while queueing and to adhere to the given boarding sequence (e.g., boarding in zones) [8]. Prior to cabin entry, a mandatory temperature screening or even a COVID-19 test can be required [9]. All these measures can increase the turnaround time of a typical aircraft by up to 15 min [10].

This situation has triggered research interest in different aspects of the boarding process and their implication on the infection risk of passengers. For example, different suggestions regarding the boarding strategy have emerged. These include back-to-front [10] or window-to-aisle boarding [11], a new reverse pyramid boarding [12] as well as restrictions to hand luggage and a safe distance between passengers [13].

The target for the new movement model of PAXelerate is to enable a more precise boarding process simulation, taking into account the aircraft cabin geometry and other passengers in proximity. Furthermore, it aims to provide a better control over the passenger walking behavior itself. These aspects will advance the new movement model compared to the previous implementation and can enable PAXelerate to contribute to the understanding of a future pandemic-proof boarding scenario.

2. Literature Review

Passenger flow simulations are the preferred tool for assessing and optimizing congested and crowded areas with regard to fast movement and safety. The following is an examination and assessment of various simulation frameworks regarding their implementation of passenger walking speed calculation depending on the environment and other people. Depending on the research purpose of the simulations, the concepts differ strongly. This is especially true concerning the modelling of human movement within the environment. Therefore, simulations focusing on the boarding of aircrafts are covered first.

An overview of existing passenger flow simulation frameworks focusing on an aircraft and terminal application is listed in Table 1 below. Aspects of the simulation by Fuchte as well as MASim by Richter found application in the development of PAXelerate.

Table 1. Overview of existing passenger flow simulation frameworks [14].

Name	Author	Year	Purpose	Simulation Type
Cast Cabin	Airport Research Center [15]	-	Aircraft boarding	Agent-based
MASim	Richter [16]	2007	Aircraft boarding	Agent-based
PEDS	Marelli et. Al. [17]	2008	Aircraft boarding	Cellular automaton
-	Schultz [18]	2010	Airport terminal	Discrete event simulation
TOMICS	DLR [19]	2011	Aircraft boarding	Agent-based
-	Fuchte [20]	2014	Aircraft boarding	Agent-based
airExodus	Fire safety engineering group [21]	2015	Aircraft evacuation	Agent-based
PAXelerate	Bauhaus Luftfahrt [14]	2016	Aircraft boarding	Hybrid agent-based

All of these simulation frameworks are either not published or only available for commercial use, thus preventing any examination of their passenger movement model. However, there are other simulation types that include a detailed human movement model such as for simulations of crowded places, e.g., events or public transport infrastructures. Depending on the particular situation, different approaches to simulate humans interacting with the environment find application. Passenger flows at events such as concerts, where many people move towards a common exit, are often modelled following gas kinetic analogies, whereas simulations for public transport facilities use approaches with a focus on the simulation of passengers as individuals [22]. In addition, these simulations vary in their modelling of the environment and the shape of the human body [23]. Finally, video recordings are applied to derive parameters or attributes of human individuals and surroundings from real-life tests [24].

2.1. Categorization

In general, simulation models can be categorized depending on their characteristics. This includes distinguishing between macroscopic and microscopic as well as discrete and continuous models [25]. Rather than modelling the behavior of each individual, macroscopic simulations model the passenger flow as a whole [22]. In contrast, microscopic models calculate no overall states of passenger flows, but rather simulate every passenger as an individual. An overview of the different model types can be seen in the following Table 2.

Table 2. Overview of different passenger simulation types.

	Macroscopic Simulation	Microscopic Simulation		
		Social Force Model	**Cellular Automaton**	**Agent-Based**
Foundation		Individual modelling of passengers and characteristics. Models differ in modelling of interactions between individuals and the environment as well as the level of spatial discretization and modelling depth. [26]		
	Partial differential equations describe variables such as density, speed and flow [26].	Passenger behavior is modelled by motion equations [27]. The individual is acting based on external forces. The walking speed depends on the sum of all external forces [22].	Discretization of space (rectangular or hexagonal cells) and time. Walking behavior is defined by rules and states for each cell [26]. Movement directions are limited. All passengers are modelled the same.	Autonomously acting agents with specific characteristics such as gender or age that interact with one another. Pathfinding and movement are derived by unique characteristics. Agents often have a cost minimization target. [22]
Human Model	Analogies from fluid dynamic particle flows with circular body shape.	Modelled in a continuous space in the shape of flat disks.	Modelled as single or multi cell sized. Shape can be rectangle or hexagon if multi cell shaped. [22]	Passenger models are not bound to specific concepts of spatial discretization or continuous spaces.

2.2. Obstacle Interaction

There are different existing concepts for passenger-obstacle interactions in passenger flow simulations [28]. In social force models, the interaction between passengers and obstacles (e.g., seats, galleys or lavatories) is modelled using forces that act on the passenger. The driving forces for a desired walking speed and the attraction towards a goal are superimposed by braking forces of nearby obstacles.

Regarding cellular automata, a cell is only blocked by one passenger or obstacle at a time. Thus, a passenger cannot make a step on a cell occupied by an obstacle. Passengers can however access a cell next to an obstacle. To prevent this, an influence area around obstacles can be introduced by combining a cellular automaton and a social force model [22]. Here, forces that would act on a passenger are converted into potential fields around obstacles that decline with increased distance. The passenger walking speed and direction can then be adapted according to this potential.

Other models use velocity-based approaches adapted from computer graphics and animations to anticipate collisions between passengers and obstacles by using speed vectors to recognize possible conflicts and adapt the vectors accordingly. Other approaches calculate three-dimensional distances, sometimes only in the field of vision of a passenger, to detect obstacles and avoid collisions. These methods require more computing power than the aforementioned methods [28].

To achieve the most realistic results, the behavior of passengers close to obstacles can also be assessed using real-life experiments. This may include the reaction time from the perception of an obstacle in a certain distance to initiating an avoidance maneuver. To simplify test setups, these assessments could also be conducted with the aid of virtual reality [29].

2.3. Passenger Interaction

Passenger interaction is based on the same concepts as obstacle interaction. However, forces, potentials, or three-dimensional distances are not only calculated for static objects but for moving passengers. This requires a recalculation after each location change as well as in the integration of novel aspects such as overtaking. Additionally, interactions between two passengers are not restricted to collision avoidance. Weidmann [30] discusses the influence of passenger density on walking speed. The relation between walking speed and density can be explained by the two effects of maintaining personal space and speed as function of step length and frequency.

Passengers demand personal space free of other individuals (and objects) for comfort [29]. To keep this personal space free, passengers tend to evenly spread over a given space and prefer to reduce their walking speed rather than compromising their personal space.

The walking speed is a function of individual step length and frequency of steps [30]. With increasing density, the available space for each passenger decreases. The step length and/or frequency will be reduced and walking speed will decrease. Above a certain density, the remaining space for every passenger is just sufficient to stand, but it is no longer possible to walk.

3. PAXelerate

After the introduction of different crowd simulation types and modelling approaches, this chapter introduces the open-source passenger flow simulation PAXelerate [31]. It is based on a 2D agent-based modelling approach with aspects of a cellular automaton (see Section 3.2). It has been developed by Bauhaus Luftfahrt as a boarding assessment tool and provides fast and easy results for the boarding performance of both conventional and novel cabin layouts. The underlying algorithm for the simulation is called the cheapest path A Star and operates in a grid-based cabin representation [14,32].

3.1. General

For the AVACON project, support for the CPACS file format [33] has been integrated into PAXelerate. This enables a fast import and integration of various aircraft concepts into the tool via an interface. CPACS, which can "hold data from a variety of disciplines considered in an aircraft design process" [34] is a common language for aircraft design and can contain information on the shape and structure of the fuselage as well as the size and positioning of monuments and seats within the cabin.

PAXelerate comprises the CAFE cabin designer and simulation module (Figure 2), with the latter one being the focus of this publication. It enables the batch simulation of cabin scenarios generated with Monte Carlo based passenger and boarding properties and also contains movement and pathfinding models. The cabin input required for the simulation is delivered by the cabin configurator module, which allows for an import, modification and export of CPACS files. Additionally, common design rules and CS25 conformability can be assessed.

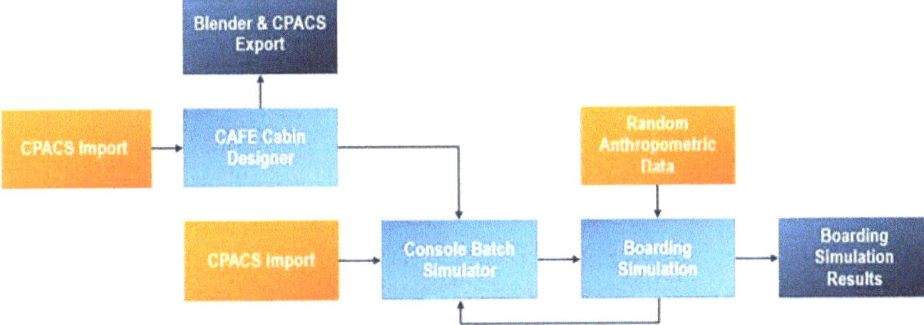

Figure 2. PAXelerate module structure.

3.2. Current Movement Model

The currently implemented movement model of PAXelerate is based on a cellular automaton approach with a grid of nodes, where each node contains various properties. The cabin is discretized in the x-y-plane and the distance between neighboring nodes is currently set to 10 centimeters. Other distances are configurable as well. This delivers a node grid with a 10 by 10 nodes per square meter resolution of the cabin area along its width and length axis.

Each node contains properties defining its location comprising an x and y value as well as various other parameters such as whether the node represents an obstacle or a passenger is currently located on the node. An obstacle represents an object within the cabin that interacts with a passenger during the pathfinding and walking processes. This can be a galley, a lavatory, a seat or any other object located on the cabin floor. Additionally, each node contains a value providing information on the difficulty for a passenger to walk on the node. This is called the potential and makes the PAXelerate movement model a hybrid of a social force model and a cellular automaton. Each object in the cabin is surrounded by this potential with a linear gradient, creating a map. This motivates the passenger to keep a distance to objects, as the difficulty of using these nodes is higher compared to nodes with a larger distance to obstacles. The cheapest path, meaning the path with the lowest total sum of cost potential in all nodes being used for walking, is determined by an A star algorithm operating with the cheapest path condition.

The walking behavior of the passengers itself is modelled in a simplistic way, where the speed is derived by an age-speed-distribution. After a delay with a duration of the theoretical time needed to reach the next node position, the passenger object is moved one node ahead. No external factor has any influence on the walking speed, but the size and amount of luggage a passenger is carrying decreases the walking speed by a predetermined factor. A minimum distance to other passengers in combination with a check if the next scheduled node plus the minimum distance is already occupied delivers the decision to stop walking and wait for the preceding passenger to clear the path. Lastly, an acceleration and deceleration procedure consisting of a basic linear equation influences the walking speed when approaching the minimum distance mentioned above.

Summing up, the movement model in PAXelerate consists of an agent-based approach for the modelling of passengers and their characteristics, while the movement itself is based on a cellular automaton approach with a grid of nodes and discretized steps within the simulation bounds. The obstacle avoidance aspects are derived from a social force model, creating a hybrid model, as explained in Section 2.2.

4. Novel Movement Modelling Approach

For the AVACON project, a new approach advances the movement model previously implemented in PAXelerate. The new design aims to calculate the walking speed of a passenger within the cabin as a function of its surroundings and other passengers, as seen in Figure 1. Two challenges have been identified in this regard, with the first being the modelling of the impact of three-dimensional objects on the walking speed of passengers at every moment and every position within the cabin. The second challenge is the modelling of the influence of other, already seated and walking, passengers within the cabin environment on the walking speed. The following sections will give an overview of both aspects of the novel model.

4.1. Methodology

Ideally, the distance between each geometry surface of every element within the cabin and the passenger itself has to be calculated at every time step of the simulation in order to get an exact measurement of the surroundings of a passenger. The problem therein is the immense computational effort required to calculate 3D distances of every object surface to every passenger as well as all passengers amongst each other. As the procedure has to be repeated for every time step, there is a need for a more simplified approach, which can reduce the calculation effort by increased model abstraction, while yielding sufficient information for a reasonably accurate simulation. The solution to this problem lies in the structure of the PAXelerate boarding simulation, which consists of a discretized cabin area built out of a grid of nodes. As these nodes represent a specific location, each node can also contain the distance information to the closest cabin monument. It is thus possible to pre-calculate the distance of every node within the cabin to the closest monument in advance of a batch simulation, thereby reducing the required number of calculations to just one. This can be done given that no cabin monument changes position throughout the simulation loops.

Concerning the model abstraction, an additional measure has been taken. Looking at the human body, four distinctive elements of the human body were defined for modeling the influence on a human's walking behavior in a constricted space. Those body parts are the knee, hip, shoulder and head, each positioned at a different height and each exposed to a different layout of the cabin at the respective height. As listed in Table 3 below, the cabin model is thus in a first step divided into four layers, with each layer comprising of a different grid of nodes and each representing the cabin and the containing objects at a specific height above the cabin floor. The lower three layers contain information on the position and size of cabin monuments at the specific height, whereas the upper ceiling layer contains information about the cabin ceiling height as well as the size and position of the overhead bins.

Table 3. Division of the cabin into multiple layers.

Layer Name	Start Height [m]	End Height [m]	Purpose
Ground layer	0.0	0.65	Passenger and object modelling
Mid layer	0.65	1.10	
Top layer	1.10	Cabin height	
Ceiling layer	-	-	Ceiling contour

The selection of the different heights is based on four neuralgic body parts of the human body during walking. This includes the knee in the ground layer, the waist in the mid layer as well as the shoulders and head in the top layer. The height of the layers is designed to surround the specific neuralgic point, based on average human body dimensions as seen in Table 4 and in relation with the given cabin elements such as seat objects.

Table 4. Neuralgic human body points [35].

Body Point	Mean Height Female [m]	Mean Height Male [m]	Layer
Knees	0.46	0.50	Ground
Waist	0.84	0.89	Mid
Shoulders	1.33	1.44	Top

4.2. Modelling of Passengers

Inside the grid-based cabin layout of the PAXelerate boarding simulation, passengers are modelled as discretized rectangular shapes. At each of the different heights of the three modelling layers, the passenger is represented with a different size within the grid. As can be seen in Figure 3 below, the top layer body shape consists of the shoulder dimensions, the mid layer consists of the waist width and depth and the ground layer contains the knee width and depth. All values are based on the data seen in Table 5 below.

Table 5. Neuralgic human body points [35] (* estimated).

Body Layer		Width [m]		Depth [m]	
		Female	Male	Female	Male
Knees *	Mean	0.34	0.34	0.22	0.24
	deviation	0.02	0.02	0.02	0.02
Waist	Mean	0.3427	0.3418	0.2271	0.2486
	deviation	0.0224	0.0203	0.021	0.0207
Shoulders	Mean	0.4301	0.4912	0.2394	0.2432
	deviation	0.023	0.026	0.0211	0.0215

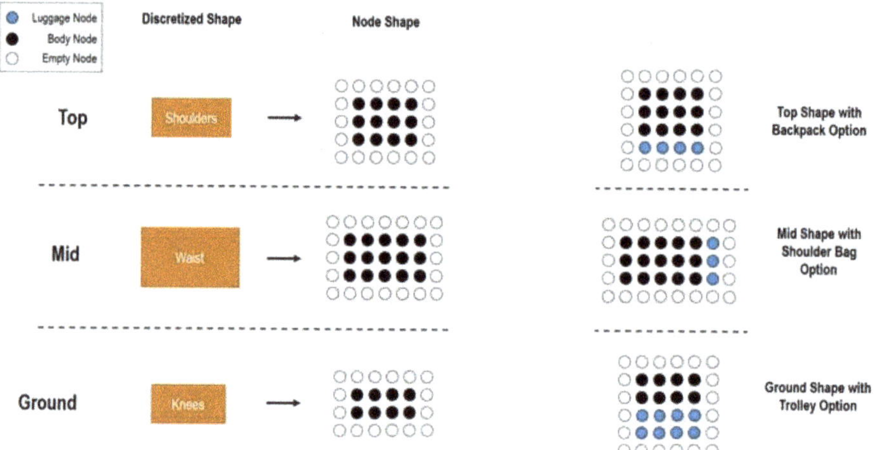

Figure 3. Representation of the human body and luggage within the simulation grid.

The values used in the simulation are, as PAXelerate uses a Monte Carlo approach for all anthropometric data, generated individually for each passenger from the mean and deviation values of a normal distribution.

A potential carry-on luggage is amended to the human body at respective height layer (see luggage nodes in Figure 3 above). The luggage options currently implemented contain a backpack extending the shoulder and waist depth and a shoulder bag extending waist width. Additionally, the options contain a cabin trolley extending the knee depth backwards if pulled or in front if pushed. Lastly, a combination of multiple carry-on luggage items is also taken into account.

During the course of the boarding simulation, the carry-on luggage is considered an integral part of the human body up to the point of luggage stowing. Additionally, every luggage carried by the passenger decreases the walking speed by a certain factor.

4.3. Modelling of Cabin Monuments

The following Figure 4 highlights the seat dimensions used for the discretized shape of cabin seats at the different height layers. As the width of a seat is almost identical at the armrest and floor level, the same dimension is used for the lower two layers. The top width is smaller due to the slimmer backrest dimensions. Considering the depth of the seats, each layer has a different dimension, beginning at the floor level with the base footprint size, continuing to the middle layer, where the armrest dimension is critical and ending at the top width of the seat, consisting of the backrest depth.

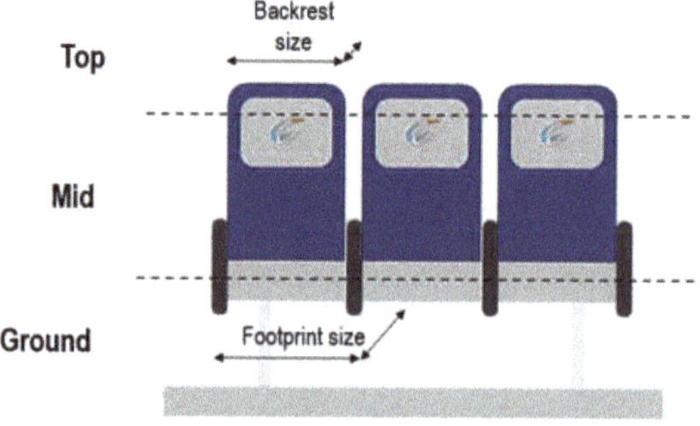

Figure 4. Dimensions of a typical aircraft seat used for the different layers.

Figure 5 depicts the transition of the cabin seat into a discretized rectangle inside the PAXelerate node grid for each of the different layers. All other cabin monuments such as lavatories or galleys are modelled in the same way. However, monuments with a vertical wall are modelled with the same dimensions in each of the layers. The overhead bins are not considered cabin monuments and are thus not included in this modelling approach but rather dealt with in a separate fashion as described in a later section.

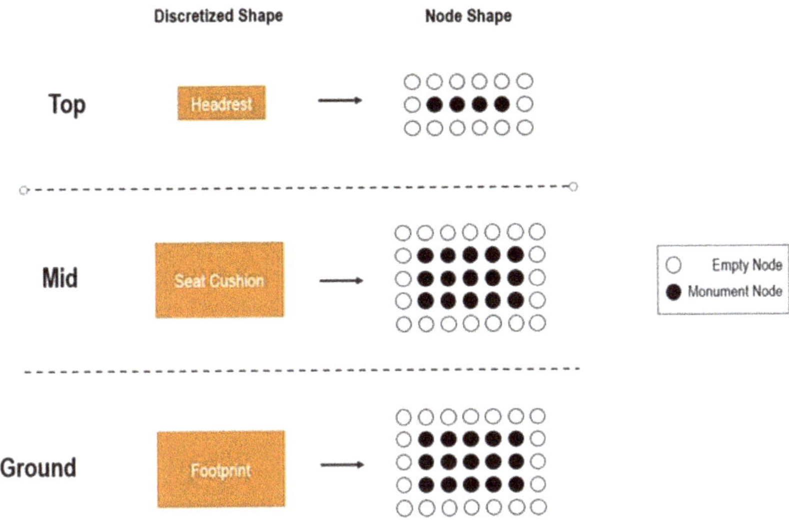

Figure 5. Transition to discretized seat dimensions for each layer.

4.3.1. Influence Area of Cabin Monuments

In order to estimate the effect of the distance of objects to a passenger, each cabin monument inside the simulation grid is surrounded by a potential φ_{node}, creating a so-called influence area (see Equation (1)). This is the core method of reducing the computation effort by pre-calculating the linear distance between objects for each simulation step. This is feasible, as each node knows its own position to the closest object in advance. A passenger walking through the nodes within this influence area will decrease their walking speed depending on the node's properties and the pre-calculated distance value $d_{closest}$ stored within the node. The potential correlates to the linear interpolation between φ_{min} and φ_{max} in dependence of the inverse distance to the object.

$$\varphi_{node}(d) = f(d_{closest}, \varphi_{min}, \varphi_{max}) \quad (1)$$

The value of φ_{min} and φ_{max} were chosen in order to comply with boarding times published by Airbus [36] or parameters extracted form video sequences by Steiner and Philipp [37]. In fact, it requires real-life experiments to determine the detailed impact of different obstacle types on passenger walking speed.

As can be seen in Figure 6 below, the potential is cut off at a specific distance $d_{threshold}$ of the object. This is based on the assumption that after surpassing a given distance, no effect on the walking speed can be detected and thus a computation and storing of these values is obsolete and would reduce the computation performance.

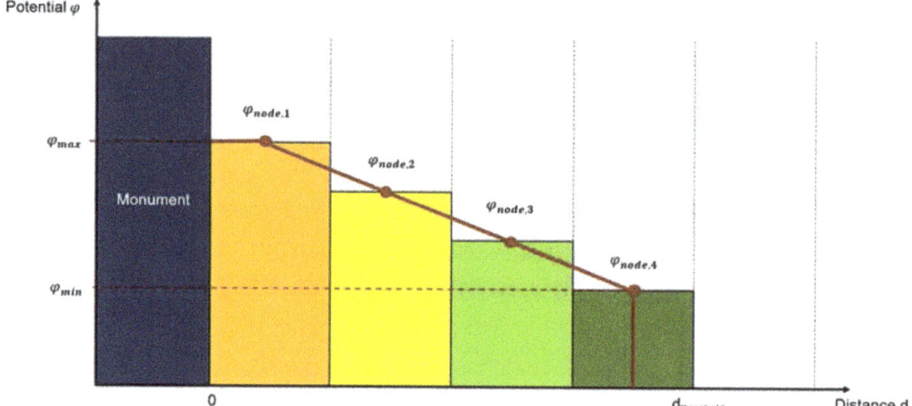

Figure 6. View of the potential, representing the impact on passing objects.

Figure 7 shows the potentials surrounding a group of seats at two different heights, one representing the lower parts of the seats and one representing only the backrests and their surrounding potential area.

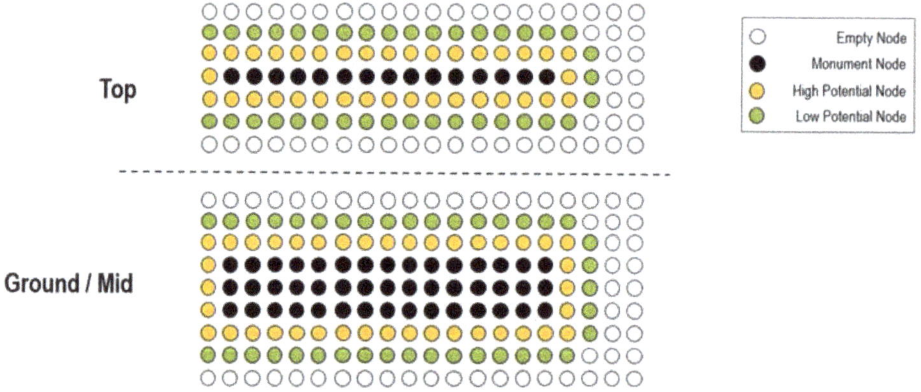

Figure 7. Exemplary potential around a group of seats.

The only time this node grid changes throughout a simulation is when a passenger is seated onto a seat. This increases the potential surrounding the respective seat by raising the φ_{max} value, thus integrating effects on the walking behavior of passengers by others already seated.

4.3.2. Modelling of the Overhead Bins and Ceiling

The ceiling layer is similar to other layers but further includes the ceiling height at a certain location as highlighted in Figure 8. The head and shoulder clearances are thus additionally stored at each node covered by the ceiling-layer.

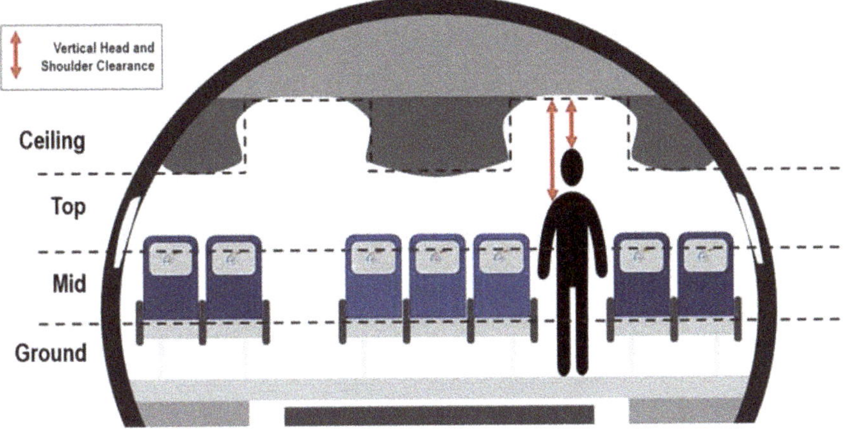

Figure 8. Depiction of the ceiling and overhead bin layer. Shown here is the cross-section of the AVACON research baseline (ARB) [5].

4.3.3. Calculating the Effect of Close Objects on the Walking Speed

At first, the sum of all potentials that act on a passenger while walking over a node is calculated for each step. All nodes occupied by the passenger, as seen in Figure 3, represent the current position of the passenger. Each node can have a certain potential φ_{node} in each layer from nearby objects (see Figure 6). These potentials are first summed up over all occupied nodes of one layer and then over all layers from $i = 0 \ldots n$, where n is the amount of layers in the model. Finally, the mean potential $\varphi_{Obstacle}$ is calculated by dividing this sum by the total number of nodes $N_{obstacle}$ occupied in all layers (see Equation (2)). This results in a potential $\varphi_{Obstacle}$ averaged over all nodes in all layers occupied by a passenger.

$$\varphi_{Obstacle} = \frac{\sum_{i=0}^{n} \left(\sum \varphi_{node,i}\right)}{\sum_{i=0}^{n} N_{obstacle.i}}. \tag{2}$$

The same equation covers the low ceiling and overhead bin effects, but additionally the vertical distance between head, shoulders, and ceiling is considered within the function for $\varphi_{Ceiling}$ as well. The sum of $\varphi_{Obstacle}$ from Equation (2) and the potential of the ceiling $\varphi_{Ceiling}$ is divided by the maximum potential φ_{max} defined in Section 4.3.1 (see Equation (3)). This value is then subtracted from 1 and multiplied with the free walking speed v_o (see Section 4.4.1). This results in the walking speed with obstacle interference $v_{obstacle}$.

$$v_{obstacle} = v_o * \left(1 - \frac{\varphi_{Obstacle} + \varphi_{Ceiling}}{\varphi_{max}}\right). \tag{3}$$

4.4. *Impact of Passengers*

4.4.1. Fundamental Relations

The modelling of the interaction between passengers and their impact on the walking speed is derived from the social force and cellular automaton hybrid approach explained in Section 3. In general, the walking speed of a human in a crowded area—where free, unrestricted walking is not possible—depends on the crowd density. This approach is based on the so-called fundamental diagram by Weidmann [30] as described in Section 2.3. The diagram is based on the following Equation (4), where v_o = 1.34 m/s is the free walking speed and $\rho_{human,\ max}$ = 5.4 m^{-2} is the density of humans at which a standstill occurs.

$$v\left(\rho_{human}\right) = v_0 * \left\{1 - exp\left[-1.913 * \left(\frac{1}{\rho_{human}} - \frac{1}{\rho_{human,\ max}}\right)\right]\right\}. \tag{4}$$

The resulting graph, depicting the walking speed in dependence of the density of people per square meter, can be seen in Figure 9. As depicted, the walking speed reaches zero at a density of 5.4 people per square meter, resulting in a standstill. Free walking speed is maintained up to a density of roughly 0.4 people per square meter, after which a steeper decrease in walking speed can be observed.

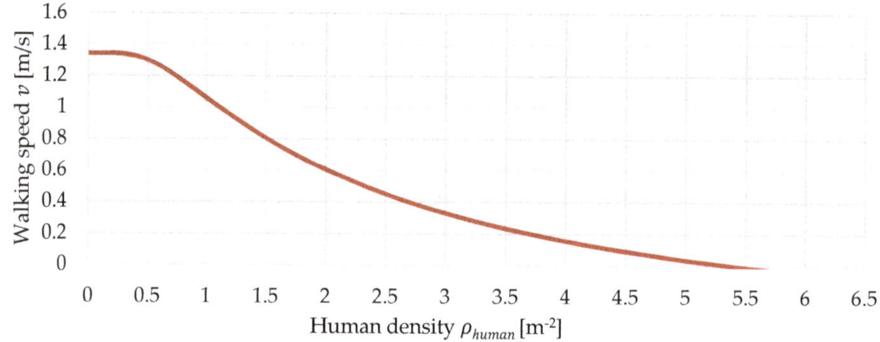

Figure 9. Walking speed in dependence of crowd density according to [30].

Generally, there are other implementations of a walking speed and human density dependency such as for example the one used by Richter [16] and seen in the following Equation (5). The model introduced in this paper however is based on the concepts introduced by Weidmann as his model has no case differentiation and consists of a single continuous equation.

$$v\left(\rho_{human}\right) = v_0 * \begin{cases} 1, & \rho_{human} < 0.8\, m^{-2} \\ \left(0.032 * \rho_{human}^2 - 0.37 * \rho_{human} + 1.27\right), & \rho_{human} \geq 0.8\, m^{-2} \end{cases} \tag{5}$$

4.4.2. Transition to Model for PAXelerate

As shown in Equation (4), the passenger density ρ_{human} inside the cabin is the only unknown variable that has to be determined. A widespread approach is to divide the simulated area into smaller parts and calculate the passenger density by dividing that area by the number of passengers inside [16]. However, this force-based approach of applying a walking speed depending on an external variable contradicts the principle of independence of each passenger in PAXelerate's agent-based approach.

Furthermore, today's aircraft cabins usually offer narrow corridors for walking, which are often too tight to allow overtaking or counterflow. Future cabin layouts are, excluding modifiable seat designs which can temporarily alter the aisle width [38], unlikely to increase aisle widths due to space and cost considerations. Additionally, passengers always prefer lane formation when walking in the same direction such as in aircraft cabins [27].

Due to this lane formation, a passenger is mainly influenced by the others walking directly ahead. Other passengers any further ahead or behind have no direct influence on the walking speed, as they are not directly perceived. Looking at a passenger individually, the critical area concerning crowd density effects and thus the area in which a passenger perceives others is solely the area directly in front. Consequently, the distance to the passenger in front determines the density experienced when walking in lane formation. The density can ultimately be seen as the free space in front of a passenger.

If this space decreases, a passenger has to adapt the step size and slows down to avoid approaching the passenger ahead. The area of such as personal space is equal to the reciprocal of the density as seen in Equation (6).

$$A_{personal} = \frac{1}{\rho_{human}}. \quad (6)$$

Using the model of Weidmann and his assumptions regarding the needed lateral space while walking $w_{lateral}$ = 0.71 m, a rectangular personal area can be defined together with the distance in walking direction d according to the following Equation (7).

$$A_{personal}\,(d) = w_{lateral} * d. \quad (7)$$

Using Equations (6) and (7), the following Equation (8) can be derived. This equation delivers a dependency between the human density according to Weidmann and the personal space of a passenger.

$$\rho\,(d) = \frac{1}{A_{personal}(d)} = \frac{1}{w_{lateral} * d}. \quad (8)$$

When integrating Equation (8) into Equation (4), the following Equation (9) results. The corresponding graph can be seen in Figure 10. It depicts the dependency between the distance to a preceding passenger and the resulting walking speed.

$$v_{passenger}\,(d) = v_0 * \left\{1 - exp\left[-1.913 * \left(d * w_{lateral} - \frac{1}{\rho_{human,\ max}}\right)\right]\right\}. \quad (9)$$

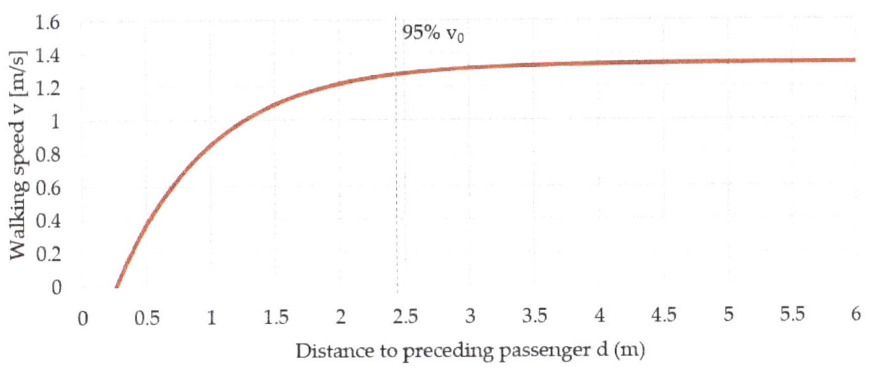

Figure 10. Walking speed in dependence of the distance to the preceding passenger.

The equation above also delivers the distance of d_{min} = 0.26 m, corresponding to a density ρ_{human} = 5.4 m^{-2} and representing the minimum distance at which a standstill occurs. The curve is of asymptotic nature for v_0 = 1.34 m/s, where 95% of v_0 are reached at d_{max} = 2.47 m and 99% of v_0 are reached at d = 3.65 m. For performance and simplification reasons, the maximum considered distance for a potential influence of preceding passengers is thus selected to be at the 95% v_0 line. The two values of maximum and minimum influencing distance result in the rectangular of Figure 11 below.

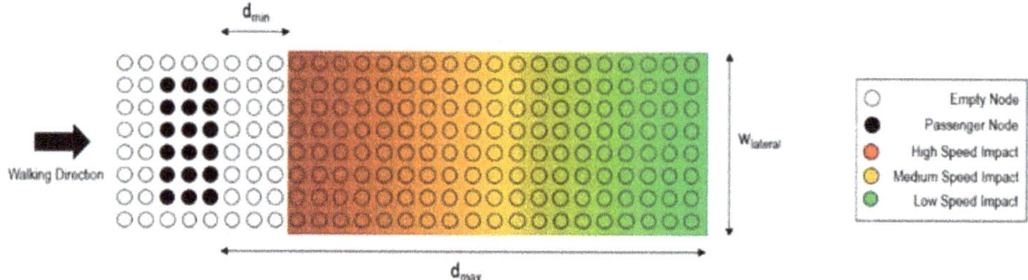

Figure 11. Schematic of the influence area of a walking passenger. The walking speed depends on the distance d to a preceding passenger and if it is within the rectangle defined as the influence area.

Combining the information from Equation (9) as well as Figures 10 and 11, a passenger is in the free walking mode and is not influenced by any preceding passengers on the same path up to a distance of 2.47 m. This is explained by the fact that up to a distance d of greater than 2.47 m, the resulting $v_{passenger}$ of Equation (9) is nearly constant. From 2.47 m down to a distance of 0.26 m, the walking speed will gradually slow down according to Equation (9). This can also be seen in the transition from a low to a high impact on the walking speed as seen in Figure 11. At a distance of 0.26 m, $v_{passenger}$ of Equation (9) equals 0 and the passenger comes to a standstill until the passenger in front increases the distance beyond the threshold.

4.4.3. Implementation of Model into PAXelerate

The shape, position, and personal influencing area of a passenger is saved in a discretized way, as seen in Figure 11. The influence area is saved by applying a potential in the same way as for obstacles to each node within the rectangle. A node can save the potential of more than one passenger. If a passenger steps on a node that is part of such an influence area of other passengers, the speed factor is calculated according to Equation (9) if both passengers are walking. If the preceding passenger is already seated, the potential is rather used for the calculation of the factor and the other passenger is regarded as an obstacle. If different passengers have an influence on the current position, the most influencing passenger is determined by algorithms searching for the highest potential and considering the current orientation of the involved passengers. The speed factor is calculated only once by calculating an average footprint shape and not for each layer separately.

4.5. Model Summary

Information about nearby obstacles and influencing passengers are saved as a property for each node. At each step, the effects of nearby objects and other passengers on the walking speed are calculated separately. Finally, as seen in Equation (10), the lower speed, hence the higher and more influential speed factor is chosen to calculate the walking speed for the next step.

$$v_{current} = min\left(v_{obstacle} \,\middle|\, v_{passenger}\right). \tag{10}$$

Generally, the approach chosen for the movement model was selected because it builds on the foundation of PAXelerate as a hybrid cellular automaton and agent-based simulation. The obstacle interaction is based on a social force approach and the interaction between passengers in the constricted aisle is based on an adapted walking speed and human density relation. Finally, the following Table 6 provides an overview of changes to PAXelerate regarding the new movement model implementation.

Table 6. Comparison of old and new passenger model in PAXelerate.

Aspect	Old Model	New Model	Section
Model dimensions	2D cabin floor	3D cabin using multiple stacked 2D layers	Section 4.1
Passenger shape	2D rectangle	Distinctive rectangle shape in each stacked cabin layer at neuralgic body points	Section 4.1
Movement model	"Stop and go" using free walking speed when path is unobstructed and stopping when preceding passengers reach a threshold distance	Adapted speed and crowd density relation by Weidmann using the personal space around individuals	Section 3.2, Section 4.3
Influence of other passengers	Collision avoidance only	Walking speed is decreased by all passengers in proximity, depending on their state (seated, standing, or walking)	Section 4.3
Obstacle influence	Considered only for collision avoidance during initial path finding process	Obstacles decrease walking speed of passengers in their proximity depending on obstacle shape in each model layer	Section 4.2

5. Validation

This chapter presents the validation of the new movement model implemented for the AVACON project and is introduced in this paper. It comprises of a default boarding simulation loop with default passenger properties. As cabin layouts, the AVACON research baseline as well as an Airbus A320 type aircraft are selected.

5.1. AVACON Research Baseline

The AVACON research baseline (ARB) is a mid-range aircraft and the foundation of the AVACON project and is based on a Boeing 767 aircraft [5]. The corresponding cabin layout can be seen in Figure 12 below. In a related paper regarding PAXelerate and the AVACON project, the PAXelerate boarding simulation has been applied to this reference aircraft with a variety of different boarding doors and strategies [39].

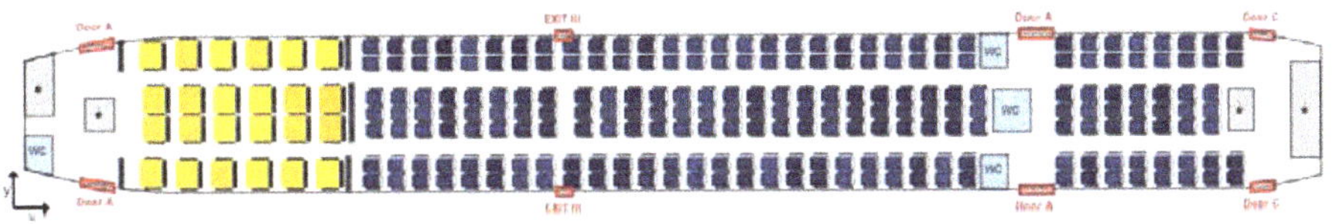

Figure 12. Cabin layout depiction of the AVACON research baseline.

Due to the Monte Carlo approach used for all passenger properties within PAXelerate (see Section 4.2), the simulation is non-deterministic and rather delivers a normal distribution of boarding time results. The following input parameters in Table 7 are used for the assessment of both the old and new movement models using the ARB aircraft.

Table 7. Overview of the input parameters used in the boarding simulations.

Parameter	Value	Unit
Passengers	255	-
Discretized cabin grid resolution	10×10	Nodes per m^2
Boarding doors	Front left	-
Number of simulation loops	100	-
Load factor	100	%
Boarding sequence	Random	-
Boarding rate	18	Passengers per minute

The following Figure 13 shows the boarding time distribution of the two models. Each dot represents a boarding time result for the respective model, whereas the bounds of the colored boxes represent the 25th and the 75th percentile, respectively. A horizontal line within the boxes represents the mean boarding time. For the old model, the execution of 100 simulation loops resulted in an average boarding time of 16:27 (mm:ss) for this twin aisle aircraft with 252 seats and the standard deviation accounted to 34 s. In comparison to this, the new version of PAXelerate incorporating the novel movement model and keeping all other settings resulted in a boarding time of 20:56 with a standard deviation of 54 s. This represents both longer boarding times compared to the previous implementation as well as an increased simulation uncertainty as represented by an increased box height in Figure 13 and an increased standard deviation. For reference, official boarding process times for an Airbus A330 (which has a comparable cabin to the ARB) suggest ~25% higher values compared to the old PAXelerate simulation [1]. The new results thus still lie within in an acceptable range. A further investigation into the cause of this increase will be performed in the future, but the fine-tuning of the model parameters for a single-aisle A320 style aircraft and the overall slowing of passenger movement are expected to have a strong influence.

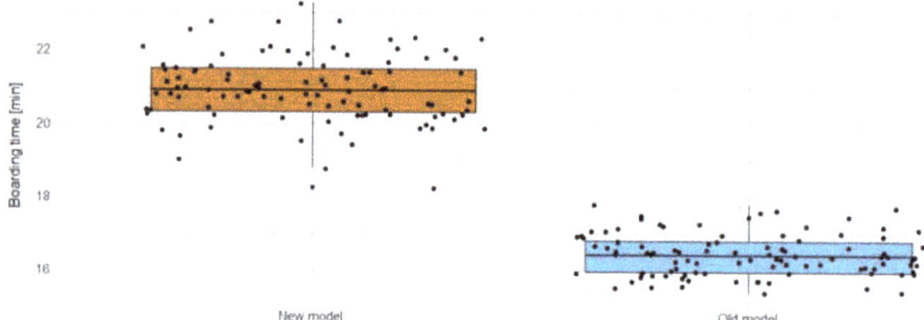

Figure 13. Boarding times of the ARB when using the different movement model implementations.

As a statistical test for the results of the two boarding simulation models, the Welch two-sample t-test has been chosen as both result samples are normally distributed (Shapiro-Wilk normality test) but differ in their variance (f-test). The resulting *p*-value of $<2.2 \times 10^{-16}$ proves that the two models and their respective boarding time results are significantly different.

5.2. Airbus A320

As mentioned, the model parameters were first calibrated and fine-tuned for an A320 style aircraft, thus potentially explaining the deviations in the AVACON research baseline results. The parameters and settings for this validation remain unchanged compared to Table 7, but the total number of passengers in the cabin layout depicted in Figure 14 accounts to 180.

Figure 14. Cabin layout depiction of an Airbus A320-200.

As highlighted by the boxes in Figure 15 below, the simulation of the previous model results in an average boarding time of 15:36 with a standard deviation of 43 s using the old model. The new model results in a boarding time of 16:26 with a standard deviation of 47 s. Using the Welch two-sample t-test for a statistical test, the *p*-value of 7.874×10^{-9} proves that the two models and their boarding time results are significantly different. The results show that the new model increases the boarding times slightly. This behavior is again explainable by the overall reduction in passenger movement speed due to the new movement model.

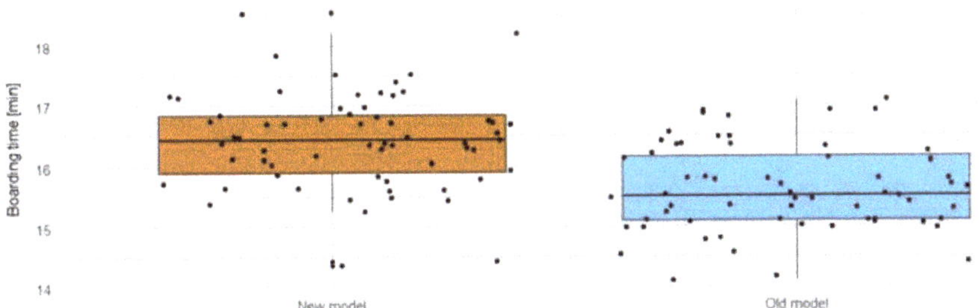

Figure 15. Boarding time distribution of an Airbus A320 with the new movement model.

5.3. Summary

The following Table 8 highlights the results of the validation for both the AVACON research baseline aircraft and the Airbus A320 type aircraft.

Table 8. Overview of the validation results.

Validation Case	Source	Mean Boarding Time (min)	Standard Deviation [min]
AVACON ARB	Old model	16:27	00:34
	New model	20:56	00:54
Airbus A330	Manufacturer data [1]	~25% higher than old model	-
Airbus A320	Old model	15:36	00:43
	New model	16:26	00:47
	Manufacturer data [36]	~19 min	-

The validation performed for this paper is able to show the applicability of the model to boarding simulations and the ability to generate better results compared to the previous implementation as the delta to official references decreases. In order to assess unconventional aircraft cabin concepts where no existing data is available, real world experiments and measurements need be performed for the determination of influence factors for the various cabin monuments. This can for example be done by recording a boarding procedure at an airport [37] or even using virtual reality to get a more detailed insight into humans moving within a virtual 3D cabin space [29].

6. Application and Implications

After the completed validation of the model for existing scenarios, PAXelerate can now be used to look at a customization of the cabin layout or passenger behavior changes. The actual scenarios assessed below include a modification of the aisle width and of the overhead bin location for an Airbus A320. Concerning the passenger behavior, a currently relevant topic is the impact of COVID-19. As a safe distance of 1.5 to 2.0 m to other people should be maintained at all times, this suggestion can also be applied to the boarding process [40]. As the new model is capable of controlling this minimum distance to preceding passengers, an initial assessment of the impact of such a regulation will be performed below using an Airbus A320.

6.1. Changes to Cabin Geometry

The following assessment is based on a default Airbus A320 configuration. The modification to the cabin consists of an increase in aisle width of 25% or 50%, respectively. This results in a corresponding shift of the overhead bin location as well as an increase in fuselage width. All model parameters except the number of passengers remain the same compared to Table 7. The resulting average boarding time of 50 iterations is 16:08 min with a standard deviation of 00:54 min for the 25% increase scenario and 15:55 min with a deviation of 00:55 for the 50% increase scenario. While the differences in the mean boarding times for the 50% wider aisle are significant (Welch two sample t-test p-value = 0.002068), the deviation in boarding times for the 25% wider aisle scenario is insignificant (p-value = 0.08598).

The slight decrease in average boarding times for the 50% change highlights the potential for an increased aisle width, as passengers might feel less constricted in a wider aisle, thereby reducing the mean boarding times by up to 3%. The effect of overtaking possibilities in combination with a wider aisle has not been considered. The true impact of this aisle widening depends on the parameters used in the potential of every node within the model. A deviation to the simulation result may thus occur after real life experiments have been executed and tweaked parameter values have been integrated. Nevertheless, the results indicate an effect of the altered cabin geometry on boarding times and lead the way for the model to evolve and improve with future versions of PAXelerate.

6.2. Changes to Passenger Behavior

In order to assess the passenger movements due to the COVID-19 restrictions, the model is adapted to include a minimum distance between passengers of d_{min} = 1.75 m, shifting the graph of Equation (9) along the positive x-axis. All other properties remain unchanged compared to the validation scenario of the A320. The decision for an Airbus A320 and against the ARB is because short distance travel seems to be the likelier option for a reemerging commercial flight schedule in the short term as intercontinental travel will be restricted for a longer period of time [41,42].

The average boarding time for 50 iterations is 27:30 min with a standard deviation of 1:19 min. The significance of this deviation is proven by a Welch two sample t-test with a p-value of $2.2 * 10^{-16}$. This represents an increase in boarding times of 67.4% compared to the reference case presented in the validation chapter. The reason for this increase is partly explainable by the reduction of simultaneous actions being performed by passengers. Fewer passengers can stow luggage or enter the aircraft per time step, as the distance between passengers is 6.5 times higher, resulting in increasing overall boarding times. As the boarding process lies on the critical path of the turnaround process for most scenarios [1], an increase in boarding times by this amount will lead to an extended airport turnaround duration and a negative economic impact. This result highlights the risks and effects of the application of this minimum distance regulation.

As the minimum distance requirement might also be enforced during flight, a suggested vacant middle seat in every seat group is another option to be considered [40]. The following assessment thus combines the boarding process with a safe distance and an empty middle seat in each seat group, limiting the time two passengers come in very close contact to each other. For a study with 50 iterations, the average boarding time results in 16:23 min with a standard deviation of 37 s. The deviation of this result compared to the reference is insignificant (Welch two sample t-test p-value = 0.8092). This shows that, with the reduced amount of passengers, the boarding sequence has on average the same duration as the reference scenario. However, as only 120 passengers can board the airplane, thus reducing the load factor to 66%, the economics of such a scenario render the application in a real world scenario questionable.

As mentioned in Section 1, there are different other suggestions for boarding procedures with reduced interaction between passengers such as a window to aisle boarding, rear to front boarding [40] and even more unconventional concepts [10–12]. These scenarios distribute the passengers throughout a boarding process in a way that reduces the total amount of contact with already seated passengers. However, they can cause increased boarding times or aisle areas with high interference between passengers, as shown in a recent publication for the AVACON research baseline [39].

6.3. Overview

The following Table 9 highlights the results of the application for both the cabin modifications and the passenger behavior adaptions performed for this publication. All assessments are conducted using an Airbus A320 cabin.

Table 9. Overview of the assessment results.

Airbus A320 Scenario	Average Boarding Time (min)	Standard Deviation (min)	Delta (%)
Reference	16:26	00:48	-
Aisle width + 25%	16:08	00:54	(−1.8)
Aisle width + 50%	15:55	00:55	−3.1
Safe distance	27:30	01:19	+67.4
Safe distance + vacant middle seats	16:23	00:37	(−0.2)

7. Summary

The integration of the newly developed model empowers PAXelerate to simulate a more detailed boarding process and enables a better understanding of the influence of cabin layout changes to the aircraft's boarding performance. This allows for the identification of critical aspects of a cabin layout such as a local narrowing of the aisle and therefore gives cabin designers the opportunity to tackle potential issues in an early design phase.

In general, the potential for a swifter boarding process by changing aspects of the cabin layout, such as altering the aisle width or relocating cabin monuments, may enable airlines to quicken the turnaround process without high effort. Passengers on the other hand may feel less crowded during the boarding procedure as potential bottlenecks can be removed in advance by rearranging monuments. Lastly, the model introduced in this paper also enables insight into potential paths for future improvements of the design of the AVACON project's next aircraft iterations and their respective cabin layouts.

Concerning the COVID-19 assessment, it can be said that the suggested measures may effectively reduce close contact to other passengers. However, considering the vast economic impact of these measures by increasing boarding times or reducing the aircraft's capacity, other mitigation methods such as an increased hygiene level or facial masks may be economically a less harmful measure to enable a safe and economic future travel scenario.

8. Future Work

The initial implementation of the new passenger movement model has proven to be a valuable addition to the PAXelerate boarding simulation. Specific aspects of the boarding simulation and the movement model in particular provide the potential for future model improvements. As mentioned previously, this includes the adjustment of the factors and parameters used in the novel movement model based on real world measurements and experiments during a boarding sequence, monitoring the impact of cabin monuments and their placement on the walking speed of real humans. This task will be dealt with during the next steps of the simulation framework enhancement.

Regarding the PAXelerate boarding simulation as a whole, the implementation of an advanced luggage handling and overhead bin storage model is of high priority. This includes a more realistic luggage model, incorporating the sizes and weights of luggage and their respective positioning inside the luggage bins by passengers. The size and position of the luggage bins play a crucial role, as there might not always be an empty luggage bin close to the passenger's seat, thus delaying or complicating the luggage stowing procedure and the flow of passengers.

Lastly, a more detailed analysis of COVID-19 related parameters such as exposure time and distance between passengers will be introduced in the future, assessing potential paths for an infection risk reduction throughout the boarding process. This may also benefit novel methods for the luggage stowing procedure as well as the de-boarding of the aircraft.

Author Contributions: Conceptualization, methodology, validation, writing and visualization, M.E.; methodology, validation and visualization, T.K.; supervision, M.H. All authors have read and agreed to the published version of the manuscript.

Funding: This research received funding as part of AVACON, a research project supported by the German Federal Ministry for Economic Affairs and Energy in the national LuFo V program. Any opinions, findings, and conclusions expressed in this document are those of the authors and do not necessarily reflect the views of the other project partners.

Conflicts of Interest: The authors declare no conflict of interest. The funders had no role in the design of the study; in the collection, analyses, or interpretation of data; in the writing of the manuscript, or in the decision to publish the results.

Abbreviations

AVACON	AdVanced Aircraft CONcepts—LuFo research project
PAXelerate	Boarding simulation developed by Bauhaus Luftfahrt
CPACS	Common Parametric Aircraft Configuration Schema
CAFE	Cabin And Fuselage Environment, module of PAXelerate
Symbols	
φ	Potential of a node
ρ	Density of humans
v	Walking speed of a passenger
d	Distance to preceding passengers on the aisle

References

1. Airbus, S.A.S. Airbus A330—Aircraft Characteristics—Airport and Maintenance Planning. Available online: https://www.airbus.com/content/dam/corporate-topics/publications/backgrounders/techdata/aircraft_characteristics/Airbus-Commercial-Aircraft-AC-A330.pdf (accessed on 13 September 2019).
2. Schultz, M.; Kunze, T.; Fricke, H. Boarding on the critical path of the turnaround. Proceedings of the Tenth USA/Europe Air Traffic Management Research and Development Seminar (ATM2013), Chicago. 2013. Available online: http://www.atmseminar.org/seminarContent/seminar10/papers/316-Schultz_0127130604-Final-Paper-4-15-13.pdf (accessed on 19 December 2020).
3. Yildiz, B.; Förster, P.; Feuerle, T.; Hecker, P.; Bugow, S.; Helber, S. A generic approach to analyze the impact of a future aircraft design on the boarding process. *Energies* **2018**, *11*, 303. [CrossRef]
4. Milne, R.J.; Kelly, A.R. A new method for boarding passengers onto an airplane. *J. Air Transp. Manag.* **2014**, *34*, 93–100. [CrossRef]
5. Wöhler, S.; Hartmann, J.; Prenzel, E.; Kwik, H. Preliminary aircraft design for a midrange reference aircraft taking advanced technologies into account as part of the AVACON Project for an entry into service in 2028. *Dtsch. Luft Raumfahrtkongress* **2018**. [CrossRef]
6. IATA. IATA Calls for Passenger Face Covering and Crew Masks. Available online: https://www.iata.org/en/pressroom/pr/2020-05-05-01/ (accessed on 2 December 2020).
7. IATA. Precautions to Take When Flying. Available online: https://www.iata.org/en/youandiata/travelers/health/precautions-to-take-flying-by-air-in-covid-times/ (accessed on 2 December 2020).
8. Flughafen München GmbH. Travelling in Times of the Coronavirus Pandemic. Available online: https://www.munich-airport.com/travelling-in-times-of-the-coronavirus-pandemic-8395611 (accessed on 2 December 2020).
9. IATA. IATA Calls for Systematic COVID-19 Testing Before Departure. Available online: https://www.iata.org/en/pressroom/pr/2020-09-22-01/ (accessed on 2 December 2020).
10. EUROCONTROL and Airport Research Center GmbH. Impact Assessment of COVID-19 Measures on Airport Performance. 2020. Available online: https://www.eurocontrol.int/sites/default/files/2020-09/eurocontrol-impact-assessment-covid-19-airport-performance-2020.pdf (accessed on 2 December 2020).

11. Milne, R.; Delcea, C.; Cotfas, L.-A. Airplane boarding methods that reduce risk from COVID-19. *Saf. Sci.* **2020**. [CrossRef]
12. Milne, R.; Cotfas, L.-A.; Delcea, C.; Craciun, L.; Molanescu, A. Adapting the reverse pyramid airplane boarding method for social distancing in times of COVID-19. *PLoS ONE* **2020**, *15*. [CrossRef] [PubMed]
13. Schultz, M.; Fuchte, J. Evaluation of aircraft boarding scenarios considering reduced transmissions risks. *Sustainability* **2020**, *12*, 5329. [CrossRef]
14. Schmidt, M.; Engelmann, M. PAXelerate—An open source passenger flow simulation framework for advanced aircraft cabin layouts. In Proceedings of the 54th AIAA Aerospace Sciences Meeting, American Institute of Aeronautics and Astronautics, San Diego, CA, USA, 4–8 January 2016. [CrossRef]
15. Airport Research Center GmbH. Cast Cabin. Available online: https://arc.de/ (accessed on 3 December 2020).
16. Richter, T. Simulationsmethodik zur Effizienz und Komfortbewertung von Menschenflussprozessen in Verkehrsflugzeugen. In *Disserataion, Technische Universität München, Institut für Luft und Raumfahrttechnik*; Verlag Dr. Hut: München, Germany, 2007.
17. Marelli, S.; Mattocks, G.; Merry, R. The role of computer simulation in reducing airplane turn time. *AERO Mag.* **1998**, *1*, 10.
18. Schultz, M. Entwicklung eines individuenbasierten Modells zur Abbildung des Bewegungsverhaltens von Passagieren im Flughafenterminal. Ph.D. Thesis, Technische Universität Dresden, Dresden, Germany, 2010.
19. German Aerospace Center (DLR). Traffic Oriented Microscopic Simulator–TOMICS. Available online: https://www.dlr.de/fw/ (accessed on 3 December 2020).
20. Fuchte, J. Enhancement of Aircraft Cabin Design Guidelines with Special Consideration of Aircraft Turnaround and Short Range Operations: Dissertation: Technische Universität Hamburg: Harburg, Germany, DLR e.V. 2014. Available online: https://elib.dlr.de/89599/1/Fuchte%20FB-2014-17%20Version%20Druck.pdf (accessed on 19 December 2020).
21. Fire Safety Engineering Group, airExodus. Available online: http://fseg.gre.ac.uk/exodus/ (accessed on 3 December 2020).
22. Kneidl, A. Methoden zur Abbildung Menschlichen Navigationsverhaltens bei der Modellierung von Fußgängerströmen. Dissertation, Technische Universität München, Lehrstuhl für Computergestützte Modellierung und Simulation, München, Germany. March 2013. Available online: https://mediatum.ub.tum.de/1131501 (accessed on 19 December 2020).
23. Rindsfüser, G.; Klügl, F. Agent-based pedestrian simulation: A case study of the bern railway station. *disP* **2007**, *170*, 2007. [CrossRef]
24. Bohari, Z.A.; Bachok, S.; Osman, M.M. Simulating the pedestrian movement in the public transport infrastructure. *Procedia Soc. Behav. Sci.* **2016**, *222*, 791–799. [CrossRef]
25. Schadschneider, A.; Klingsch, W.; Klüpfel, H.; Kretz, T.; Rogsch, C.; Seyfried, A. Evacuation dynamics: Empirical results, modeling and applications. In *Encyclopedia of Complexity and Systems Science*; Meyers, R.A., Ed.; Springer: New York, NY, USA, 2009; pp. 3142–3176.
26. Daamen, W. Modelling Passenger Flows in Public Transport Facilities. Ph.D. Thesis, Delft University, Delft, The Netherlands, September 2004. Available online: http://resolver.tudelft.nl/uuid:e65fb66c-1e55-4e63-8c49-5199d40f60e1 (accessed on 19 December 2020).
27. Helbing, D.; Molnár, P. Social force model for pedestrian dynamics. *Phys. Rev. E Stat. Phys. Plasmas Fluids Relat. Interdiscip. Top.* **1995**, *51*, 4282–4286. [CrossRef] [PubMed]
28. Kielar, P.M. Kognitive Modellierung und Computergestützte Simulation der Räumlich-Sequenziellen Zielauswahl von Fußgängern. Ph.D. Thesis, Technische Universität München, Lehrstuhl für Computergestützte Modellierung und Simulation, Munchen, Germany, May 2017.
29. Sanz, F.A.; Olivier, A.-H.; Bruder, G.; Pettré, J.; Lécuyer, A. Virtual proxemics: Locomotion in the presence of obstacles in large immersive projection environments. In Proceedings of the IEEE Virtual Reality Conference, Arles, France, 23–27 March 2015; pp. 75–80.
30. Weidmann, U. *Transporttechnik der Fussgänger: Transporttechnische Eigenschaften des Fussgängerverkehrs, Literaturauswertung*; ETH Zurich: Zürich, Switzerland, 1993.
31. Bauhaus Luftfahrte, V. PAXelerate Boarding Simulation. Available online: www.paxelerate.com (accessed on 19 December 2020).

32. Schmidt, M.; Engelmann, M. Boarding process assessment of novel aircraft cabin concepts. In Proceedings of the 30th International Congress of the Aeronautical Sciences (ICAS), Daejeon, Korea, 25–30 September 2016; p. 3568.
33. Bachmann, A.; Kunde, M.; Litz, M.; Schreiber, A.; Bertsch, L. Automation of Aircraft Pre-design Using a Versatile Data Transfer and Storage Format in a Distributed Computing Environment. In Proceedings of the Third International Conference on Advanced Engineering Computing and Applications in Sciences, Sliema, Malta, 11–16 October 2009. [CrossRef]
34. German Aerospace Center (DLR). CPACS—Common Language for Aircraft Design. Available online: https://cpacs.de/ (accessed on 3 July 2019).
35. Gordon, C.G.; Churchill, T.; Clauser, C.E.; Bradtmiller, B.; McConville, J.T.; Tebbetts, I.; Walker, R.A. *Anthropometric Survey of US Army Personnel*; Summary Statistics, Interim Report for 1988; United States Army Natick Research, Development and Englneering Center: Natick, MA, USA, 1990.
36. Airbus Customer Services. *A320-200 Aircraft Charactersitics Airport and Maintenance Planning*; Airbus S.A.S: Blagnac, France, 2019.
37. Steiner, A.; Philipp, M. Speeding up the airplane boarding process by using pre-boarding areas. In Proceedings of the 9th Swiss Transport. Research Conference, Monte Verità, Ascona, 9–11 September 2009.
38. Götz, M. Engineering Concept Study of an Innovative Sideward Retractable Aircraft Seat. Master's Thesis, Institute of Aircraft Design, Technische Universität München, Munich, Germany, May 2014.
39. Engelmann, M.; Hornung, M. *Boarding Process Assessment of the AVACON Research Baseline Aircraft*; Deutscher Luft- und Raumfahrtkongress: Darmstadt, Germany, 2019. [CrossRef]
40. IATA Medical Advisory Group. Restarting Aviation Following COVID-19: Medical Evidence for various Strategies being discussed as at 27 April 2020. 2020. Available online: https://www.iata.org/contentassets/f1163430bba94512a583eb6d6b24aa56/covid-medical-evidence-for-strategies-200508.pdf (accessed on 19 December 2020).
41. Qantas Airways Limited, International Network Changes. Available online: https://www.qantas.com/au/en/travel-info/travel-updates/coronavirus/qantas-international-network-changes.html (accessed on 5 November 2020).
42. New York Times. Coronavirus Travel Restrictions, Across the Globe. Available online: https://www.nytimes.com/article/coronavirus-travel-restrictions.html (accessed on 5 May 2020).

Publisher's Note: MDPI stays neutral with regard to jurisdictional claims in published maps and institutional affiliations.

MDPI
St. Alban-Anlage 66
4052 Basel
Switzerland
Tel. +41 61 683 77 34
Fax +41 61 302 89 18
www.mdpi.com

Aerospace Editorial Office
E-mail: aerospace@mdpi.com
www.mdpi.com/journal/aerospace

www.ingramcontent.com/pod-product-compliance
Lightning Source LLC
LaVergne TN
LVHW070139170726
843515LV00007B/1833

* 9 7 8 3 0 3 6 5 4 2 2 5 6 *